Feedstocks for the Future

ACS SYMPOSIUM SERIES **921**

Feedstocks for the Future

Renewables for the Production of Chemicals and Materials

Joseph J. Bozell, Editor
National Renewable Energy Laboratory

Martin K. Patel, Editor
Utrecht University

**Sponsored by the ACS Division
Cellulose and Renewable Materials**

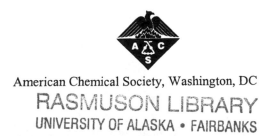

American Chemical Society, Washington, DC

Library of Congress Cataloging-in-Publication Data

Feedstocks for the future: renewables for the production of chemicals and materials / Joseph J. Bozell, editor ; Martin K. Patel, editor ; sponsored by the ACS Division of Cellulose and Renewable Materials

      p. cm. — (ACS symposium series ; 921)

    "Developed from a symposium sponsored by the Division of Cellulose and Renewable Materials at the 227th National Meeting of the American Chemical Society, Anaheim, California, March 28–April 1, 2004"—Pref.

    Includes bibliographical references and index.

    ISBN-13: 978–0–8412–3934–0 (alk. paper)

    1. Renewable energy sources.

    I. Bozell, Joseph J., 1953– II. Patel, Martin K., 1966– III. American Chemical Society. Division of Cellulose and Renewable Materials. IV. American Chemical Society. Meeting (227th : 2004 : Anaheim, Calif.). V. Series.

TJ808.F44   2005
621.042—dc22

                               2005048308

The paper used in this publication meets the minimum requirements of American National Standard for Information Sciences—Permanence of Paper for Printed Library Materials, ANSI Z39.48–1984.

Distributed by Oxford University Press

ISBN 10: 0-8412-3934-7

PRINTED IN THE UNITED STATES OF AMERICA

Foreword

The ACS Symposium Series was first published in 1974 to provide a mechanism for publishing symposia quickly in book form. The purpose of the series is to publish timely, comprehensive books developed from ACS sponsored symposia based on current scientific research. Occasionally, books are developed from symposia sponsored by other organizations when the topic is of keen interest to the chemistry audience.

Before agreeing to publish a book, the proposed table of contents is reviewed for appropriate and comprehensive coverage and for interest to the audience. Some papers may be excluded to better focus the book; others may be added to provide comprehensiveness. When appropriate, overview or introductory chapters are added. Drafts of chapters are peer-reviewed prior to final acceptance or rejection, and manuscripts are prepared in camera-ready format.

As a rule, only original research papers and original review papers are included in the volumes. Verbatim reproductions of previously published papers are not accepted.

ACS Books Department

Contents

Preface

At the conclusion of the very long (albeit very productive) week (March 28–April 1, 2004) described in this volume, I found myself ahead of schedule for the shuttle to the airport. Nonetheless, the driver was happy to meet me first, as long as I had no objection to riding for a longer time in the shuttle to pick up the passengers on his list. Not surprisingly, some of the other riders were running a little late, resulting in more than one stop at the conference hotels until all the passengers were accounted for. To pass the time, I began to watch some of the stores that were on this circular route, and since our symposium was focused on renewables, I was paying particular attention to the California gasoline prices, among the highest in the nation. During our drive, we passed a gas station exhibiting a price of $2.05/gallon for regular. Over the next 20 minutes as we continued on our route, we picked up several additional riders, and then passed the same gas station.

The price had risen to $2.09/gallon.

Of course, this was simply an amusing coincidence, but it provided an exclamation point for the week's meeting and an opportunity for further discussion of renewables among the riders on the way to the airport. The period of time since this meeting has seen continuing price pressure on our most strategic import. At this writing (early 2005), oil and gasoline prices are approaching the record (when adjusted for inflation) high prices of the late 1970s, China and the third world are rapidly increasing their demand for oil, OPEC has made the ominous announcement that for the first time in their history, oil costs are outside of their control (see http://www.msnbc.msn.com/id/7190109/), and the U.S. Senate has narrowly approved drilling in the Alaskan National Wildlife Refuge. These occurrences are of particular interest when viewed in the context of Hubbert's curve, which projects a peak in world oil production in or around 2010. According to that model, global production will gradually decline, but it is unlikely that demand will d ecline i n p arallel. Developing clean and sustainable alternative raw materials is thus increasingly vital to the world's economies.

The chapters included in this American Chemical Society (ACS) Symposium Series volume present an overview of ongoing efforts to address these global challenges. As part of the spring 2004 ACS meeting, Martin Patel and I were asked by the ACS Division of Cellulose and Renewable Materials to organize a symposium entitled *Feedstocks for the Future: Renewables for the Production of Chemicals and Materials*. We received an excellent response: the symposium spanned all 5 days of the meeting in 12 sessions, and comprised 60 oral and more than 25 poster presentations. This volume presents a selected collection of these papers best representing the wide v ariety o f t opics covered during the week. Of course, it is not our intent to position these examples as the last word in renewables development. Rather, we hope that these specific examples will spur others to think broadly about renewable feedstocks and to apply their own unique expertise and technology to the continued expansion and development of this field.

Joseph J. Bozell

National Renewable Energy Laboratory
1617 Cole Boulevard
Golden, CO 80401

Martin K. Patel

Utrecht University Department of Science, Technology and Society (STS)/Copernicus Institute
Heidelberglaan 2
NL–3584 CS Utrecht
The Netherlands

Feedstocks for the Future

Chapter 1

Feedstocks for the Future: Using Technology Development as a Guide to Product Identification

Joseph J. Bozell

National Renewable Energy Laboratory, 1617 Cole Boulevard, Golden, CO 80401

Biorefinery development addresses two goals: displacement of imported petroleum (an *energy* goal) and enabling biorefineries (an *economic* goal). These goals are simultaneously addressed if chemicals (producing an economic impact) are closely integrated with the more common production of fuels (producing an energy impact). However, incorporating chemicals is frequently impeded by the question "what product should we make?" The impact of products on biorefinery operation is much clearer when the focus is shifted to broad technology development, rather than a more fragmentary product-by-product approach. Yet conversion technology tailored for the structural features of biomass still lags far behind that available for nonrenewables. In 2004, the Cellulose and Renewable Materials division of the ACS sponsored a symposium "Feedstocks for the Future: Renewables for the Production of Chemicals and Materials", focusing on new technologies able to bridge this gap. This chapter introduces some key concepts, and overviews the information described in subsequent chapters.

Introduction

The *biorefinery* is now well established as a unifying concept for the transformation of renewable raw materials into chemicals, fuels, and power. A biorefinery should be viewed as analogous to a petrochemical refinery, in that it takes the complex raw materials of nature, separates them into simpler building blocks, and converts these building blocks into marketplace products. In its most basic form, a biorefinery will supply *carbohydrates*, in the form of cellulose, hemicellulose, and monomeric sugars, *aromatics*, in the form of lignin, and *hydrocarbons* in the form of plant oils. Each of these building blocks can feed into appropriate operating units using conversion technology tailored for their unique structural features.

Yet, the biorefinery industry is still orders of magnitude smaller than today's petrochemical industry.[1] A symposium entitled "Chemicals and Materials from Renewable Resources" was held as part of the fall 1999 National ACS meeting, and served as the prelude to the contributions presented in this book.[2] That meeting attempted to identify the barriers impeding biorefinery development, and had a specific focus on the production of chemicals. The meeting highlighted processes that offered new insight into the challenges presented by renewables, and possible solutions to meet those challenges. The context of the meeting was that the primary barrier between the concept and realization of chemicals from renewables was *technology development*, i. e., the diversity of technology available for the conversion of renewable raw materials into final products pales in comparison to the petrochemical industry.

This book overviews the follow-up symposium held at the spring 2004 ACS meeting, entitled "Feedstocks for the Future: Renewables for the Production of Chemicals and Materials". The focus and intent of the meeting remained the same as 1999, but the level of interest and participation indicated that the last five years have seen a significant upswing in activity. In 1999, the symposium covered about 1.5 days, and attracted about 15 papers. The 2004 symposium was held over all 5 days of the meeting in 12 different sessions, and attracted 60 oral and more than 25 poster presentations. Understanding of transformation processes is growing rapidly. However, technology development still remains as the primary barrier to the successful production of chemicals and materials from renewables.

The Challenge: Making a Case for Technology Development

Conversion Technology Must Be Developed Before Product Identification

Biorefinery development has two high level goals: the displacement of imported petroleum in favor of domestic raw materials (an *energy* goal) and the enabling of a robust biobased industry (an *economic* goal). The energy goal can be addressed by biorefineries producing fuel, primarily fermentation EtOH. However, since fuel is a high volume, but low value product, new, standalone fuel facilities are often burdened by a low return on investment, making their construction less desirable. Building a biorefinery based on chemical products can realize a much higher return on investment, but lacks the potential for a large energy impact. This results in attempts to identify "blockbuster" products whose energy impact might be significant. This is nonproductive, since few of these opportunities exist and chemical production accounts for only about 7-8% of our oil imports.[3]

Analyses carried out by the U. S. Department of Energy (DOE) consistently reveal that chemicals have the greatest impact when they are included in an integrated biorefinery scenario.[4] When a biorefinery produces both fuel and chemicals, the energy and economic goals can be met simultaneously. In an integrated operation, high value products become the economic flywheel supporting low value fuel, leading to an overall profitable biorefinery operation that also exhibits an energy impact.

Nonetheless, research on chemical products from renewable feedstocks lags research on materials such as fuel (i. e., EtOH or biodiesel) or biopower. A significant contributing factor to this situation is the incredible diversity of materials currently supplied by today's chemical industry. The sheer number of opportunities makes identification of an appropriate target difficult. Unfortunately, an additional, and perhaps bigger barrier to progress also arises when the first step in a research program is not an effort to clearly understand transformation processes, but rather a search among all of these potential opportunities for an answer to the question "what product should we make?" Such an approach immediately places chemicals in an unreasonably fragmented context, especially when compared to "single product" efforts on fuels and power. Moreover, the question is frequently premature, since the amount of fundamental knowledge on the types of conversion technologies most applicable to renewable building blocks is far behind that available for nonrenewables.

The example of the petrochemical industry, which provides the model for the biorefinery, offers an interesting illustration. The petrochemical industry did not start by identifying exact structures of compounds wanted from the crude oil raw material. Rather, the product slate developed as fundamental research was carried out on crude oil to find those broad technologies most applicable to the properties of the raw material, and identifying the structures most easily made *from* these technologies. Kerosene production led to thermal cracking, steam cracking and catalytic cracking.[5] In each case, the product slate from the processes changed. Only *after* the industry identified what structures were easily available from these broad technologies did they begin to identify and develop specific products. In the same way, the renewables industry needs to examine broad technologies, and find out what structures most easily result by using these technologies on carbohydrate, lignin, or oleochemical feedstocks. Product definition will naturally occur from that point. A consistent message from companies representing many large segments of the chemical industry is that the ability to make *any* single, defined structure, pre-identified or not, as long as it is inexpensive, is valuable. There is a most interesting industrial sense of "if you build it, they will come".

Moreover, the success of the petrochemical industry has evolved over many decades, and has resulted from a deep understanding of conversion processes and mechanisms at the molecular level. Reaction conditions and engineering processes then developed to reflect this understanding. Success for the biorefinery industry will require a similar approach of first understanding the types of processes best suited for the building blocks of nature, and developing engineering processes needed to move the molecular level understanding from the laboratory to commercial operation. Without this understanding, biorefinery development frequently becomes an effort to force fit renewable building blocks into the production of known chemical products. This approach is easy from an operational standpoint, because an infrastructure already exists within the chemical industry for the use and conversion of a certain set of building blocks. However, this approach frequently exhibits poor economics, because of the great structural and chemical differences between highly oxygenated renewables, and oxygen-poor nonrenewables.

Technology Development Simplifies the Complexity of Pre-identifying Products

As was the case for petrochemicals, development of appropriate technology for the biorefinery will not occur immediately. It is critical to recognize that in comparison to fuels and power, chemicals and materials are, by far, the most technically complicated of the potential biorefinery outputs. The diversity inherent in chemicals and materials accurately reflects the nature of the chemical industry itself, anticipated to be the primary customer for any technology development. The fuels and power components are *convergent*, while the chemicals component is *divergent* (**Figure 1**). Importantly, this realization begins to simplify and strengthen the case for technology development prior to product identification.

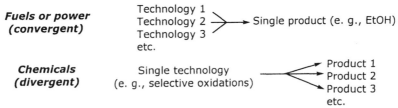

Figure 1-Fundamental Differences Between Components of the Products Platform

Research in fuels and power tends to investigate a wide number of different technologies to produce a single or very small number of outputs, for example, EtOH. Since the structure of the product is clearly defined, price targets for these single products are well known, making process analysis ideally suited for determining what technologies offer the best opportunity for research investment. These targets also immediately establish the allowable costs of process technology, and the amount of flexibility allowable in a production scheme. For a fuel process, if a technology does not meet these price targets, it is discarded in favor of more economical processes. Research related to a biorefinery fuel operation is convergent: the investigation of several technologies focused on a single product.

In contrast, research on chemicals often uses a single, broad based technology and applies it to the production of several different refinery outputs. Each product will have its own valid price targets, depending on the market and application. This flexibility results in a greater breadth of allowable conversion processes. If a technology does not meet the price targets of one product, it need not be discarded, since it may be applicable to a material with a different projected cost structure. Research on products is divergent: one frequently finds the opportunity to focus on one technology directed at several products. Such a realization also has the effect of simplifying a research effort. Instead of trying to identify an appropriate list of product targets from a huge number of possibilities, one recognizes that a biorefinery will first need to understand what types of structures result from typical refinery operations: selective reductions and oxidations, bond making/breaking processes, the identification of catalysts appropriate to biomass transformation, etc.

Product Identification Can Be Highly Useful in the Proper Context

Nonetheless, this argument must not be construed as a suggestion to eliminate ongoing activity in identifying product opportunities. Product identification is most useful when it is used as support for research activities rather than a prerequisite. Decision makers having responsibility for investing millions of dollars in new processes or facilities can use the results of technology development to reduce the multiplicity of specific product opportunities. The output of a biorefinery must offer a good opportunity for profitability because within the chemical industry, making the wrong decision on a product can mean the loss of many millions of dollars of capital investment. The development of polylactic acid as a commercial product is illustrative.[6] Original projections described a fully operating plant by 2001, and a 1 billion lb/yr capacity by 2006. The markets have not developed as expected, and the single operating facility is not yet at full capacity. Accordingly, operators have scaled back their projections and have refocused their efforts on smaller, niche markets. In the same way as decision makers need to provide strong support for technology development within a research program, those of us carrying out technology development must be able to provide good pointers for the eventual application of this technology. We are faced with a puzzle: *how can the biorefinery model the success of the petrochemical refinery, address the clear need to carry out fundamental technology development in conversion of renewable building blocks while also showing that such research can and will lead to profitability and identifiable marketplace products?*

The Opportunity: Chemicals Will Play a Key Role in Enabling the Biorefinery

The solution to this puzzle is still developing for the biorefinery. An ongoing combination of evaluation and research activities provides a very encouraging picture of how efficient biomass conversion processes will help support a developing biorefinery industry. The following examples are of particular importance.

DOE's "Top 10" Report for Sugars

In 2004, DOE released their first "Top 10" report that described an initial group of specific target molecules that could reasonably be produced from biorefinery carbohydrates.[7] The great value of this report is that it provides an appropriate balance between the need for technology development and efforts to pre-identify biorefinery outputs. This report is the first in a series that will eventually address product opportunities from each of the major biorefinery building block streams (carbohydrates, lignin, and oils). The report clearly identifies an initial group of specific chemical structures that could reasonably be made within a biorefinery (**Table 1**).

Succinic, fumaric, and malic acids
2.5-Furandicarboxylic acid
3-Hydroxypropionic acid
Aspartic acid
Glucaric acid
Glutamic acid
Itaconic acid
Levulinic acid
3-Hydroxybutyrolactone
Glycerol
Sorbitol
Xylitol/arabinitol

Table 1 - The "Top 10" Chemicals from Carbohydrates

But importantly, the report is equally clear in stating that the structures of **Table 1** are not the last word in product identification. The report is not an attempt to assign structures for manufacturing or a business plan for chemical producers. Those decisions must be left to industry. This report shows that by identifying specific structures reasonably prepared from sugars, a simultaneous identification of broad technology development needs within the biorefinery can be made. Thus, the report acknowledges the need to have potential targets for any planned research activities ("what product should we make?"), and also sets the stage for fundamental development and understanding of technologies (oxidations, reductions, bond forming/breaking, etc.) tailored for renewable building blocks. The ideal research activity would lead to partnerships between research institutions, using their expertise to develop the necessary technology, and industry, using their expertise to identify what specific chemicals could be made from biomass using this technology. The list of products is also dynamic. It is clear that the biorefinery of 2050 may or may not be making, say, itaconic acid as a product. However, it will certainly still be using the broad technologies developed for an early generation itaconic acid process.

Can the Impact of Technology Development be Quantified?

Decision makers also frequently search for information regarding the impact of investing in new technology. Ongoing process evaluation at the

National Renewable Energy Laboratory has attempted to provide a measure of the impact of including chemicals within the biorefinery. The evaluation looks at the product-independent effect of replacing nonrenewable carbon with renewable carbon. As a simple example, we have investigated the impact of producing aromatic chemicals from lignin. The analysis is intended to provide high-level pointers as to the potential of such transformations, and does not assume that technology exists for these transformations. In many cases it may not. The results of such an analysis can help determine whether technology development would lead to greater success for the biorefinery. **Table 2** shows the three highest volume aromatic chemicals (BTX) produced by the chemical industry.

Chemical	Production (10^9 lb)
Benzene	20.4
Xylene	13.8
Toluene	11.1
Total	**45.3**

Table 2 – The Top Five Commercial Aromatic Chemicals

A simple stoichiometric calculation indicates that production of this amount of aromatic carbon would require about 91×10^9 lb of lignin. This scenario also offers a striking energy perspective for an integrated biorefinery making both fuels and products. Lignin always offers a parallel source of carbohydrates since it will be part of a lignocellulosic raw material. An energy impact can be projected by assuming that all of these sugars would be available for EtOH production. If wood is assumed to be the lignocellulosic feedstock, and wood is assumed to be 25% lignin, 91×10^9 lb of lignin would also supply 272×10^9 lb of sugar, equivalent to almost 21×10^9 gallons of EtOH. Given that the U. S. consumes 136×10^9 gallons of gasoline/year, this amount of EtOH would be about 15% of the current national gasoline use.[8] In other words, *technology development to make aromatic products in the biorefinery that meet the current aromatic carbon needs of the U. S. could enable EtOH production equivalent to almost 15% of our current supply.*[9]

An important question closely related to technology development is embedded in these data. Terephthalic acid and phenol are also large volume aromatics produced in the petrochemical refinery, as derivatives of p-xylene and benzene, respectively. If the biorefinery develops a process to BTX from lignin, then inclusion of terephthalic acid and phenol in this assessment is "double counting". However, the uncertainty in the best technology to employ for lignin conversion highlights the importance of research and development in moving the biorefinery concept forward. The structure of lignin suggests alternate approaches that could produce benzene, xylene, phenol and terephthalic acid (or their functional equivalents) as separate, independent streams, which would raise the EtOH impact of these calculations by about 20%. Only after the most appropriate conversion processes are identified will this uncertainty be resolved.

Identification of Emerging Recent Research and Evaluation Activities

Finally, the contributions in this book present an overview of ongoing research activities whose success will provide answers for the challenges described in this chapter. Evaluation and projection is only successful when validated by research success. The papers collected in this book were chosen from the symposium to provide the reader with the broadest overview of current research activities, and those most representative of the theme of the symposium. Within this volume, the reader will find the following topics.

The Promise of Renewable Feedstocks

Several contributions illustrating the promise of chemicals from renewables open the book. The first paper of the symposium, submitted by Metzger, provides an overview of the concept of sustainable development, illustrated by examples from his laboratory. Wilke builds on this topic, and describes how advances in selective plant genetics can lead to improved feedstocks. An elegant example of how renewable building blocks can be exploited by industry is provided by Manzer, who describes DuPont's experience in the use of levulinic acid as a pre-identified renewable feedstock. Gallezot's contribution presents an overview of broad technology development, describing the use of catalytic oxidations for the conversion of biomass.

Oleochemical Building Blocks

Oleochemicals were called out as topics of particular interest for this symposium. As the biorefinery evolves, oleochemical technology could serve to drive a new operating unit comprising biodiesel as a fuel, and oleochemicals as hydrocarbon-based chemical products. The contribution from Larock's laboratory offers syntheses of new biobased polymers from oleochemical starting materials. The paper by Hutchings describes a new process for the selective oxidation of glycerol, addressing the broader issue of new and inexpensive glycerol supplies resulting from the advent of biodiesel. The contribution from van Haveren describes the development of new, oleochemical phthalate esters.

New Technologies for the Conversion of Renewables

One session of the symposium was devoted to technologies not normally associated with renewable feedstocks, but which offer new concepts for potential use within the biorefinery. The paper by Coates describes new opportunities for the formation of polymers using CO_2 as a C_1 feedstock. McElwee-White describes a process for the catalytic electrochemical oxidation of MeOH. Given the huge impact that catalysis has on the operation of today's petrochemical refinery, it is anticipated that the biorefinery will use catalysts designed to accommodate the structures of biomass. New transformation technology using transition metal catalysis is described in the papers by Gable, Horwitz and Macquarrie.

Bioconversions

Biotechnology continues to be an important contributor to the biorefinery, especially for the conversion of carbohydrates. The paper by Richard describes a new approach for the fermentation of C_5 sugars, providing methodology for more efficient conversion of biomass carbohydrates to EtOH. The contribution from Nakas discusses the bioproduction of polyhydroxyalkanoates using levulinic acid as a carbon source. Stipanovic describes new approaches for using hemicellulose as a chemical feedstock.

Life Cycle Analysis

A focus area for this meeting was a discussion of new approaches to life cycle analyses of biobased processes is described in three papers. Anex reports an evaluation of processes for the production of 1,3-propanediol. Niederl describes an analysis of biodiesel production from tallow, while the paper from Bohlmann reports on life cycle issues surrounding polyhydroxylalkanoate production.

Novel Products from Renewables

Finally, several papers describing the development and use of renewable building blocks are included. Fitzpatrick describes the use of levulinic acid as a chemical feedstock, as well as its continued development. New nanostructural materials based on biopolymers are discussed by van Soest. Robinson presents an overview of a biorefinery sugar platform leading to the production of polyols. A new approach for the extraction of heteropolysaccharides is given by

Karlsson, and Guilbert describes the use of protein as a component of polymers, composites and other green materials.

Conclusions

Renewables have the potential to provide a new and sustainable supply of basic chemical building blocks. Nature produces as much as 170×10^9 tons of biomass annually, with over 90×10^9 tons present as carbohydrate in the form of cellulose.[10] The corn industry processes $8 - 10 \times 10^9$ bushels of corn/yr, each containing about 33 pounds of carbohydrate as starch, and equivalent to almost 500×10^6 barrels of crude oil.[11] The pulp and paper industry consumes over 100×10^6 metric tons/yr of wood, which is about 25% aromatics as lignin,[12] and other biomass industries (oleochemical, sugar cane, etc.) process about 2×10^9 tons of oil crops annually.[11]

In light of a highly successful petrochemical industry producing thousands of products, it is natural to first look for those opportunities that are structurally identical to existing materials. As exemplified by the papers collected in this volume, the greatest success in biorefinery development will arise when technology is developed to convert this supply of raw material into products that take the best advantage of the unique structural features of biomass.

Acknowledgements

We would like to thank the Cellulose and Renewable Materials division of ACS for sponsoring this symposium, and hope that such meetings will continue to be part of the division's plans. We also greatly appreciate the effort of all the authors that participated, including those not represented in this volume. It is through their efforts and those of many others that the potential of chemicals from renewables will be realized.

References

1. There are exceptions. Both corn wet milling and the pulp and paper industry are examples of biorefineries, and may eventually serve as the starting locations for broader biobased chemical production.
2. Bozell, J. J., ed., *Chemicals and Materials from Renewable Resources*, ACS Symposium Series **2001**, 784.

3. Donaldson, T. L.; Culberson, O. L. *Energy* **1984**, *9*, 693; Lipinsky, E. S. *Science* **1981**, *212*, 1465; Hanselmann, K. W. *Experientia* **1982**, *38*, 176.
4. See the U. S. Department of Energy Multiyear Technical Plan for products, available at http://www.eere.energy.gov/biomass/.
5. Green, M. M.; Wittcoff, H. A. *Organic Chemistry Principles and Industrial Practice*, Wiley, New York (2003).
6. Tullo, A. *Chem. Eng. News* **2005**, Feb. 28th, 26.
7. *Top Value Added Chemicals from Biomass Volume I. Results of Screening for Potential Candidates from Sugars and Synthesis Gas*; U. S. Department of Energy, **2005**, available at http://www.nrel.gov/docs/fy04osti/35523.pdf.
8. Displacement of gasoline by EtOH is actually about a 1.1:1 (EtOH/gasoline) substitution. For the purposes of this calculation, a 1:1 substitution was assumed.
9. Statistics available from the Energy Information Agency at http://www.eia.doe.gov/neic/quickfacts/quickoil.html.
10. Eggersdorfer, M.; Meyer, J.; Eckes, P. *FEMS Microbiol. Rev.* **1992**, *103*, 355.
11. Varadarajan, S.; Miller, D. J. *Biotechnol. Prog.* **1999**, *15*, 845; "The World of Corn", National Corn Growers Association, viewable at http://www.ncga.com.
12. *North American Pulp and Paper Factbook*; Miller Freeman, San Francisco.

Chapter 2

Sustainable Development and Renewable Feedstocks for Chemical Industry

Jürgen O. Metzger and Ursula Biermann

Department of Chemistry, University of Oldenburg, Oldenburg, Germany

The principles of the United Nations Conference on Environment and Development (UNCED), held in June 1992 in Rio de Janeiro, the World Summit on Sustainable Development, held in August 2002 in Johannesburg, and Agenda 21, the comprehensive plan of action for the 21th century, adopted 12 years ago by more than 170 governments, address the pressing problems of today and also aim at preparing the world for the challenges of this century. The conservation and management of resources for development are the main focus of interest, to which the sciences will have to make a considerable contribution. The encouragement of environmentally sound and sustainable use of renewable natural resources is one aim of Agenda 21. In this contribution we investigate innovations in chemistry for such a development focusing exemplarily on chemical uses of fats and oils as renewable feedstocks. Since base chemicals are produced in large quantities and important product lines are synthesized from them, their resource-saving production is especially important for a sustainable development. New processes based on renewable feedstocks are significant here. Most products obtainable from renewable raw materials may at present not be able to compete with the products of the petrochemical industry, but this will change as oil becomes scarcer and oil prices rise. In the long run, renewable resources could replace fossil raw materials. The competition of the cultivation of food and of feedstocks for industrial use can be met by a global program of reforestation of areas wasted in historical time by human activities.

Introduction

First of all, we should have a thorough and very precise understanding what **sustainable development** may be. In principle that is quite simple; in practice it is much more difficult (*1,2*).

For example: may the international space station contribute to a sustainable development of mankind as claimed by the German ministry of Science and Technology? "One key principle of German space-programme policy is that space programmes should be seen and used as important means of developing sustainable policy world-wide." (*3*)

Will the creation of the "carbon sequestration leadership forum" and the stimulation of world wide research into how CO_2 produced through burning fossil fuels can be captured at source and stored deep underground promote sustainable development (*4*)?

Can the global energy consumption, increasing continuously by more than 50% from 2001 to 2025 be sustainable? And can the distribution of the energy consumption expected in 2025 - 1 billion people in the industrialized countries consume 50%, 8 billion people in the developing countries 50% as well – be assigned as sustainable? (*5*)

And finally: Could it be possible that the results of synthetic organic chemistry performed with fatty acids obtained by saponification of the seed oil of some plants eventually become a contribution of chemistry to a sustainable development (*6,7*)?

Sustainable Development

Sustainable development is being understood as the implementation of the Rio process, as developed in the Rio Declaration and Agenda 21, (*8*) including its on-going advancements such as the Johannesburg Declaration and Plan of Implementation of the World Summit on Sustainable Development in 2002 (*9*). Science as well as industry will best contribute to promote sustainable development being devoted to the implementation of the Rio process. The 27 principles of the Rio Declaration were made concrete in the forty chapters of Agenda 21, the comprehensive plan of action for the 21st century which was adopted by more than 178 governments in Rio. We should keep in mind that these documents should be our compass to assess sustainability (*10*). Here we want to discuss some of the many unsolved problems on the way to a sustainable development outlined in Agenda 21 and some aspects of renewable feedstocks for the chemical industry exemplarily referring to Chapter 4 "Changing Consumption Patterns" focusing "on unsustainable patterns of production and consumption" and "national policies and strategies to encourage changes in unsustainable consumption patterns" (8).

In 1995, the energy consumption in Germany was about 14 exajoule. That is approximately the same as in 2002. The chemical industry used 1.7 exajoule, about 12 % of the total and 45 % of the energy consumed in all manufacturing processes. The main energy consumption was invested in the production of the organic chemicals. Approximately 51% of the total energy was used as feedstock for chemical products - the non-energetic consumption - and 49 % as process energy. A potential saving of about 15% of the total energy was identified by reduction of the process energy (11).

"Reducing the amount of energy and materials used per unit in the production of goods and services" is an important aim of Agenda 21. However, a reduction factor of 4 (12) or even 10 (13) is thought to be necessary on the way to a sustainable development. In addition to the demand for food, the demand for other goods will grow substantially with an increasing adoption of the standard of living in developing countries. That is a real challenge for chemists and chemical engineers to realize this goal in chemical processes and products. What can chemists do?

Chemistry can contribute to the conservation of resources by the development of

- more efficient and environmentally more benign chemical processes to reduce the energy consumed in chemical industry.

- chemical products that are environmentally more benign and enhance significantly the efficiency of production processes and products in the other manufacturing areas; to reduce the energy consumed in all manufacturing processes and finally and most important

- products that allow the consumer to use resources more efficiently to reduce the energy consumed in daily life (14).

Base chemicals

Base chemicals, chemicals that are each produced worldwide in more than one million tons per year, are most important to discuss with respect to the conservation of resources. A large part of the energy in chemical industry is used for base chemicals and most energy may be saved by potentially improving their production processes. The gross energy requirements – the requirements of fossil energy – used for the production of some important base chemicals are given in Figure 1. It is evident that all petrochemical base chemicals have a much higher gross energy requirement – the sum of process energy and feedstock energy – than base chemicals derived from biomass such as ethanol and rape seed oil having zero feedstock energy. Here, the differences in the fossil resource consumption are so high that it can be assumed that products based on renewables must clearly be more sustainable than petrochemical products. Most interesting is the fact that the GER of ethanol derived from corn is only one third of that of petrochemical ethanol. Thus, clearly, we should begin to substitute petrochemical base chemicals by renewables. Let us consider in more detail and exemplarily propylene oxide, the base chemical with the highest GER.

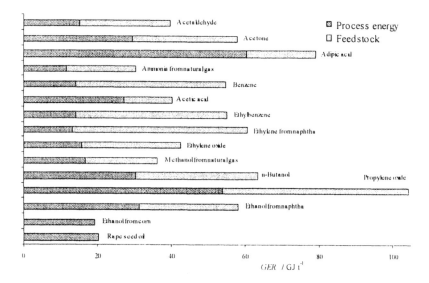

Figure 1. Gross energy requirements (GER) for important base chemicals. Data are taken from Ref. 11.

Propylene oxide is one of the top 50 chemicals. More than 4 million t/a are produced worldwide. It is reacted via polyetherpolyols to form polyurethanes and via propylene glycol to form polyesters (*15*). Obviously, the polyol and diol functionalities are used to form with diisocyanates derived from diamino compounds polyurethanes and with diacids polyesters, respectively. These functionalities are available from renewable feedstocks: carbohydrates, oils and fats, proteins and lignins.

Without any doubt, alternatives to propylene oxide and also to other oxidized base chemicals should be possible to be developed based on renewable feedstocks. We want to focus on the examples of a possible contribution of synthetic organic chemistry to a sustainable development given on vegetable oils, the renewable feedstock we are working with (5). Vegetable oils can easily be epoxidized (Figure 2). That is a well known industrial process. Obviously, this process should be much improved i. e. by direct catalytic oxidation using oxygen from the air (1).

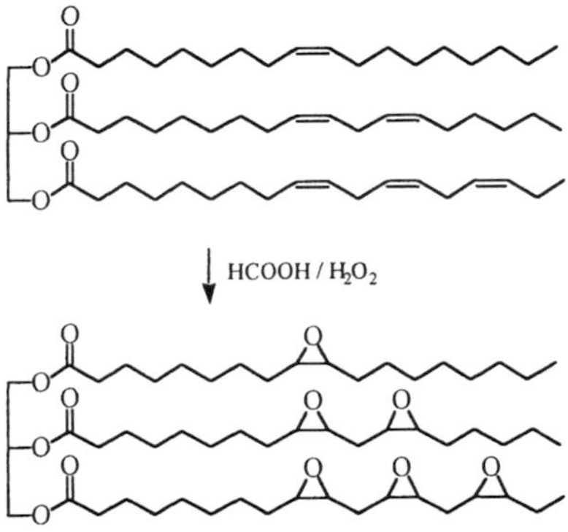

Figure 2. Epoxidation of a vegetable oil.

We introduced a UV-curable coating with linseed oil epoxide as binder that has the additional advantage that it is processed free of organic solvents thus avoiding VOCs (1). The initiator of this cationic curing process is, for example, a sulfonium hexafluoroantimonate (16). The best green as well as most economical solvent is no solvent at all in chemical reactions and especially in the processing of chemical products mostly performed by the non-chemical industry (17).

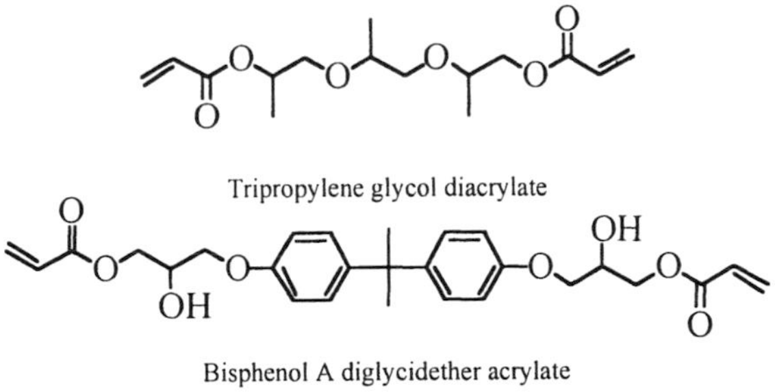

Tripropylene glycol diacrylate

Bisphenol A diglycidether acrylate

Figure 3. A mixture of tripropylene glycol diacrylate and of Bisphenol A diglycidether acrylate is an important petrochemical UV-curable coating (1).

However, there are of course also petrochemical UV-curable coatings available - for example a 1: 1 mixture of the diacrylates given in Figure 3 Tripropylene glycol diacrylate is produced from propylene oxide (*15*).

Agenda 21 calls for "criteria and methodologies for the assessment of environmental impacts and resource requirements throughout the full life cycle of products and processes", from the cradle to the grave. In other words, we have always to answer the question: Which alternative is greener (*18*) or more sustainable? Patel et al. studied the environmental assessment of bio-based polymers and natural fibres (*19*). Thus, we used Life Cycle Assessment (LCA) and compared the UV-curable coating with linseed oil epoxide as binder with an often used 1:1 mixture of the diacrylates. The results showed clearly the advantages of the renewable feedstock – centralized as well as decentralized processing. The gross energy requirements as well as the CO_2 and other emissions are almost one order of magnitude smaller when linseed oil epoxide was used.(*1*) This result is most important: It is indeed feasible to reduce the gross energy requirement for a product by one order of magnitude using renewable feedstocks.

Thus, the encouragement of the environmentally sound and sustainable use of renewable natural resources is an important aim of Agenda 21.

Renewable Feedstocks

The biomass cycle (Figure 4) shows the advantages of renewable feedstocks. Biomass is formed by photosynthesis; extraction gives renewable feedstocks such as vegetable oil, starch and others. These are processed to give the renewable base chemicals such as fatty acids, glycerol, glucose etc. Further processing gives the useful products that, after utilization, can be biologically degraded to give again carbon dioxide and water. That is the ideal type of biomass cycle giving no additional carbon dioxide. Of course, some process energy is needed for the farmer, the fertilizer, the extraction and processing. However, the chemical energy accumulated in the product comes completely from the sun. In contrast, petrochemicals are derived completely from fossil feedstocks. Most importantly, when renewables are used as base chemicals for organic synthesis, nature's synthetic input is used to obtain in one or only very few chemical reaction steps those complex molecules which petrochemically are only accessible in multistep reaction sequences, thus reducing the process energy as well.

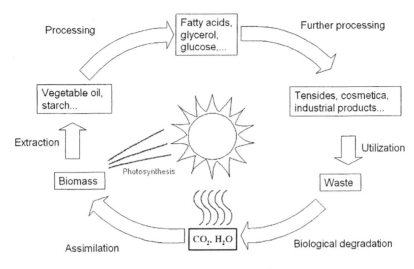

Figure 4. Biomass cycle

Dicarboxylic Acids Derived from Renewable Feedstocks

Adipic acid is a base chemical that is produced from benzene in 2.3 million t/a worldwide with a GER of about 80 GJ/t. There are important alternatives based on renewables for the production of dicarboxylic acids available.

Ozonization of oleic acid gives azelaic and pelargonic acid (*20*). However, ozone is much too expensive and energy consuming. It should be a real challenge for chemists to develop a catalyst that enables this reaction to proceed quantitatively with oxygen from the air. This reaction applied to unsaturated fatty acids with the double bond in other positions of the alkyl chain will open access to diacids of different chain length, such as adipic and lauric acid from petroselinic acid. Tridecane diacid will be obtained from erucic acid (Figure 5).

Long chain diacids can also be obtained by metathesis of ω-unsaturated fatty acids, e. g., 10-undecenoic acid (Figure 6). ω -Unsaturated fatty acids of different chain length can be produced by metathesis of e. g. oleic acid with ethylene (*7*).

Cognis developed a metabolically engineered strain of *Candida tropicalis* to oxidize a terminal methyl group of an alkyl chain. Reaction of fatty acids (e. g. oleic acid) gives the respective C 18 diacid (*21*). Remarkably, adipic acid can be synthesized in a combined biotechnological-chemical process via muconic acid that is hydrogenated to give adipic acid (*22*). One problem is that two molecules of the energy expensive hydrogen are needed for the hydrogenation step.

An assessment of the sustainability of these different processes to produce diacids based on renewables would be most interesting.

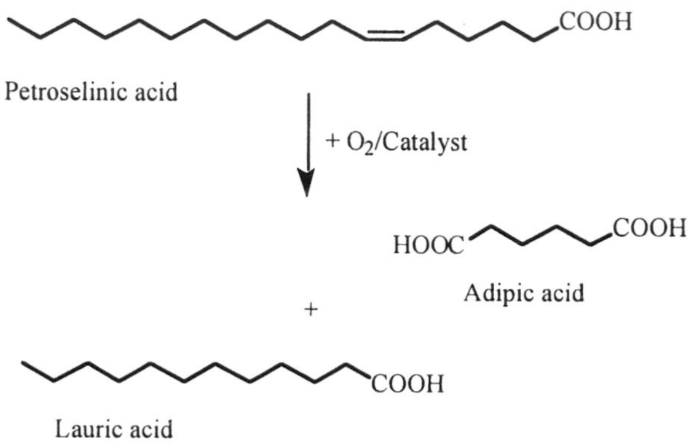

Figure 5. *Oxidative scission of unsaturated fatty compounds.*

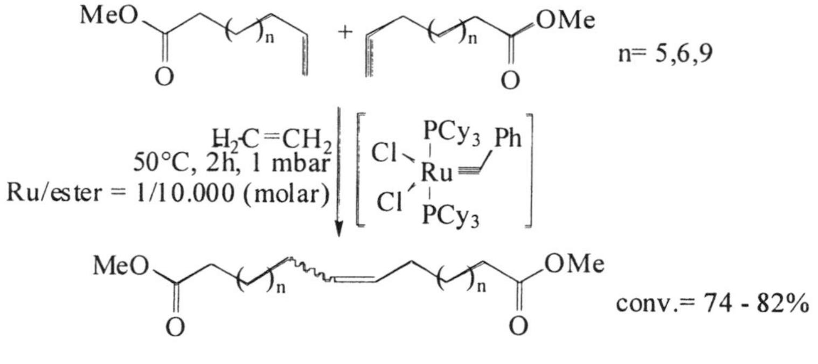

Figure 6. *Metathesis of ω-unsaturated fatty esters to give long-chain diacid diesters.*

Increasing the Agricultural Biodiversity

We can expect that the increasing usage of renewable feedstocks will also enlarge agricultural biodiversity. Petroselinic acid from *Coriandrum sativum* was already mentioned as a possible feedstock for adipic acid. *Calendula officinalis* offers a most interesting C18 fatty acid with a conjugated triene system, being available on industrial scale (*23*). Reaction of calendic acid methyl ester with maleic anhydride gives the Diels-Alder adduct with high regio- and stereoselectivity (Figure 7). The X-ray structure analysis shows the remarkable all *cis* configuration of the four substituents of the cyclohexene derivative (*24*). α-Eleostearic acid is the main fatty acid of tung oil and an isomer of calendic acid. Tung oil is commercially available on industrial scale. Diels-Alder reaction with maleic anhydride gives the respective Diels-Alder product again with high regioselectivity and stereoselectivity (*24*). These Diels-Alder adducts may potentially substitute petrochemical diacids such as phthalic acid anhydride in the production of alkyd resins.

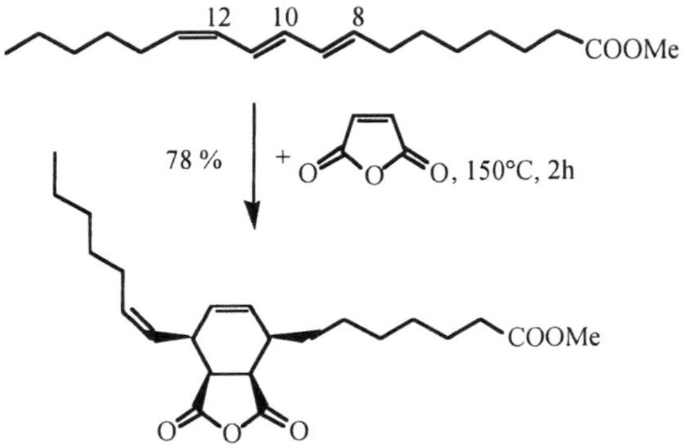

Figure 7. Diels-Alder reaction of calendic acid methylester and maleic anhydride and X-ray structure analysis of the product.

Vernonia oil is a naturally occuring epoxidized vegetable oil containing the chiral enantiomerically pure vernolic acid, thus offering most interesting synthetic possibilities. *Vernonia galamensis* originates from tropical and subtropical Africa. Nowadays it is cultivated in Zimbabwe, Kenya, Ethiopia and in parts of South America. As an example, the synthesis of an enantiomerically pure fat derived aziridine is given (Figure 8). In the first step, nucleophilic attack of azide gives the azido alcohol and an interesting fat derived pyrrol. The azido alcohol can be reduced to give the enantiomerically pure α-aminoalcohol. Reaction with triphenyl phosphine gives the aziridine (*25*).

Biodiesel, Rapeseed Asphalt, Insulating Fluids

The production and consumption of biodiesel in Germany will reach more than one million tons this year. In the European Union 2.2 Million tons are expected in this year. Biodiesel can be used as chemical feedstock as well .

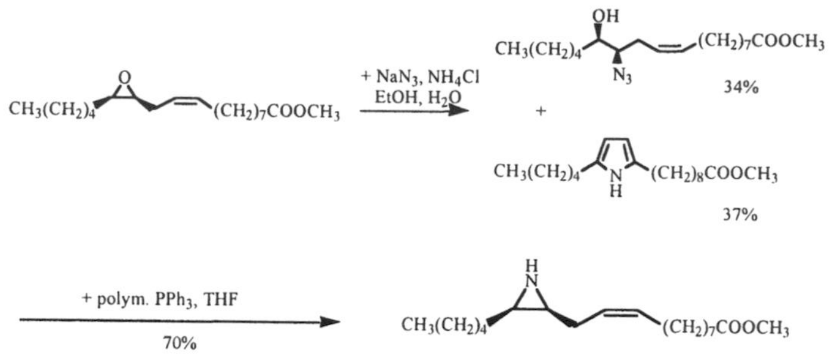

Figure 8. Methyl vernolate as substrate for the synthesis of the aziridine methyl (9Z,12S,13R)-12,13-epimino-9-octadecenoate.

A new and interesting application of rapeseed oil is rapeseed asphalt. Asphalt containing four percent of rapeseed oil is perfectly suited for the surface treatment of roads, causes an increase in surface density, and prolongs service life of the roads. Each square metre of rapeseed asphalt contains the rapeseed oil of one square metre of a rapeseed field (26).

Another important application is the use of plant oils as insulating fluids for electric utilities showing remarkable advantages compared to the respective petrochemical products (27).

Industrial Use of Renewables and Food Production.

One could ask which quantities of oils and fats and of other renewable feedstocks are available to displace petrochemicals. In 2002 the chemical industry in Germany used approximately 900,000 t of oils and fats and 1×10^6 t of renewables such as cellulose, starch, sugar and others. That is about 8 % of total feedstock consumption of about 20×10^6 t petroleum equivalents (1). This percentage could be doubled or even tripled considering the above mentioned increasing consumption of biodiesel in Germany and in the European Union. The 'Biomass Technical Advisory Committee' presented the 'Vision for Bioenergy and Biobased Products in the United States' (28). In 2030, 25% of the production of organic chemical products is expected to come from renewable feedstocks. However, it seems to be most difficult to substitute a higher percentage or finally the total of fossil feedstocks because an important problem most narrowly connected with the industrial use of renewables is the competition of the cultivation of food on the limited available agricultural area

(*29*). Also food demand and consumption will increase dramatically. The world population will rise from the present 6 billion to about 9 billion in 2050 (*30*).

For this reason, the UN programs to combat deforestation and desertification are most important. In short, "efforts should be undertaken towards the greening of the world", to enlarge the terrestrial biosphere (*8, 9*). Unfortunately, very little has been done to implement these programs, especially because the industrialized countries have evidently not been interested in contradiction to their obligations accepted in Agenda 21 (*31*). They have been more interested in other things, for example in the planet Mars! However, there is nothing new on Mars as one can see comparing the images taken in 2004 with the first image taken in 1976. It can be assumed that in a hundred or a thousand years there will be no difference (*10*). Obviously, it would be much better for a sustainable development to use the money to regenerate wasted areas on earth, for example in northern Africa the ancient granary of the Roman Empire or in Sinkiang or in other regions of our world wasted by human activities in historical times and unfortunately still today (*32*).

The sciences and especially chemistry and chemical industry could make enormous contributions to those programs. For example, hydrogels are important products of chemical industry. These hydrogels are able to bind a four hundred fold of water, reversibly. Hydrogels were used by Hüttermann (*33*) and Chinese scientists to cultivate semiarid regions in China (*34*).

CO_2-Sequestration

The UN programs to combat deforestation and desertification face another problem. The capture and storage of CO_2 is becoming necessary with the upcoming coal based economy in the coming decades of this century (*35*). A respective program was called in Germany COORETEC (CO two reduction technologies) and in the U.S. Future Gen. We mentioned already the international program to establish "a carbon sequestration leadership forum" (*4*) The costs for separation of CO_2 are estimated to 18 – 60 €/t and for transport and deposition to 10 – 24 €/t. However, the most efficient system of CO_2 sequestration, validated over millions of years, is the terrestrial biosphere. Costs for sequestration of CO_2 by reforestation are estimated to be 1 – 5 €/t. Important additional advantages of a reforestation would be the generation of sufficient biomass for chemical and energetic use, the improvement of the global climate and the improvement of water retention. Furthermore employment would be generated in many developing countries (*36*).

Conclusions

In conclusion, we can see that the CO_2 sequestration program is obviously non sustainable as is the international space station, and global energy consumption and distribution. In contrast, synthetic organic chemistry with biomass as renewable feedstock including fats and oils may contribute eventually to a sustainable development. A global reforestation program of areas wasted by human activities in historical time would give sufficient biomass as renewable feedstock for chemical and energetic use without competition to food production. The greening of the world can be realized if we want to do that.

References

1. Eissen, M.; Metzger, J. O.; Schmidt, E.; Schneidewind, U. *Angew. Chem. Int. Ed.* **2002**, *41*, 414-436.
2. Boschen, S.; Lenoir, D.; Scheringer, M. *Naturwissenschaften* **2003**, 90, 93-102.
3. Federal Ministry of Education and Research (BMBF), Public Relations Division, Ed., *Facts and Figures Research 2002*, Bonn, **2002**, p. 244, http://www.bmbf.de/pub/facts_and_figures_research_2002.pdf
4. Carbon Sequestration Leadership Forum, http://www.fe.doe.gov/programs/sequestration/cslf/
5. Energy Information Administration/ *International Energy Outlook* **2003**, http://www.eia.doe.gov/oiaf/ieo/
6. Biermann, U.; Friedt, W.; Lang, S.; Lühs, W.; Machmüller, G.; Metzger, J. O.; Rüsch gen. Klaas, M.; Schäfer, H. J.; Schneider, M. P. *Angew. Chem. Int. Ed.* **2000**, *39*, 2206-2224
7. Biermann, U.; Metzger, J. O. *Top. Catal.* **2004**, *38*, 3675-3677.
8. *Report of the United Nations Conference on Environment and Development*, Rio de Janeiro, 3-14 June 1992; http://www.un.org/esa/sustdev
9. *Johannesburg Declaration on Sustainable Development* and *Johannesburg Plan of Implementation*; http://www.un.org/esa/sustdev
10. Metzger, J. O. *Green Chem.* **2004**, *6*, G15 – 16.
11. Patel, M. *Closing carbon cycles*, Proefschrift Universiteit Utrecht, 1999; http://www.library.uu.nl/digiarchief/dip/diss/1894529/inhoud.htm; Patel, M. *Energy* **2003**, *28*, 721 – 740.
12. Von Weizsäcker, E. U., Lovins, A. B., Hunter Lovins, L. *Factor Four: Doubling Wealth - Halving Resource Use. A New Report to the Club of Rome.* St Leonards (NSW), Australia, 1997.
13. Schmidt-Bleek, F. *Das MIPS-Konzept, Weniger Naturverbrauch- mehr Lebensqualität durch Faktor 10*, Droemer & Knaur, Munich, 1998.

14. Metzger, J. O.; Eissen, M. *C. R. Chimie,* **2004,** *7,* 569-581.
15. Kahlich, D.; Wiechern, K.; Lindner, J. in: Elvers; B.; Hawkins, S.; Russey, W.; Schulz, G. (Eds.) *Ullmann's Encyclopedia of Industrial Chemistry,* Fifth, Completely Revised Edition Wiley-VCH, 1993, Vol. A22, p. 239 – 260.
16. Crivello, J. V.; Narayan, R. *Chem. Mater.* **1992,** *4,* 692 – 699.
17. Metzger, J. O. *Chemosphere* 2001, *43,* 83-87.
18. Curzons, A. D.; Constable, D. J. C.; Mortimer, D. N.; Cunningham, V. L. *Green Chem.* **2001,** *3,* 1-6.
19. Patel, M.; Bastioli, C.; Marini, L.; Würdinger, E.: *Life-cycle assessment of bio-based polymers and natural fibres.* Chapter in the encyclopaedia *"Biopolymers",* Vol. 10, Wiley-VCH, 2003.
20. Fayter, R. G. in *Perspektiven nachwachsender Rohstoffe in der Chemie;* Eierdanz, H., Ed.; VCH, Weinheim, 1996, pp.107 – 117.
21. Craft, D. L.; Madduri, K. M.; Eshoo, M.; Wilson, C. R. *Appl. Environ. Microbiol.* **2003,** *69,* 5983-91.
22. Draths, K. M.; Frost, J. W. In *Green Chemistry, Frontiers in Benign Chemical Syntheses and Processes;* Anastas, P. T.; Williamson, T. C., Eds.; Oxford University Press, Oxford **1998,** pp. 150–177.
23. Janssens, I. R. J.; Vernooij, W. P. *Inform* **2001,** *12,* 468 – 477.
24. Biermann, U.; Eren, T.; Metzger, J. O. in *Book of Abstracts,* 25th World Congress and Exhibition of the ISF, *12–15 October 2003, Bordeaux, France.*
25. Fürmeier, S; Metzger, J. O. *Eur. J. Org. Chem.,* **2003,** 649-659
26.. http://www.carmen-ev.de/dt/industrie/projekte/rapsasphalt.html
27.. http://www.cooperpower.com/library/pdf/98077.pdf
28. Biomass Research and Development Technical Advisory Comitee, *Vision for Bioenergy & Biobased Products in The United States,* , October 2002.
29. Wackernagel, M.; Schulz, N. B.; Deumling, D.; Linares, A. C.; Jenkins, M.; Kapos, V.; Monfreda, C.; Loh, J.; Myers, N.; Norgaard, R.; Randers, J. *Proc. Natl. Acad. Sci. USA* **2002,** *99,* 9266-9271.
30.. Bongaarts, J.; Bulatao, R. A., *Beyond Six Billion: Forecasting the World's Population* , Panel on Population Projections, Committee on Population, National Research Council, Washington, D.C., 2000; http://www.nationalacademies.org.
31. *UN Program to Combat Desertification,* http://www.unccd.int/main.php ; *The sixth session of the Conference of the Parties (COP) of the United Nations Convention to Combat Desertification,* Havana, Cuba, 25 August - 5 September 2003; http://www.unccd.int/cop/cop6/menu.php
32 Lal, R. *Science* **2004,** *304,* 1623-1627.
33. Hüttermann, A.; Zommorodi, V; Reise, K. *Soil Tillage Res.* 1999, *50,* 295 – 304.

34. Ma, C H. Nelles-Schwelm, C.E. *Application of Hydrogel for Vegetation Recovery in Dry-Hot Valley of Yangtze*, Yunnan Science and Technology Press, Kunming, **2004.**
35. *The Carbon Dioxide Dilemma, Promising Technologies and Policies,* National Academies Press, Washington, D.C., **2003**; http://www.nap.edu
36.. Hüttermann, A.; Metzger, J. O., *Nachr. Chem.* **2004,** *52,* 1133 – 1136.

Chapter 3

Starting at the Front End: Processes for *New* Renewables as Feedstocks of the Future

Detlef Wilke[1,2] and Yuri Gleba[1]

[1]Icon Genetics AG, Munich, Germany
[2]Dr. Wilke & Partner Biotech Consulting GmbH, Wennigsen, Germany

The need for additional "new renewables" and the potential contribution of molecular plant genetics to the breeding of new industrial crops producing such novel plant ingredients are being discussed. Efforts to add new chemical candidates derived from existing plant-based feedstocks have so far met with limited success. It therefore seems necessary not only to invest in new derivatization technologies utilizing existing plant-based raw materials, but also to genetically engineer crops capable of producing new renewables. Such renewables should be inexpensive, convertible into bulk derivatives and chemical intermediates, and the conversion processes, or at least their early generations, should be compatible with the established infrastructure of the chemical industry. Icon Genetics AG is developing 'switchable amplification' production platforms which overcome the problems of limited yield and biotoxicity of recombinant proteins and enzymes made in plants. These expression processes, magnICON[TM] and rubICON Seed[TM], are also a promising way for metabolic engineering of new compounds in vegetative plant biomass or in seed. Both systems allow for separation of growth/biomass formation and plant propagation, on one side, and product formation, on the other. Ingredient synthesis can therefore, at least theoretically, reach its biological maximum. Host plant species can be chosen from among non-food crops that have a necessary pool of biochemical precursors for overproduction of the compound of interest.

Renewables – Present Status

Physical, chemical and biochemical conversion of plant-based oligo- and polymers into industrial bulk products as well as into specialties is a well established technology, which favourably extends the product spectrum obtained through petrochemical synthesis. Volumewise, less than 50 million tons p.a. of chemical intermediates and end-products are derived from renewable plant-based raw materials and constitute less than 15% of about 400 million tons provided by petrochemistry. This imbalance is not so much a result of a relative shortage or lack of availability of renewables compared to petrochemical feedstocks, rather, it reflects the versatility of ethylene, propylene and olefin based chemical synthesis in meeting the product and product range requirements of an industrialized world.

Carbohydrates

Roughly 80 % of the intermediates and end-products based on renewables use carbohydrate feedstocks of agricultural and forestral origin: starch, enzymatic starch hydrolysates (oligosaccharides and glucose), sugar (sucrose) and cellulose. These carbohydrates are converted through physical processing, chemical derivatization, and biochemical/ biotechnological conversion steps into specialty and bulk compounds. These polymers or small molecules conserve their distinctive property, the oxygenated carbon moiety. Whether organic acids, bio-ethanol, modified starch, cellulose or various speciality intermediates, these bio-based compounds compete with petrochemical alternatives in terms of cost performance and functionality, which in part are functions of its unique chemical structure like chirality or other kinds of stereospecificity.

Oleochemicals

The high stereospecific oxygen content of carbohydrate feedstocks versus petrochemicals represents a competitive advantage in certain market segments, but it restricts the use of these cheap and abundant plant-based feedstocks for other applications. The restriction is twofold: not only are most synthetic polymers like polyethylene and polypropylene structurally quite dissimilar to carbohydrates and derivatives, but in addition, such highly oxidized biochemicals do not properly fit into the processing routes established by the chemical industry for olefin feedstocks. Vegetable oils are structurally closest plant molecules to fossil hydrocarbons and are used as feedstocks for the production of long chain fatty alcohols and bio-diesel, but outside of such specialized market segments, they still cannot compete with petrochemicals.

Rationale for an Increased Use of Bio-Based Products

Though bio-based products are a reality, their overall contribution to the chemical marketplace is modest, from complete absence in some segments, to only 10 – 20 % share in others, with one notable exception being an outstanding share of 50 % in the EU surfactants market (1). There are also a few industrial markets like the pulp & paper industry which are dominated by renewables, but these markets are typically not considered as constituents of the chemical industry at all.

Whereas present day production and consumption of renewables is mainly based on cost performance and unique functionalities, other parameters such as biocompatibility, biodegradability and sustainability are less important. Particularly, the shrinking worldwide crude oil reserves have never been taken as a serious issue by the chemical industry, because the bulk consumption of fossil feedstocks by energy generation and transportation industries gives the impression that even with depleting crude oil reserves, there is still enough fossil feedstock for a secure supply of the chemical industry's demands. Rather, the chemical industry was occasionally evaluating renewables as cost competitive alternatives to olefins when crude oil prices were rising. However, there is an additional motive nowadays to consider an increased usage of carbon derived from agricultural production, namely, as a strategic tool to net contribution to CO_2 emission reduction. According to some estimates, "white biotechnology", i. e. industrial biotechnology converting agricultural feedstocks into intermediates and end-products, may finally account up to 20 % of the savings of greenhouse gas emissions under the Kyoto protocol (2). Physico-chemical processing and chemical conversion of renewables is further contributing to this task. It is a matter of fact that plants and photosynthetic microbes are the only biological and industrially applicable systems which can remove CO_2 from the atmosphere through the CO_2 fixation. Worldwide CO_2 production from fossil carbon sources exceeds 20 billion tons p.a. The significance of agricultural and forestral CO_2 sequestration from the atmosphere and its physical capacity limitations, photosynthetically productive terrestrial surface, is illustrated by recent ecological overshoot models (3).

Providing that energy saving conversion and processing routes are in place, carbon feedstocks derived from annual crops, when converted into short-lived industrial and consumer products, are emission neutral, and when converted into durable products, can even lead to a net CO_2 saving (sequestration). Therefore, governments and governmental agencies get more and more arguments to foster the industrial use of renewables and products derived from agricultural feedstocks. USDA´s proposed Federal Purchase Regulations favoring bio-based products is a good example.

Engineering New Industrial Crops

Need for New Plant-Based Feedstocks

If plant based renewables in the future are to play a more significant role as feedstocks for industrial and consumer products, several issues must be addressed:

further agronomic improvements of existing food and industrial crops used as feedstocks for non-food market ends;

further improvement of physical, chemical and biotechnological conversion processes to extend the product spectrum similar to the versatility of present day petrochemistry;

creation of novel plant ingredients which better fit the intermediate and end-product requirements of the chemical and other processing industries.

During the past 100 years, tremendous progress has been made in yield improvement (per hectare productivity) of major food plants like corn, cereals, sugar cane and soybean, which all feed both nutritional as well as non-food markets, and also in the creation of new crops and crops with new ingredients, such as sugar beet and canola (rapeseed). This progress resulted from a combined continuing effort of agricultural technology, agrochemistry (fertilizers and pesticides), as well as plant breeding. However, there is a general belief that the yield improvement potential of the first two contributors (technology and agrochemicals) is already past its peak, and that the improvement by conventional plant breeding, although made faster and more effective (through introduction of modern methods such as use of molecular markers) is reaching its biological limits.

Carbohydrate conversion into new industrial intermediates and end-products like propanediol, methane or syngas needs either net energy input ("reduction equivalents") or decarboxylation, which results in a significant and costly yield loss and lowers the CO_2 fixation net gain. This balance loss is also a conceptual draw-back of present fuel ethanol fermentation from sucrose and starch hydrolysates. So far, the only bulk intermediates with no or minimal carbon loss are fermentative lactic acid and the oleochemicals (with its compulsory by-product glycerol).

Conventional plant breeding and first-generation plant genetic engineering does not easily lead to new industrial crops with ingredients which may be more suitable feedstocks for chemical processing. The established breeding strategies depend on full biocompatibility of the target compound with plant growth, fertility and seed germination. Breeding of new industrial crops ("designer crops") is therefore a compromise between ingredient production and healthy plant growth and development, and typically the content of novel plant ingredients in such engineered varieties is rather low, resulting in poor production economics (hectare yields of crop harvest, and weight per weight ingredient content of the harvested crop). Most ingredients designed to better fit existing chemical conversion routes are expected to exhibit a physiological incompatibility or even toxicity when present in a cell at economically feasible concentrations, and would therefore interfere with plant development, biomass formation and seed set/viability.

A remarkable peculiarity of today's plant biotechnology industry developing genetically modified plants is the absence of potential "blockbuster" products in R&D pipelines in the 'output traits' category. Except perhaps for omega-3 fatty acids, all other traits either have too small potential markets to appeal to large companies, or their development has been met with serious technical challenges. Examples of the latter type include various modified oils, poly-ß-hydroxybutyrate (biodegradable plastics), collagen, and silk protein, which may be considered as bulk product candidates, and proteins like phytase, human serum albumin etc., which in terms of the chemical industry are clear specialty products. In most of these cases, expression of the trait of interest is not high enough to allow for a competitive production process. Moreover, in those cases where high level of expression can be achieved, agronomic performance/yield of the crop is also severely affected. The conflict between high level of expression of a gene of interest, on one side, and growth and development of the plant expressing this gene, on the other side, is a well-known phenomenon. The problem can be overcome by designing new generation transgenics in which the 'growth phase' is separated from the 'production phase', the technology route typically employed by microbial secondary metabolite fermentation. In plants, some native or experimental proteins and molecules are being produced at very high levels, with up to 35% of total protein in a leaf (large subunit of ribulose-1,5-bisphosphate carboxylase/oxygenase), and up to 35% of total protein in a seed (storage proteins), and there are all reasons to believe that the same should be possible with commercially useful products. Such high productivity will however most likely require more sophisticated expression controls than just using a strong constitutive promoter, as is the case with the first generation transgenic crops today.

Icon Genetics has developed new Transgene Operating Systems™ which address several of these issues. Our primary focus so far was on the safe, high-yield production of recombinant proteins and enzymes. However, the new expression systems shall also be applicable for engineering plants producing small molecules through co-expression of several genes which control (initially short) metabolic pathways of up to five additional enzymes, thus yielding transgenic plant hosts and processes for new metabolites. In principle, industrial crops with the existing precursor pool of primary and secondary metabolites and its derivatives, whether monomers or polymers, may all be modified through such pathway engineering.

Transient Gene Expression in Leaf Tissue

magnICON™ is a new generation expression platform based on activation of proviral amplicons in plants (*4,5*). The plant host is non-transgenic, and the foreign genetic information is delivered to the plant cell as a T-DNA vector by *Agrobacterium tumefaciens* transfection. All genetic engineering is performed in *E. coli* and transferred to *Agrobacterium* which is then sprayed on or infiltrated into the leaves of mature host plants. Preferred model plants are two tobacco species, *Nicotiana tabacum* (commercial tobacco) or *N. benthamiana*, the crop plants that can yield up to 100 tons of leaf biomass per hectare outdoors, and significantly more in a greenhouse. Upon *Agrobacterium* transfection ("infection") of the leaf tissue, the DNA pro-vectors are converted into RNA amplicons that spread throughout the whole plant by cell-to-cell or systemic movement, and the transgene(s) of interest is transcribed using a viral RNA-dependent RNA polymerase and translated into a functional protein by the plant cell machinery. This results in a fast reprogramming of the whole plant's protein biosynthesis in favor of the expression of the recombinant protein(s) or enzyme(s) of interest. The whole process, from *Agrobacterium* delivery to the harvest of the expression products, takes no longer than 5 to 12 days. This transient expression system is fully compatible with specialty protein production on a greenhouse scale, but obviously it is less appropriate for large scale bulk production. Due to its easiness and speed, however, magnICON™ is an ideal 'desktop' platform for optimization of genetic constructs as well as for functional screening of large numbers of enzymes derived from *in vitro* mutagenesis or directed evolution, and for rapid engineering of new metabolic pathways. Since up to 5 bacteria all carrying different DNA vectors can be directed to the same leaf cell, multiple genes and gene constructs, respectively, can be administered and tested in the same experiment.

Separation of Growth and Production in Transgenic Plants Using Switches

A system that gives a reliable control over transgene expression requires at least two or more components. Depending on the configuration of the components, the system acquires the possibility to be in two states: 'on' and 'off'. These configurations have been designed and implemented through the use of molecular genetics tools, and proper engineering of their interaction as a part of actual plant development or life cycle. Switching can be achieved in a variety of ways: by hybridization of two plants (bringing the two genetic components together in the hybrid cell) or by delivery of a message/component from outside the cell (a small molecule activator for a chemical switch; or a virus/bacteria that delivers a more complicated effector molecule to plant cells). As a transient expression system magnICON^{TM} is particularly suited for the synthesis of enzymes which as such, or through their metabolic reaction products, have a negative impact on plant growth and propagation. Based on the *Agrobacterium* transfection of mature plants and its rapid harvest a few days later, biomass and product formation are entirely separated, and even toxic enzymes and metabolites, respectively, can be synthesized without interfering with plant's optimal growth and progeny seed formation. This advantage can be preserved in a future scale-up version of magnICON^{TM}, in which the proviral amplicon is stably integrated into the transgenic host, but placed under the control of a strictly regulated "non-leaky" inductor. We are presently designing such an inducible system which would allow for growth and seed production of the transgenic crop because of a "zero level" uninduced expression level, but would switch on the gene expression and product synthesis upon administration of a small molecule chemical inductor ("switch") in the production field shortly before harvesting the crop.

Such a stably transformed inducible system based on the magnICON^{TM} amplification cassette can virtually be designed for any crop variety. It is particuarly suited for those new plant ingredients which are best synthezised in vegetative plant tissue like leaves, stems, tubers, and roots. Compared to product synthesis in seeds, a production in vegetative tissue has a clear advantage, since hectare yields of harvested biomass can easily reach 100 ton per hectare and above and are therefore up to hundred times higher than those of seed harvests, which range from 1.5 to 8 tons per hectare depending on the respective crop variety. This hectare yield advantage of fresh biomass may compensate for eventual lower contents in kg of compound per ton of harvested crop as well as for the general advantage of seeds being a storable material.

Onset of Product Synthesis in Hybrid Seeds Only

Production of new ingredients in seeds avoids campaign operation and is therefore more cost economic with respect to capital investment and labor force. When using cross-pollinating species as production hosts, one can stably transform the female line with an inactive magnICON™ amplification cassette, which is then activated by a genetic switch (for example by a site-specific recombinase that "flips" part of the amplicon DNA thus rendering it operational) supplied by the pollen of the male parent. Both male and female parent lines can be grown and propagated separately without any functional transgene expression and ingredient synthesis, respectively, and the product is formed exclusively in hybrid seeds after the activation of the magnICON™ amplification cassette. Various genetic switches could be employed for such an approach, like transcriptional trans-activators encoded by the genome of the male parent (6). Another approach is a gene 'fragmentation' followed by genetic complementation in a hybrid, such as our rubICON™ technology (7); in the case of our platform, the two complementing gene fragments are provided as two gene constructs incorporated in isoallelic positions on two homologous chromosomes residing in two parent lines. Functionally active gene product is then formed in hybrid seeds either on RNA level through ribozyme-mediated trans-splicing or on protein level by intein-based trans-splicing.

For the majority of crops which are not strict cross-pollinators, hybrid seed formation itself is a technical problem. We have developed a universal hybrid seed system based on the rubICON™ gene fragmentation technology which allows for the breeding of male sterile female lines as fertile crossing partners (8). This hybridization system can easily be combined with a switchable gene expression onset in cross-pollinated seeds.

New Plant Ingredient Candidates

A transgenic expression system targeting biosynthetic processes in seeds looks particularly attractive for agricultural production of new starches and oils, which are typically accumulated in seeds as storage material used during subsequent seed germination. If such an expression system is restricted to the first generation hybrid seeds (F_1) and allows for metabolic engineering without any impact on plant growth, fertility and seed viability of the parental lines, it may be used to make modified and entirely new carbohydrate polymers, as well as modified oils and other acetyl-CoA derivatives for which seeds are the most appropriate biosynthetic plant organs.

By contrast, overproduction of new low molecular compounds like sugars and derivatives secreted into the plant vacuole, and overproduction of technical enzymes, protein based polymers and other performance proteins may be more suitable in leaves, stems, tubers, and roots, which biomass can be obtained at much higher yields per hectare.

So far, efforts to add further product candidates with bulk potential derived from cheap and abundant plant-based feedstocks have met with limited success. From the point of view of genetics and physiology, three kinds of new feedstocks should be considered: yield-improved existing or modified plant ingredients such as carbohydrates and oils; new ingredients in a given crop plant but of plant kingdom origin like indigo or long chain microalgal hydrocarbons; and new ingredients of microbial genetic origin, such as poly-ß-hydroxyalkanoates (PHAs).

Many yield improvement approaches involve metabolic engineering steps that are devoted to existing plant specialty ingredients like vitamin E (9,10), carotenoids (11,12) and flavors and aromas (13). Pathway engineering thereby may involve both plant enzymes as well as prokaryotic and eukaryotic microbial enzymes. These nutritional and dietetic specialties have no bulk feedstock potential, but improvements in terpenoid and isoprenoid biosynthesis may pave the way for the future production of industrial polyisoprenoids like latex and rubber in crops (14).

Crops improved in their carbohydrate metabolism are probably closest to market introduction, with the transgenic potato producing high-amylopectin starch for the pulp and paper industry representing the most advanced case (15,16). Modification of central metabolism resulted in starch yield increase of potato, and similar improvements have been accomplished in other carbohydrate accumulating crops like cerials (17,18).

Microalgal long-chain fatty acids and alcanes like those derived from Botryococcus braunii (19), when produced in annual field crops, may extend the established oleochemistry for market ends beyond surfactants, lubricants and bio-diesel. Although a cloning strategy for pathway engineering starting from acetyl-CoA up to long chain hydrocarbons looks very straightforward and theoretically involves no more than three or four genes/enzymes, practical experience suggests precaution. Metabolic engineering of vegetable oils with the aim of changing the fatty acid composition turned out to be more difficult than expected (20).

Microalgal hydrocarbons are an example where molecular genetic improvements and transgenic expression can also be tested in the natural host or its close phylogenetic relatives (21). Each feedstock production and conversion system, whether based on plants, microbes, or chemistry, has its pros and cons, and there is no one universal platform for foreseeable future (22). Host-compound choices must therefore be taken on a case-by-case basis. Poly-ß-hydroxyalkanoates as biodegradable polyesters are bacterial storage polymers and can be produced by fermentation on cheap starch hydrolysate with high volume and biomass yield (23). Nevertheless, commercialization of the compound as a bulk polymer so far has not been accomplished because the overall production costs and process steps such as isolation and purification of the bacterial intracellular PHA granules, have still to be improved.

On the other hand, plants are established bulk polymer sources with robust, large scale processing technology in place, and agricultural production of commodity products is considered to provide a clear cost advantage compared to any submersed cultivation whether it is an algal cultivation in open ponds or a microbial fermentation in stainless steel tanks. First poly-ß-hydroxybutyrate production in the transgenic model plant *Arabidopsis thaliana* was achieved in 1992 (*24*). Productivity was low (up to 0.01 % of leaf fresh weight), and plant growth and seed production were reduced when higher amounts of the biosynthetic enzymes had accumulated. Targeting of the biosynthetic enzymes from cytoplasm to the chloroplasts considerably increased productivity (up to 14 % of leaf dry weight, equivalent to approximately 1.4 % of fresh weight – see *25*). Growth and fertility of the transgenic plants were unaffected. Others could further increase the poly-ß-hydroxybutyrate content in *Arabidopsis* leaves to more than 4 % of fresh weight, equivalent to approximately 40 % of leaf dry weight, but on the expense of stunted growth and a loss of fertility of the high-producing transgenic lines (*26*). Monsanto Co. (*27*) has estimated the break-through productivity for economic production of PHAs in plants to be at 15 % dry weight, but could not reach more than 13 % content of dry mass in homozygous *Arabidopsis* plants (leaf biomass), and 7 % of seed weight in seeds of heterozygous canola plants (*Brassica*). Expression in leaf chloroplasts of the fodder plant alfalfa (*Medicago sativa*) gave very low yields (up to 0.18 % of leaf dry mass), and there was no deleterious effect on growth and fertility (*28*). Clearly, we are not there yet.

Conclusions

If the share of renewables used by the chemical industry is to significantly increase in the future, it may become necessary not only to invest in new conversion and derivatization technologies for existing plant-based raw materials but also to engineer *new* renewables. We are convinced that molecular plant genetics will play an important role in accomplishing this task by creating new designer crops and by providing practical solutions to such parameters as synthesis of desired precursor, precursor yields, biomass content and agronomic characteristics of the designer crops. We have implemented transient expression systems as fast development tools for testing multi-component genetic constructs in biochemical pathway engineering, which can shorten typical experimental protocols based on stable chromosomal transformation of host plants, from typically one to three years, down to a few weeks.

Transient expression and its transgenic counterpart, switchable gene expression in vegetative biomass or in hybrid seeds, allow separation of plant biomass formation from product synthesis. This platform could be particularly attractive for manufacturing new plant ingredients with non-physiological properties or interfering with proper plant growth and propagation, as has been recognized for high-yield expression of PHAs and certain oils in transgenic plants.

Hectare yields of vegetative plant biomass are much higher than harvests of seeds. With exception of the three major grain crops (corn, wheat and rice) that have a 3 to 8 tons per hectare average yield, most other seed crops yield between 1.5 and 3 tons seed per hectare. Yields of leaf of vegetative biomass, on the other hand, can easily reach 100 tons per hectare. Depending on the biochemical precursor availability in different tissues, development strategies may therefore focus on the most economic production route taking into account both the plant physiology and the agronomic parameters. For safety reasons, preference may be given to non-food crops, unless other measures can be implemented which secure a strict segregation of the crop streams dedicated to food, on one side, and industrial market ends, on the other.

So far, a complex metabolic engineering in plants is still a vision, and cloning of the genes which code for the essential enzymatic steps of an engineered metabolic pathway may not be sufficient for engineering high-yield varieties. Successful metabolic engineering of new designer crops may therefore need a lot more testing of simultaneously expressed genes of isozymes or even biochemically unrelated regulatory genetic elements. In case of bacterial PHA synthesis, for instance, a special protein, phasin coded by the *phaP* gene, covers the surface of the PHA granules (*29*). The presence of phasin in transgenic plant hosts overexpressing the three enzymes necessary for PHA production may or may not be advantageous, but complex genetic set-ups as revealed from genomics can be taken as a hint that successful metabolic engineering may involve more empirical efforts than a simple transfer of just the biochemical pathway into a plant.

New plant expression systems leading to new functioning pathways in crop varieties are essential elements of these contemplated novel industrial processes, but there is also a need for concerted action of all players along the commercialization chain. Conventional plant breeding and agronomic optimization are important, as is recovery and purification process engineering. In order to become an accepted new feedstock, additional enzymatic conversion technologies as well as chemical derivatization and polymerization processes, may need to be developed. Examples of entirely new bio-based end-products such as polylactic acid teach us how important it is to start application and market development work as early as possible. The overall process from design to product can easily take 20 years and more.

Ideally, the combined efforts should lead to new industrial crops that produce a single bulk ingredient (plus value adding by-products) at a market price between $200 and $700 per ton. Starch, vegetable oils and sucrose are all the commodity products that fall in this cost range, and they successfully compete in certain segments of the chemical business with petrochemical derivatives based on crude oil prices of $20 to 30 per barrel. At a crude oil price of $50 and above (something we already experienced, albeit temporarily), the manufacturing costs of the renewables will also go up, but to a lesser extent than petrochemical cost, and competitiveness of a plant-derived feedstock will improve. Due to the long development times necessary for engineering optimized designer crops, in the meantime, biorefinery concept may be exploited (30). The new plant ingredient, still low in absolute yield, may be produced as an add-on component, whereas the basic production economics will rely on the carbohydrate, oil, protein, fiber, and energy value of the refined crop, until the industrial learning curve has been completed. It took the (petro)chemical industry half a century to establish its paramount position, and a bio-based processing industry may need as much time (31).

References

1. Ehrenberg, J. Current situation and future prospects of EU industry using renewable raw materials 2002, http://europa.eu.int/comm/enterprise/environment/reports_studies.htw
2. Van Arnum, P. *Chemical Market Reporter* **2003**, *June 16, FR 6*
3. Wackernagel, M.; Schulz, N.B.; Deumling, D.; Callejas Linares, A.; Jenkins, M.; Kapos, V.; Monfreda, C.; Loh, J.; Myers, N.; Norgaard, R.; Randers, J. *PNAS* **2002**, *99*, 9266-9271.
4. Gleba, Y., Marillonnet, S. & Klimyuk, V. *Curr. Opin. Plant Biol.* **2004**, 7, 182-8.
5. Marillonnet, S., Giritch, A., Gils, M., Kandzia, R., Klimyuk, V., Gleba, Y. *PNAS* **2004**, *101*, 6852-7.
6. Mascia, P.N., Flavell, R.B. *Curr. Opin. Plant Biol.* **2004**, 7, 189-195.
7. Gleba, Y. In: Genomics for Biosafety and Plant Biotechnology, J. Nap et al., eds., pp.77-89, IOS Press, 2004.
8. DE 103 25 814.0, A process for high yield and safe production in hybrid seeds. June 6, **2003**, to Icon Genetics AG
9. Cahoon, E.B., Hall, S.E., Ripp, K.G., Ganzke, T.S., Hitz, W.D., Coughlan, S.J. *Nat. Biotechnol.* **2003**, *21*, 1082-1087.
10. Ajjawi, I, Shintani, D. *Trends Biotechnol.* **2004**, 22, *104-107*.
11. Ye, X., Al-Babili, S., Kloti, A., Zhang, J., Lucca, P., Beyer, P., Potrykus, I. *Science* **2000**, *287*, 303-305.
12. Mann, V., Harker, M., Pecker, I., Hirschberg, J. *Nat. Biotechnol.* **2000**, 18, *888-892*.

13. Lewinsohn, E., Schalechet, F., Wilkinson, J., Matsui, K., Tadmor, Y., Nam, K.H., Amar, O., Lastochkin, E., Larkov, O., Ravid, U., Hiatt, W., Gepstein, S., Pichersky, E. *Plant Physiology* **2001**, *127*, 1256-1265.
14. Mooibroek, H., Cornish, K. *Appl. Microbiol. Biotechnol.* **2000**, *53*, 355-365.
15. Kuipers; A.G., Soppe, W.J., Jacobsen, E., Visser, R.G. *Plant Mol. Biol.* **1994**, *26*, 1759-1773.
16. Lloyd, J.R., Landschütze, V., Kossmann, J. *Biochem. J.* **1999**, *338*, 515-521.
17. Regierer, B., Fernie, A.R., Springer, F., Perez-Melis, A., Leisse, A., Koehl, K., Willmitzer, L., Geigenberger, P., Kossmann, J. *Nat. Biotechnol.* **2002**, *20*, 1256-1260.
18. Smidansky, E.D., Clancy, M., Meyer, F.D., Lanning, S.P., Blake, N.K., Talbert, L.E., Giroux, M.J. *PNAS* **2002**, *99*, 1724-1729.
19. Banerjee, A., Sharma, R., Chisti, Y., Banerjee, U.C. *Botryococcus braunii*: *Crit. Rev. Biotechnol.* **2002**, *22*, 245-279.
20. Drexler, H., Spiekermann, P., Meyer, A., Domergue, F., Zank, T., Sperling, P. Abbadi, A., Heinz, E. *J. Plant Physiol.* **2003**, *160*, 779-802.
21. León-Banares, R., González-Ballester, D., Galván, A., Fernández, E. *Trends Biotechnol.* **2004**, *22*, 45-52.
22. Wilke, D. *Appl. Microbiol. Biotechnol.* **1999**, *52*, 135-145.
23. Braunegg, G., Lefebvre, G., Genser, K.F. *J. Biotechnol.* **1998**, *65*, 127-161.
24. Poirier, Y., Dennis, D.E., Klomparens, K., Somerville, C. *Science* **1992**, *256*, 520-523.
25. Nawrath, C., Poirier, Y., Somerville, C. *PNAS* **1994**, *91*, 12760-12764.
26. Bohmert, K., Balbo, I., Kopka, J., Mittendorf, V., Nawrath, C., Poirier, Y., Tischendorf, G., Trethewey, R.N., Willmitzer, L. *Planta* **2000**, *211*, 841-845.
27. Valentin, H.E., Broyles, D.L., Casagrande, L.A., Colburn, S.M., Creely, W.L., DeLaquil, P.A., Felton, H.M., Gonzalez, K.A., Houmiel, K.L., Lutke, K., mahadeo, D.A., Mitsky, T.A., Padgette, S.R., Reiser, S.E., Slater, S., Stark, D.M., Stock, R.T., Stone, D.A., Taylor, N.B., Thorne, G.M., Tran, M., Gruys, K.J. *Int. J. Biol. Macromol.* **1999**, *25*,303-306.
28. Saruul, P., Srienc, F., Somers, D.A., Samac, D.A. *Crop Science* **2002**, *42*, 919-927.
29. Pötter, M., Madkour, M.H., Mayer, F., Steinbüchel, A. *Microbiology* **2002**, *148*, 2413-2426.
30. Ohara, H. *Appl. Microbiol. Biotechnol.* **2003**, *62*, 474-477.
31. Wilke, D. *FEMS Microbiol. Rev.* **1995**, *16*, 89-100.

Chapter 4

Biomass Derivatives: A Sustainable Source of Chemicals

Leo E. Manzer

DuPont Central R&D, Experimental Station, Wilmington, DE 19880–0262

Ongoing use of petroleum for transportation fuels, chemical feedstocks and energy is not sustainable. It is imperative that new catalysts for existing processes and new products are developed based on renewable sources, to reduce US dependence on foreign oil. The US is well suited to be a major player in this area due to a large supply of biomass. DOE has identified 12 chemicals that can be produced from sugars as key feedstocks for future biorefineries. Catalytic transformation of these future feedstocks provides new market opportunities. Levulinic acid is one of DOE's top 12 bioderived feedstocks and it can be obtained directly by cellulose hydrolysis. A number of high value derivatives such as pyrrolidones, monomers for the synthesis of high Tg amorphous polymers and fuel additives can be prepared. This paper summarizes some of our early results.

Introduction

The worldwide demand for petroleum continues to rise at a rapid pace while supplies are finite. On a global basis (*1*) the US produces only 2% of the world petroleum products while consuming about 26% of global production (Table I). As developing nations increase their needs for hydrocarbon products (primarily transportation fuels), supplies will become even tighter. It is therefore important that the US develop alternative forms of energy and feedstocks based on renewable resources.

Table I. 2003 Oil Producing and Consuming Nations (*1*)

Oil Producers (% of World Production)		Oil Consumers (% of World Production)	
Saudi Arabia	26	USA	26
Iraq	11	Japan	7
Kuwait	10	China	6
Iran	9	Germany	4
UAE	8	Canada	4
Venezuela	6	Russia	3
Russia	5	Brazil	3
Libya	3	S. Korea	3
Mexico	3	France	3
China	3	India	3
Nigeria	2	Mexico	3
USA	2	Italy	2

The US is particularly well suited to develop independence on foreign oil from biomass due to the large resources as shown in Table II. The US has 40% or the world corn, 28% of the world oilseeds and 46% of the world soybean production yet only 7% of world petroleum production.

The US Department of Energy has recently issued a report (*2*) outlining the top 12 building block chemicals that can be produced from sugars via biological or chemical conversions. They are shown in Table III. Levulinic acid is one of the key building blocks and has been the focus of some of our research at DuPont. This paper will outline some of our research to illustrate the potential for new chemical transformations of these biomass building blocks.

Table II. 1999 US vs Global Production of Crude and Renewable Products

	Worldwide (billions lbs)	US (billions lbs)
Corn	1300	530
Raw Sugar	300	18
Oilseeds	650	185
Soybean	346	159
Crude Oil	6800	445

Table III. US DOE top 12 biomass derived chemical building blocks.

US DOE Future BioBuilding Blocks	
1,4-diacids (succinic, fumaric and malic)	itaconic acid
2,5-furan dicarboxylic acid	levulinic acid
3-hydroxy propionic acid	3-hydroxybutyrolactone
aspartic acid	glycerol
glucaric acid	sorbitol
glutamic acid	xylitol/arabinitol

Levulinic Acid Chemistry

Levulinic Acid Production

The acid catalysed hydrolysis of cellulose-containing biomass to levulinic acid (LA) has been known for a number of years but yields are low and tars are formed. However, by careful control of the temperature and pressure (*3*), Biofine Inc., has been able to develop a high yield process in a pilot plant. Commercial scaleup at several locations is in development (*4*). Although the literature contains numerous references to LA chemistry, the scope is limited. We now report a number of new reactions of levulinic acid.

Pyrrolidones from aryl and alkylamines

Pyrrolidones are widely used as solvents and surfactants. Their preparation is traditionally derived from the reaction of butyrolactone with alkyl- or aryl-amines. We have now discovered (*5*) that levulinic acid (or its esters), hydrogen and arylamines, in the presence of various catalysts will produce N-aryl(cycloalkyl)-5-methyl-2-pyrrolidones in very high yield under moderate pressures. Typical reaction temperatures are 150-225°C and pressures of about 500 psig. Depending on the choice of catalyst, an arylamine can produce either the N-aryl- or N-cycloalkyl- pyrrolidone. Pt-based catalysts are preferred to give the aryls while Rh is preferred to give the ring saturated derivates (Figure 1). Use of substituted arylamines provides the opportunity to produce a wide variety of cycloalkyl derivatives. Alkylamines provide direct routes to the N-alkyl pyrrolidones.

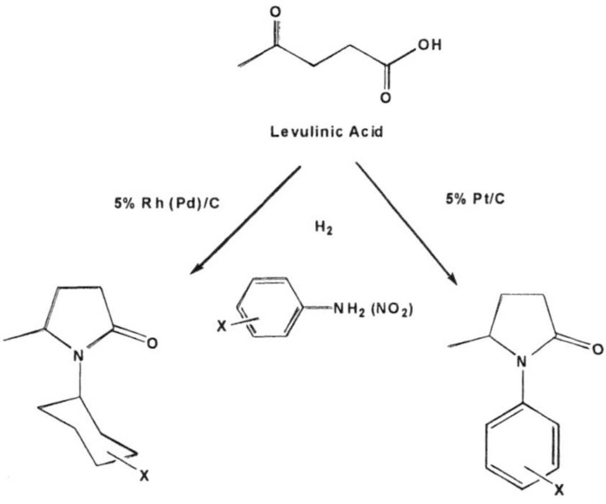

Figure 1. Preparation of Pyrrolidones from Arylamines

Pyrrolidones from nitroalkyls and aryls

A major cost in the production of existing pyrrolidones is the cost of the amine. Therefore, we moved a step back in the production chain and found (*6*) that alkyl and aryl-nitro compounds (precursors to the amine) can be readily converted to their corresponding pyrrolidones as shown in Figure 1 for nitrobenzene. Nitropropane, levulinic acid and hydrogen react in the liquid phase at 150°C, 500 psig, to give N-propylpropyl pyrrolidone in high yield.

Pyrrolidones from nitriles

Aryl and alkyl nitriles (*7*), in the presence of levulinic acid, hydrogen and a catalyst can be converted to N-alkyl pyrrolidones (Figure 2). Preferred catalysts for this reduction include Ir/SiO$_2$, Ru/Al$_2$O$_3$ and Pd/C. Excellent results are obtained with low cost nitriles such as 2-and 3-pentenenitrile which are intermediates or byproducts in the production of nylon intermediates (Invista) and should be available at very low cost. Aryl nitriles such as benzonitrile can also be used to prepare N-benzyl and N-cyclohexylmethyl pyrrolidones.

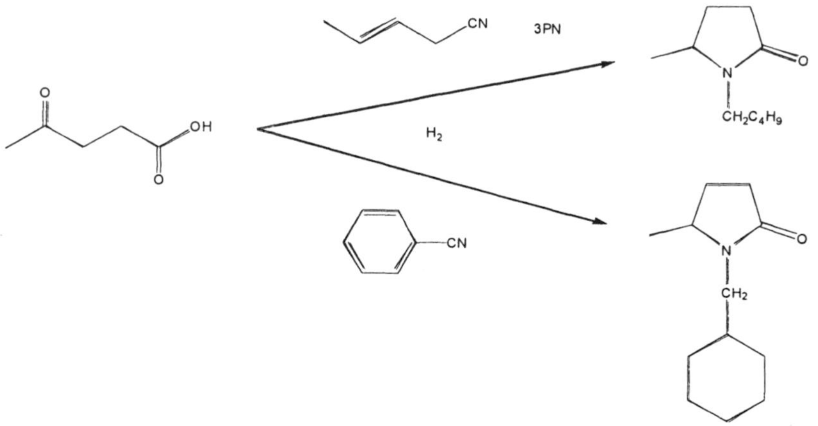

Figure 2. Preparation of Pyrrolidones from Nitriles

High Tg Acrylic Polymers

α-Methylene-γ-valerolactone (MeMBL) (Figure 3) is an acrylic monomer that has been known for a number of years (*8*) and has been of interest due to its similarity in structure to methylmethacrylate (MMA). By incorporating the lactone structure into the polymeric chain, the glass transition temperature of the homopolymer is over 100°C higher than that of poly-methylmethacrylate. However, a commercially attractive synthesis has not been available.

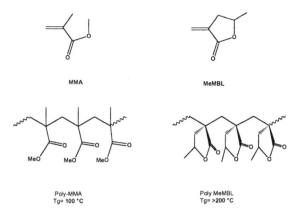

Figure 3. Methylmethacrylate comparison to α-methylene-γ-valerolactone

Synthesis of MeMBL

We report a two-step process (Figure 4) for its synthesis from levulinic acid. The first step (*9*) is the hydrogenation of levulinic acid or its esters to γ-valerolactone (GVL) in nearly quantitative yield using a Ru based catalyst at 150°C and 500 psig.

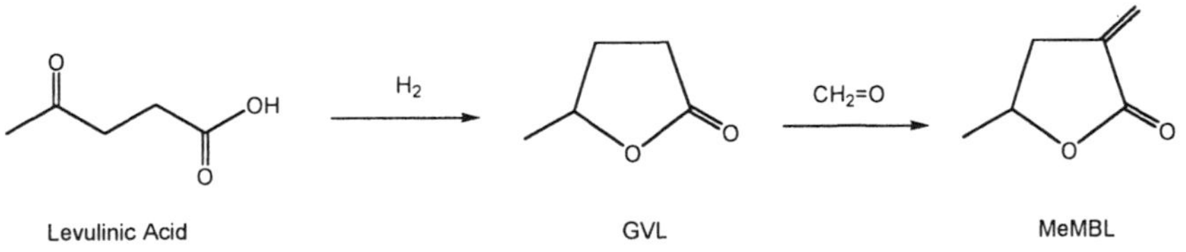

Levulinic Acid   H₂   GVL   CH₂=O   MeMBL

Figure 4. Synthetic route to MeMBL from levulinic acid

The second step involves a heterogeneous, gas phase catalytic condensation of formaldehyde with GVL over basic catalysts. Preferred catalysts are prepared by depositing Group 1 and 2 metal salts on silica (Figure 5). Catalysts prepared using Ba, Rb, and K are preferred. While the reaction is very selective to MeMBL, the process suffers from rapid catalyst deactivation. Initial conversion of GVL can be greater that 90% but within hours the conversion drops to 30-40%. Regeneration of the catalyst by heating in air at >450°C readily returns the catalyst to its initial activity. However, surprisingly, we found that regeneration can be accomplished under relatively mild conditions.

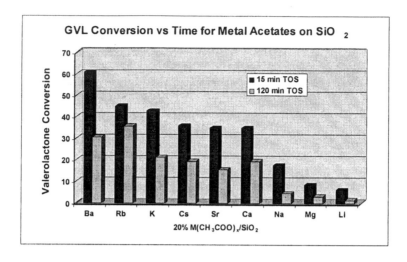

Figure 5. GVL conversion vs time on stream (TOS) for Group 1 and 2 acetates on SiO$_2$ at 340°C, using a formaldehyde to GVL ratio of 4:1

The graph in Figure 6 shows a 20+ hour run over a Ba-based catalyst. The conversion drops from >70% to less than 35% within 24 hours. The feed was stopped and nitrogen was passed over the catalyst at the reactor temperature of 250°C for two hours and then feeds were restarted. Interestingly, the catalyst was regenerated and deactivated at a similar rate. This result suggested that deactivation was not caused by coke (requiring a high temperature burn-off) but by a high boiling organic compound that was removed by flushing with hot nitrogen. This result was confirmed by microbalance reactor studies performed by the Barteau group (*10*). The low temperature regeneration which presumably removes higher molecular weight organics, suggested that if we could continuously remove these compounds during the run, longer lifetime might be obtained.

The solvating properties of supercritical fluids are well documented so we (Dr. Keith Hutchenson) ran the reaction in supercritical carbon dioxide and achieved the anticipated results shown in Figure 7. Conversion of GVL remained stable over a 5 hour run and selectivity to MeMBL was very high. Further optimization of the MeMBL synthesis in a supercritical media will be reported at a later date.

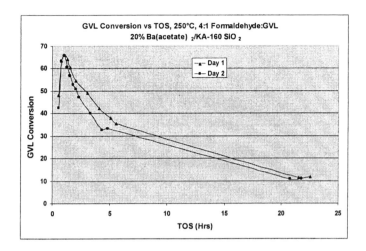

Figure 6. Effect of a 2 hour nitrogen flush on GVL conversion at 250°C following 22 hours of continuous operation

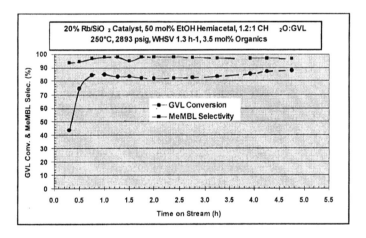

Figure 7. Conversion of GVL and Selectivity to MeMBL in a Supercritical CO₂ Reactor

Fuel Additives

Levulinic acid esters have been studied extensively (11) as fuel additives. Ethyllevulinate meets or exceeds the ASTM D-975 diesel fuel standard and numerous other qualifiers. It can be prepared directly from levulinic acid and ethanol with removal of water. Ester formation is equilibrium limited and therefore will require something like a catalytic distillation reactor to drive the reaction to completion. While researching this area we discovered an old patent (12) claiming the formation of α-angelicalactone (AL). AL was formed by heating levulinic acid to a temperature of between 150-175°C, and a pressure of 17-50 mm in the absence of a catalyst (Figure 8)

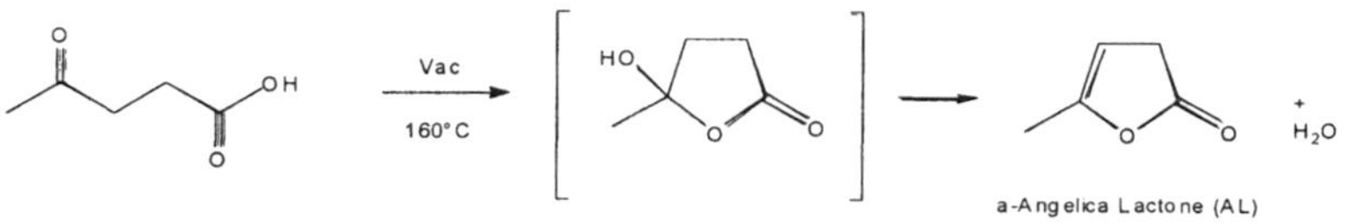

Figure 8. Preparation of a-Angelicalactone

We have discovered that AL is very reactive towards alcohols (Figure 9) and in the presence of either acid or basic catalysts high conversions of levulinate esters can be obtained (Table IV). It should be pointed out that these reactions are not optimized. Solid acid or basic catalysts are very effective. Choice of catalyst is very important as noted for the reaction of phenol with AL. Using Ba(acetate)$_2$/SiO$_2$, high conversion and selectivity to the phenylester are obtained while with a solid acid catalyst, under this one set of conditions, low yields were obtained.

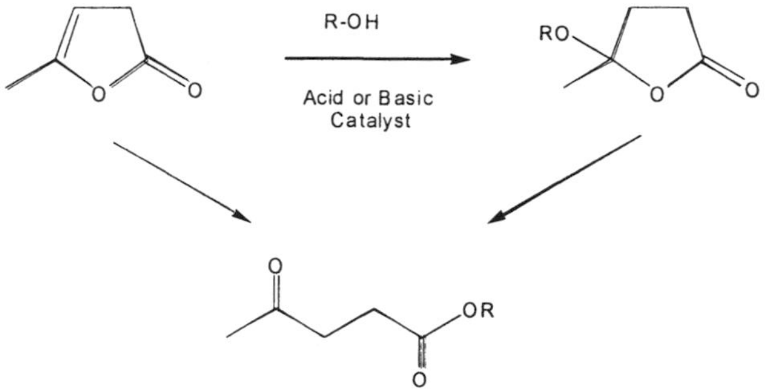

Figure 9. Levulinic Acid Esters from Angelicalactone

Table IV. Levulinate Esters from Angelicalactone

ROH	Catalyst	Time Hrs	Temp °C	AL Conv	Ester Sel
EtOH	Amberlyst-15	1	100	99.3	95.6
EtOH	20% Li(acetate)/SiO$_2$	1	150	97.3	71.9
n-BuOH	Amberlyst-15	1	150	99.8	98.6
CyOH	Amberlyst-15	1	25	85.0	87.8
MeOH	Na$_2$CO$_3$	3	100	97.5	82.4
Phenol	20% Ba(acetate)$_2$/SiO$_2$	1	150	92.2	71.0
Phenol	Amberlyst-15	1	150	100	<5

Esterification of α-Angelicalactone with olefins and water

We have also discovered another novel reaction. AL reacts with olefins and water in the presence of an acid catalyst to give alkyllevulinate esters (Figure 10). Mixtures of linear and branched products are obtained. Water is stoichiometrically required and therefore the major product, other than ester is levulinic acid. However LA can be readily recycled so overall yield can be quite high. As an example, 1-hexene reacts with AL and water at 120°C using Amberlyst 36 catalyst to give hexyllevulinate at 98.3% conversion and 83.1% selectivity to esters.

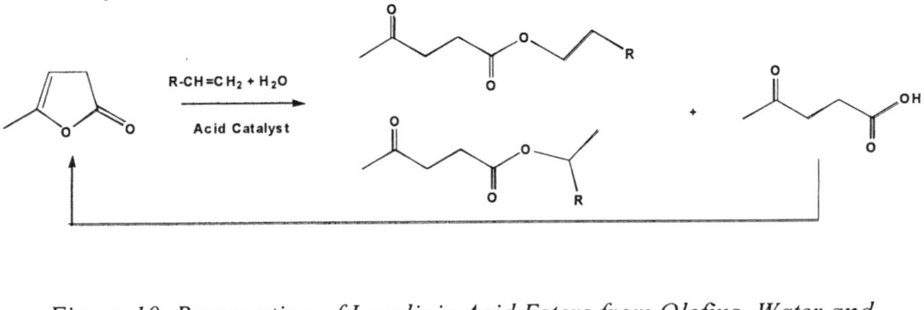

Figure 10. Preparation of Levulinic Acid Esters from Olefins, Water and Angelicalactone

Angelicalactone and formic acid

Production of levulinic acid from cellulose produces one mole of formic acid for every mole of levulinic acid produced on a stoichiometric basis. Large volume production of levulinic acid would yield large quantities of formic acid. We have found that AL and formic acid will react with olefins and water in the presence of acid catalysts to yield both formate and levulinate esters. All these products are attractive fuel additives.

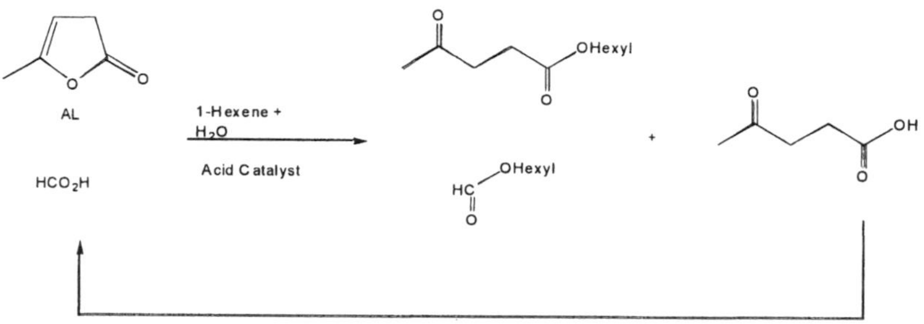

Figure 11. Preparation of Levulinic and Formic Acid Esters from Olefins, Water and Angelicalactone

Conclusions

The top 12 building block molecules from biomass, identified by the DOE study (2) present a significant new opportunity for catalytic conversion to a wide variety of specialty and commodity chemicals. Using levulinic acid as an example we have shown that rich new chemistry is possible. Pyrrolidones, reactive acrylic monomers and fuel additives are readily obtained.

Many of the platform or building block molecules are very highly functionalized, i. e., glucaric acid, so research is needed to learn how to selectively transform one or more of these functional groups into a useful molecule for society. Understanding and developing the rules for these new catalytic transformations presents a new and exciting challenge for catalysis.

Acknowledgements

I would like to recognize the many helpful discussions from Drs. Charles Brandenburg, Paul Fagan, Katerina Korovessi, Konstantino Kourtakis and Keith Hutchenson and the experimental assistance from Dr. Lien Kao and Mr. Charles Bellini.

References

1. *International Energy Annual 2001 (EIA),* Tables 11.4 and 11.10 (Updated March 2003), http://www.eia.doe.gov/iea/contents.html.
2. Werpy, T.; Petersen, G.; *Top Value Added Chemicals from Biomass: Volume I -- Results of Screening for Potential Candidates from Sugars and Synthesis Gas*; DOE/GO-102004-1992; August 1, 2004.
3. Fitzpatrick, Stephen W., US Patent 5,608,105, 1997.
4. Fitzpatrick, Stephen W., *personal communication.*
5. Manzer, L. E. US Patent 6,743,819 2004; US patent application 2004192938; US patent application 2004192937; Manzer, L. E., Herkes, F. E. US patent application 2004192933.
6. Manzer, L. E. US patent application 2004192934; US patent application 2004192936.
7. Manzer, L. E. US Patent 6,841,520 (2005); US patent application 2004192935; US patent application 2004192932.
8. McGraw, W. J. US Patent 2,624,723 (1953).
9. Manzer, L. E. US 6,617,464 (2003).
10. Angeliki A. Lemonidou, Liza López, Leo E. Manzer and Mark A. Barteau; *Dynamic Microbalance Studies of RbOx/SiO$_2$ Catalyst: Deactivation/Regeneration for Methylene Valerolactone Synthesis, Appl. Catal. A.* **2004**, *272*, 241-248.
11. Thermochemical Conversion of Biomass to Produce Value Added Fuels and Chemical Intermediates, Fitzpatrick, Stephen W., 2003 National AIChE Meeting, November 18, 2003.
12. Leonard, R. H. US patent 2,809,203, (1957).

Chapter 5

Catalytic Transformations of Carbohydrates

**Pierre Gallezot, Michèle Besson, Laurent Djakovitch,
Alain Perrard, Catherine Pinel, and Alexander Sorokin**

Institut de Recherches sur la Catalyse-CNRS, 69626 Villeurbanne, France

Catalytic processes for converting starch, starch derivatives and glycerol to valuables chemicals or polymeric materials are considered. A first approach was aimed at using molecularly pure feedstocks such as glucose or glycerol to convert them by selective catalytic reactions to specialties or fine chemicals. A second approach consisted of converting native starch via one step catalytic processes to a mixture of products fulfilling required physicochemical specifications to be incorporated in the formulation of end-products such as paint, paper or cosmetics. All the catalytic routes considered follow the principles of green chemistry.

Introduction

Biosynthesis in plants and trees using sun radiation, atmospheric carbon dioxide, water, and soil nutrients produces huge amounts of biomass estimated up to 200 Gt/y, a figure to be compared to 7 Gt/y of extracted fossil fuels. Increasing use of biomass for energy, chemicals and material supply is one of the key issues of sustainable development because bio-based resources are renewable and CO_2 neutral unlike fossil fuels. Presently, only 7% of the annual biomass is harvested for food, feed and non-food applications. Food and feed will remain priority number one, but improved agricultural techniques and genetic modification of crops will increase yields substantially. Renewables dedicated to non-food applications could come from specialized crops or forestry products giving much higher dry matter per cultivated area and a more reproducible and specific content of useful molecules for chemical applications. Moreover, large streams of crop and forestry wastes can be subject to upgrading.

Most of the non-food applications of bio-based resources fall into four categories: (i) traditional uses in timber, paper, fiber, rubber, fragrance industries, etc., (ii) thermal power generation (bio-power) by direct combustion or after catalytic or fermentative gasification, (iii) biofuels, such as ethanol produced by fermentation of carbohydrates, bio-diesel by transesterification of vegetable oïls, and hydrogen by steam reforming/WGS of biomass, (iv) bio-products, i. e., chemicals or materials produced by chemo-catalytic and/or enzymatic conversion of carbohydrates, triglycerides, and terpenes. According to the US roadmap for biomass technologies-2020 vision goals (*1*), bio-power will meet 5% of the total industrial and electric generator energy, bio-fuels 10% of the transportation fuels, and bio-based chemicals will attain 18% of the US market.

The economy and ecology of the various processes to produce bio-fuels (hydrogen, ethanol, biodiesel) should be precisely evaluated for each particular situation. In contrast, producing chemicals from renewable feedstocks generally represent a more sound and sustainable approach. The molecules extracted from bio-based resources already contain functional groups so that the synthesis of chemicals generally requires a lower number of steps than from alkanes. Synthesis could be achieved by alternative processing routes adapted to bio-based feedstocks combining catalytic and enzymatic steps, rather than by employing conventional flow sheets from hydrocarbons. Synthesis of polylactate (Dow-Cargill) and 1,3-propanediol (DuPont-Genencor) from carbohydrates are successful industrial examples of this approach. Ideally, biorefineries should produce chemicals in the first place and fuels as by-products. At any rate, research in chemistry, biochemistry and engineering is needed to decrease the cost of processing bio-based resources to produce cheaper chemicals.

This article deals with various catalytic processes that were investigated during the last few years in our laboratory. Two strategies were pursued namely: (i) highly selective catalytic conversions of pure glucose and glycerol to pure compounds that can be used as building blocks for chemistry or incorporated in the formulation of end-products and (ii) chemical modifications by one-pot catalytic reaction of native starches extracted from various cereals to obtain mixtures of polysaccharides with hydrophilic or hydrophobic properties suitable to meet specification for direct incorporation in marketable end-products. Figure 1 shows the different carbohydrate transformations investigated.

Selective conversion of glucose and derivatives

Hydrogenation of glucose

In view of the large amounts of sorbitol **3** produced batchwise by hydrogenation of glucose **2** on Raney-nickel catalysts, continuous processes

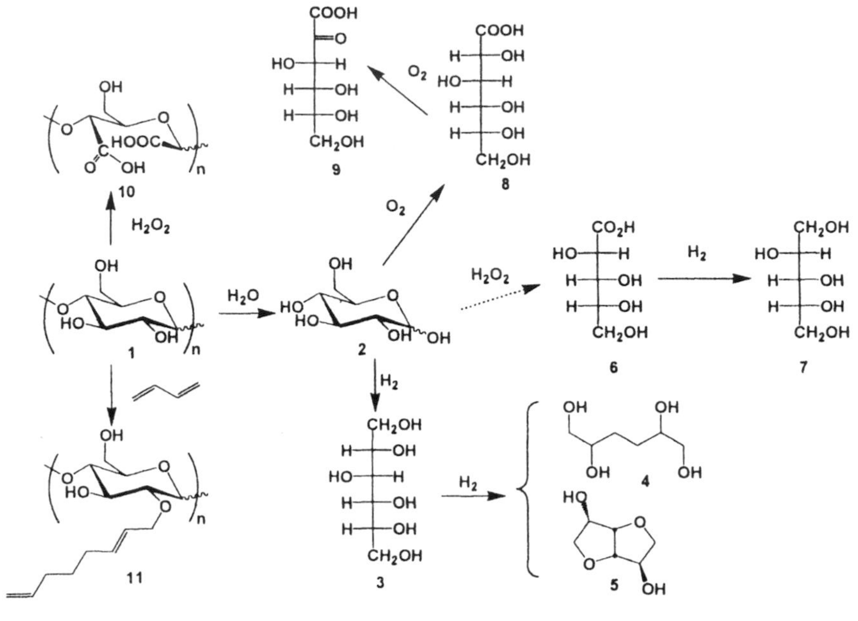

Figure 1. Catalytic conversion of starch and glucose

would be preferable and Raney-nickel would be advantageously replaced by ruthenium catalysts which are more active, selective and less prone to leaching.

Hydrogenation of 40 wt% aqueous solutions of glucose were carried out in a trickle-bed reactor on Ru/C catalyst obtained by impregnation of Norit carbon extrudates (*2*). A 99.3 % yield of sorbitol was obtained even after 596 h on stream. No leaching of ruthenium was detected. The hydrogenation was also conducted on Ru-Pt/C bimetallic catalysts of different composition prepared by co-exchange of Pt and Ru amino cations (*3*). Interestingly, the activity passed through a maximum at 1470 mmol/h/g_{Ru} for the specific atomic composition $Ru_{56}Pt_{44}$. Pt-Ru catalysts were also more selective to sorbitol because the sorbitol epimerization to mannitol decreased. The contact time with the catalyst can be increased without loss of selectivity thus allowing operation at total conversion of glucose and more than 99% selectivity over a large domain of liquid flow rates. The bimetallic catalysts loaded in the trickle-bed reactor gave a productivity of 2 tons/day/kg$_{Ru}$ of sorbitol at 99.5% purity.

Hydrogenation of arabinonic acid

There is a great interest to convert C_6 carbohydrates available in large supply from starch or sucrose into C_5 and C_4 polyols that are little present in biomass but find many applications in food and non-food products. Thus, glucose can be converted to arabitol 7 by an oxidative decarboxylation of glucose to arabinonic acid 6 followed by a hydrogenation step. The main pitfall is to avoid dehydroxylation reactions leading to deoxy-products not compatible with the purity specifications required for arabitol. Aqueous solutions (20 wt%) of arabinonic acid were hydrogenated on Ru catalysts in batch reactor (4). The selectivity was enhanced in the presence of small amounts of anthraquinone-2-sulfonate (A2S) which decreased the formation of deoxy by-products. A2S acted as a permanent surface modifier since the catalyst was recycled with the same selectivity without further addition of A2S. The highest selectivity to arabitol was 98.9% at 98% conversion with a reaction rate of 73 mmol h^{-1} g$_{Ru}$$^{-1}$ at 80°C.

Hydrogenolysis and dehydroxylation of sorbitol

This study was aimed at converting starch into mixtures of polyols that could be used in the manufacture of polyesters, alkyd resins, and polyurethanes employed in paints, powder coatings, and construction materials. Deoxyhexitols consisting of C_6 diols, triols, and tetrols 4 are suited to replace polyols derived from petrochemistry such as pentaerythritol. Sorbitol, which is easily derived from starch, was taken as starting feedstock for the hydrogenolysis studies on metal catalysts (5). To improve the selectivity of sorbitol hydrogenolysis towards deoxyhexitols, catalysts and reaction temperatures were optimised to favour the rupture of C-OH bonds (dehydroxylation reactions) rather than C-C bond rupture. Copper-based catalysts, which have a low activity for hydrogenolysis of C-C bonds, were employed to hydrogenolyse a 20 wt% aqueous sorbitol solution in the temperature range 180-240°C. Reactions carried out in the presence of a 33% CuO-65% ZnO catalyst at 180°C under H$_2$-pressure gave a 73% yield of C$_4$$^+$ polyols, and more specifically, 63% of deoxyhexitols.

In contrast, operating in the presence of palladium catalysts at 250°C under 80 bar hydrogen pressure cyclodehydration reactions of sorbitol and mannitol occurred with formation of cyclic ethers (isosorbide 5, 2,5-anhydromannitol, 2,5-anhydroiditol, and 1,4-anhydrosorbitol) (6). Up to 50% and 90% yield of isosorbide were obtained from sorbitol and mannitol, respectively. These mixtures of polyols were effectively employed by ICI paints to synthesize alkyd resins and make decorative paints which performed comparably to the commercial ones.

Oxidation of glucose to gluconate

Gluconic acid **8**, used as a biodegradable chelating agent or as an intermediate in the food and pharmaceutical industry, is produced by enzymatic oxidation of glucose. An alternative route employing oxidation with air in the presence of palladium catalysts, was investigated (7). Unpromoted palladium catalysts were active in glucose oxidation, but the rate of reaction was low because of the over-oxidation of Pd-surface, and side oxidation reactions decreased the selectivity. The beneficial effect of bismuth on the activity and selectivity was clearly demonstrated with Pd-Bi/C catalysts of homogeneous size and composition (5 wt% Pd, Bi/Pd = 0.1) prepared by deposition of bismuth on the surface of 1-2 nm palladium particles. The rate of glucose oxidation to gluconate was 20 times higher on Pd-Bi/C (Bi/Pd = 0.1) than on Pd/C catalyst, and the selectivity at near total conversion was high on the fresh and recycled catalysts (Table 1).

These results were interpreted in terms of bismuth acting as a co-catalyst protecting palladium from over-oxidation because of its stronger affinity for oxygen. This oxidation reaction is a nice example of green chemistry (one pot catalytic conversion of renewables, mild conditions, water as solvent, air as

Table 1 Product distribution in glucose oxidation

Catalyst[a] (run)	Conversion[b] (%)	Yield /mol%				Selectivity (%)
		8	**9**	**12**	**13**	
PdBi/C (1st)	99.6	99.4	<0.4	<0.4	0.2	99.8
PdBi/C (2nd)	99.7	98.9	<0.4	0.6	0.2	99.1
PdBi/C (3rd)	99.8	98.5	0.4	0.8	0.2	98.7
PdBi/C (4th)	99.9	98.5	0.4	0.7	0.2	98.6
PdBi/C (5th)	99.9	99.1	<0.4	0.6	0.2	99.2
Pd/C	82.6	78.1	1.4	2.3	0.7	94.6

a: 4.7 wt% Pd; Bi/Pd=0.1. b: after 155 min on PdBi/C and 24 h for Pd/C; **8**: gluconate; **9**: 2-ketogluconate; **12**: 5-ketogluconate + glucarate, **13**: fructose; reaction conditions: 1.7 mol L^{-1}, T= 313 K, pH 9; [glucose]/[Pd] = 787; air at atmospheric pressure.

oxidizing agent, no waste by-products, recyclable catalysts) and giving comparable selectivity and higher productivity than enzymatic glucose oxidation.

Catalytic modifications of native starch

Due to its large availability and low cost, native starch has been used for a long time in the preparation of different end-products. To obtain specific properties, native starch has to be chemically or enzymatically modified. Because native starch is an insoluble, partially crystallized solid polymer, chemical modifications are difficult to achieve and require the use of soluble catalysts. We have achieved two catalytic modifications of native starch: (i) selective oxidation to obtain carboxyl and carbonyl functions, thus making starch more hydrophilic (8), and (ii) telomerisation of butadiene with the hydroxyl groups of the glucoside units, thus providing more hydrophobic material (8,9).

Oxidation by H_2O_2 to make starch hydrophilic

Hydrophilic starch obtained by partial oxidation is widely used in paper and textile industries and can be potentially applied in a variety of applications, e.g., for the preparation of paints, cosmetics, and superabsorbents. The oxidation occurs at the C_6 primary hydroxyl group or at the vicinal diols on C_2 and C_3 involving a cleavage of the C_2-C_3 bond to give carbonyl and carboxyl functions (Figure 2).

Current industrial methods of chemical modification of starch by oxidation are based mainly on oxidizing agents like NaOCl to introduce carboxyl groups or NaIO$_4$ to obtain aldehyde functions. 2,2,6,6-tetramethyl-1-piperidinyloxy (TEMPO) was applied in combination with NaOCl/NaBr or with peroxide reagents (10) to oxidise selectively primary hydroxyl groups in polysaccharides. While these oxidations are chemically efficient they lead to inorganic wastes.

Several transition metal catalysts based on Fe, Cu or W salts (0.01 – 0.1 mol%) have been proposed to activate H_2O_2 which is a well suited oxidant from an environmental and economical point of view. However, the concentration of metal ions was quite high and because oxidised starch has good complexing properties heavy metals were retained by carboxyl functions in the modified starch.

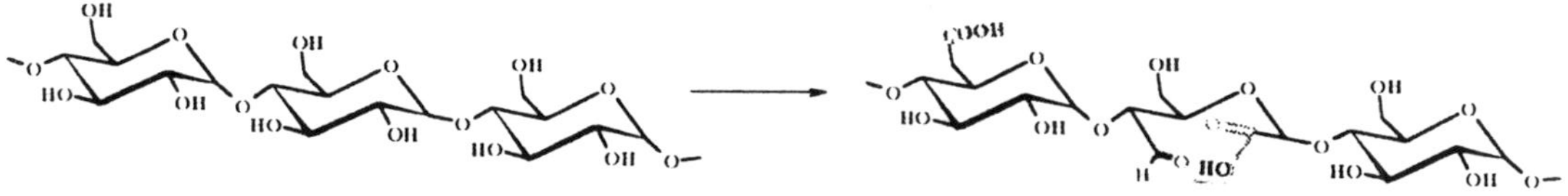

Figure 2. Oxidation of starch chains

Efficient catalytic methods for native starch oxidation with H_2O_2 in the presence of iron tetrasulfophthalocyanine (FePcS), cheap and available at industrial scale, were proposed (8,11). Starches from different origin (potatoes, rice, wheat, corn) were oxidized by H_2O_2 following two operating modes: (i) starch suspended in a water solution containing the dissolved catalysts, and (ii) starch powder wetted by small volumes of aqueous solution containing required amounts of FePcS (incipient wetness method).

Oxidation of native starch in suspension

Potato starch suspended in 400 mL of water was oxidized by adding the solution of H_2O_2 at a low continuous rate. The oxidized material was isolated after addition of ethanol followed by filtration. Table 2 gives the yield of carboxylic and carbonyl functions per 100 anhydroglucose units (AGU) formed in the oxidation process as a function of the substrate to catalyst molar ratio.

Table 2. Oxidation of starch in aqueous suspension with H_2O_2 in the presence of iron phthalocyanine. Effect of substrate/catalyst ratio.

AGU/Fe	DS_{COOH}[a]	$DS_{C=O}$[a]
25800 : 1	0.70	3.20
12900 : 1	2.00	10.40
6450 : 1	2.00	9.00

Reaction conditions: 58°C; pH: 7; reaction time: 7 h; molar ratio H_2O_2 : AGU = 1 : 2.1; [a] Degree of substitution expressed per 100 anhydroglucose units (AGU)

The best yields were obtained with a molar ratio 12900/1 (0.0078 mol %), but the oxidation was still quite efficient with 0.0039 mol % of catalyst (25800/1 AGU/catalyst ratio). The very small amount of catalyst needed for successful oxidation of starch represents a very important advantage of this catalytic system compared to oxidation methods based on metal salts that need at least 0.2 mol % (12). In addition to economical and environmental considerations this small charge of catalyst yields purer product. Comparison of FePcS with conventional $FeSO_4$ catalyst showed that oxidized starch prepared with $FeSO_4$ had a significant undesirable level of residual iron whereas the oxidised starches obtained with FePcS catalyst had practically the same final Fe-content as the native potato starch. High metal content prevents the use of oxidized starch in applications where brightness and/or low metal content are required.

Oxidation of native starch by incipient wetness method

A small volume of water containing the dissolved catalysts was added to starch under continuous powder mixing, followed by addition of hydrogen peroxide to the impregnated solid under mixing. After H_2O_2 consumption the product was washed with 2:1 ethanol-water mixture and dried. With a substrate/catalyst ratio of only 25800/1, the oxidation of potato starch by the incipient wetness method was efficient providing 1.5 carboxyl and 5.6 carbonyl functions per 100 AGU. This process was applied with success to the oxidation of starches of different physical and chemical properties (amylose/amylopectin ratio, granule size, temperature of gelatinisation) obtained from different crops (potato, wheat, rice, corn).

The SEM images of oxidised materials indicate that starch oxidised by dry method retains this granular structure. The surface of native starch grains is smooth, practically with no defects (Figure 3A). As the oxidation of starch proceeds, surface defects appear suggesting that the oxidation process starts from the granule surface ultimately forming pores through starch grains (Figure 3B, 3C).

In conclusion, the selective oxidation of native starch by hydrogen peroxide in the presence of metallophthalocyanine catalysts provided hydrophilic materials which, from primary tests, look promising for their incorporation in paints and cosmetics. These catalysts are not expensive, readily available at industrial scale and employed in very low amounts. The catalytic system was very flexible because by simple modifications of the reaction conditions it was possible to prepare oxidized starches with the desired level of carboxyl and carbonyl functions that are suitable for different applications. No wastes were formed because the process did not involve any acids, bases or buffer solutions.

 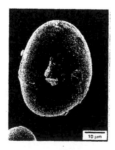 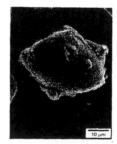

Figure 3 . SEM pictures of individual granules of potato starch: initial (A) ; oxidized potato starch with $DS_{CHO} = 2.84$, $DS_{COOH} = 0.27$ (B); oxidized potato starch with $DS_{CHO} = 8.50$, $DS_{COOH} = 4.60$ (C).

Grafting hydrocarbon chains to make starch hydrophobic

Our objective was to prepare more hydrophobic starches to incorporate them in latex preparation for decorative paints so that substrates derived from fossil fuel can be replaced by modified starches derived from renewable resources. Partial substitution of starch with acetate, hydroxypropyl, alkylsiliconate or fatty-acid ester groups was described in the literature for the synthesis of more hydrophobic starch. An alternative route was employed consisting of grafting octadienyl chains by butadiene telomerization (*8,9*). This reaction (Figure 4) was catalyzed by hydrosoluble palladium-catalytic systems prepared from palladium diacetate and trisodium tris(m-sulfonatophenyl)phosphine (TPPTS). Starch octadienyl ethers are expected to be much less sensitive towards hydrolysis compared to the esterified starches.

The reaction was first conducted with success on sucrose (*13*), but the transposition of this reaction to starch was challenging because this substrate is insoluble in water at room temperature and gelatinizes at temperature higher than *ca* 70°C. The degree of substitution (DS) should be kept low enough because modified starch should not be too hydrophobic, also for obvious economical reason the catalyst/starch ratio should be kept low.

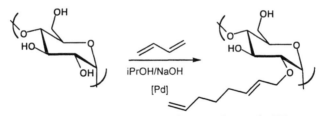

Figure 4. Telomerisation of butadiene with starch OH-groups

Figure 5 gives the results obtained in the following reaction conditions: starch 12g, butadiene ca 40g, catalyst : Pd(OAc)₂/TPPTS: 1/3, aqueous NaOH (0.1M) 50ml/iPrOH 10ml, 3h. At 90°C, with 0.5 wt.% of Pd(OAc)₂, the DS reached almost 0.6 while less than 0.3 was achieved with 0.12 wt% catalyst. At 50°C, the DS was lower but much less affected by the catalyst/starch ratio, thus the DS decreased from 0.15 to 0.08 as the amount of catalyst decreased from 0.5% to 0.05%. This is because at 90°C Pd/TPPTS complex is unstable and decomposes progressively into palladium metal whereas at 50°C the catalyst is stable enough. When the reaction was performed in the presence of 0.25% palladium acetate, 0.08 wt% residual palladium was detected by chemical analysis of modified starch after reaction at 50°C. In contrast, no palladium was detected in the modified polymer when the reaction was conducted in the presence of 0.05% palladium. An optimum DS of 0.06 suitable for application in latex preparation for decorative paints was obtained with 0.03% palladium at 50°C.

The etherified starch was further transformed by hydrogenation of the double bonds to yield the corresponding linear octyl groups using

[RhCl(TPPTS)₃], a catalyst that is soluble in EtOH/H₂O mixtures (*14*). Complete hydrogenation was obtained at 40°C under 30 bar of H₂ after 12h using 0.8 wt% Rh-catalyst. Other catalytic transformations such as double bonds oxidation and olefin metathesis could possibly be used to prepare other modified starches for various applications.

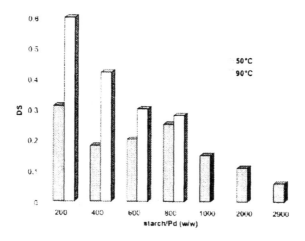

Figure 5. Modification of starch by butadiene telomerization. Influence of the catalyst/substrate ratio on degree of substitution (DS)

Catalytic conversion of glycerol

Glycerol is a highly functionalized molecule now available in large amounts as by-product of the transesterification of triglycerides to obtain biodiesel. It is important for the economy of the whole process to find new outlets for glycerol, particularly towards high value added chemicals. We have studied the catalytic oxidation of glycerol to various oxygenated derivatives of great potential interest and the hydrogenolysis of glycerol to 1,2- and 1,3-propanediols which are valuable chemicals particularly for the preparation of polymers. Figure 6 gives the targeted molecules.

The oxidation of glycerol (GLY) to glyceric acid (GLYAC) was carried out with air on palladium and platinum catalysts at basic pH (*15*). The selectivity to glycerate was 70% at 100% conversion. In contrast, using bismuth-promoted platinum catalyst the selectivity shifted to dihydroxyacetone.

β-Hydroxypyruvic acid (Figure 6), an important chemical intermediate for the preparation of different chemicals, particularly l-serine, was obtained by oxidation of glycerate with air on 5% Pt-1.9%Bi/C catalyst without pH regulation (*16*). The maximum yield was 64% at 80% conversion at acidic pH. In contrast, on the same catalyst under basic conditions (pH 10-11) an 83% yield of tartronate was obtained at 85% conversion of glycerate. This confirms the importance of acidic pH to obtain a selective complexation of the promoter with the carboxyl and α-hydroxyl group of the glyceric acid leading to a selective oxidation of the secondary alcohol function. In a similar investigation, Abbadi and van Bekkum (*17*) conducted the oxidation of sodium glycerate on 5%Bi-5%Pt/C catalyst without pH regulation. The pH decreased from 5.7 to 4.1 and 93% selectivity to β-hydroxypyruvic acid at 95% conversion was achieved.

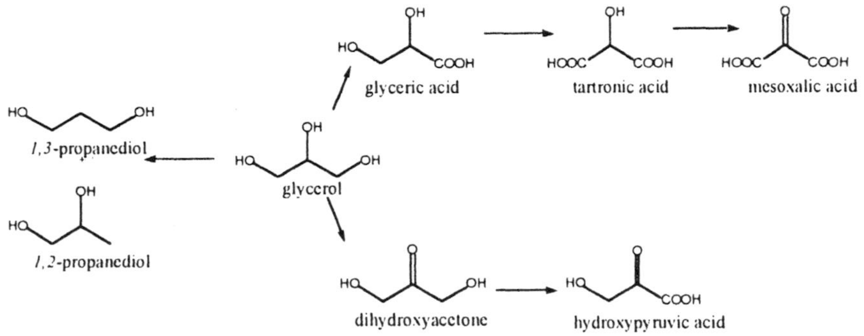

Figure 6. Products obtained by hydrodehydroxylation or by oxidation of glycerol with air on metal catalysts

Mesoxalic acid (Figure 6) is a good chelating agent and potentially a valuable synthon for organic synthesis. Fordham et al (*18*) studied the preparation of mesoxalic acid by oxidation of sodium tartronate on PtBi/C catalyst at 60°C without pH control; the maximum yield was 65% at 80% conversion. A total conversion of tartronic acid was obtained at 80°C giving 50% yield of mesoxalic acid without other by-products because they were totally oxidized into CO_2.

Hydrogenolysis of glycerol to 1,2- and *1,3*-propanediols

Our objective was to study the selective hydrogenolysis (dehydroxylation) of glycerol in the presence of heterogeneous catalysts to produce 1,2-propanediol (1,2-PDO) and 1,3-propanediol (1,3-PDO). 1,3-PDO is copolymerised with terephthalic acid to produce the polyester SORONA® from DuPont, or CORTERRA® from Shell, that are used in the manufacture of carpet and textile fibres, and exhibit unique properties in terms of chemical resistance, light stability, elastic recovery, and dyeability (*19,20*). 1,3-PDO is currently produced from petroleum derivatives such as ethylene oxide (Shell route) or acrolein (Degussa-DuPont route) by chemical catalytic routes (21,22). These diols can be produced by an alternative route involving selective dehydroxylation of glycerol. In 1985, Celanese patented the hydrogenolysis of glycerol water solution under 300 bar of syngas at 200°C in the presence of a homogeneous rhodium complex ($Rh(CO)_2(acac)$) and tungstic acid. 1,3-PDO and 1,2-PDO were produced with 20% and 23% yield, respectively (23). Taking advantage of previous work on sorbitol hydrogenolysis (5), the catalytic hydrogenolysis of glycerol was conducted over heterogeneous catalysts and reaction conditions chosen to minimize carbon-carbon bond rupture and improve the rate and selectivity towards 1,2-PDO and 1,3-PDO (24).

Aqueous solutions of glycerol were hydrogenolysed at 180°C under 80 bar H_2-pressure in the presence of various metal catalysts (Cu, Pd, Rh), supports (ZnO, C, Al_2O_3), solvents (H_2O, sulfolane, dioxane) and additives (H_2WO_4). The catalysts were tested to improve reaction rate and selectivity to the target molecules. The best selectivity (100%) to 1,2-PDO was obtained by hydrogenolysis of water solution of glycerol in the presence of CuO/ZnO catalysts. To improve the selectivity to 1,3-PDO the reaction was conducted with rhodium catalysts with tungstic acid added to the reaction medium. The best result in terms of conversion and selectivity to 1,3-propanediol (1,3-PDO / 1,2-PDO = 2) was obtained by operating in sulfolane. The presence of iron dissolved in the reaction medium was also beneficial for the selectivity to 1,3-PDO. A mechanism was proposed to account for the effect of these different parameters.

Concluding remarks

The results reported above illustrate two different approaches of producing chemicals and materials from renewable feedstocks, viz:

(i) The first approach involves molecularly pure feedstocks such as glucose and glycerol, that are easily obtained in large amounts from low priced renewables materials, e.g., starch and vegetable oils, respectively. Valuable chemicals are then derived from these platform molecules by successive selective catalytic steps leading to final pure end-products. This approach is similar to the usual route of chemical synthesis from hydrocarbon feedstocks, except that many fewer steps are required from the highly functionalized molecules derived from biomass. However, this approach may lead to costly specialties or fine chemicals and comparatively low amounts of renewables are employed.

(ii) The second approach consists of converting raw renewables feedstocks such as starch or cellulose by one step catalytic processes to a mixture of products that can be incorporated directly into final products. This approach is exemplified by the hydrogenolysis of sugar to diols, triols and tetrols useful for the preparation of polymers for decorative paints, the oxidation of native starch to hydrophilic polymeric materials suitable for incorporation in paints paper and cosmetics, and the telomerization of butadiene on native starch to hydrophobic products useful as emulsifiers. Starting from cheap renewables and one step catalytic processes with cheap co-reactants (H_2, O_2, small amounts of butadiene) one can expect to process high amounts of renewables that would replace products derived from fossil fuels.

Importantly, all the catalytic conversion processes described in the investigations reported above follow green chemistry requirements.

Acknowledgments

Roquette Frères Co. is acknowledged for supporting glucose hydrogenation and oxidation studies. We thank Rhodia Co. and Agrice agency for their support of glycerol hydrogenolysis. Investigations on the hydrogenolysis of sorbitol and catalytic modifications of native starch were conducted within EC programmes STARPOL (FAIR CT95-0837) and HYDROSTAR (G5RD-CT-1999-00051), respectively.

References

1.   Roadmap for Biomass Technologies in the United States, Biomass R&D Technical Advisory Committee, December 2002, http://www.bioproducts-bioenergy.gov
2.   Gallezot, P.; Nicolaus, N.; Flèche, G. Fuertes, P.; Perrard, A. *J. Catal.*, **1998**, *180*, 51.
3.   Gallezot, P.; Perrard, A. unpublished results.
4.   Fabre L.; Gallezot, P.; Perrard, A. *J. Catal.* **2002**, *208*, 247.
5.   Blanc, B.; Bourrel, A.; Gallezot, P.; Haas, T.; Taylor, P. *Green Chem.* **2000**, 89.
6.   Bourrel, A.; Haas, T. *Deutsch Patent* 19,749,202.9, (14/11/97,)
7.   Besson, M.; Gallezot, P.; Lahmer, F.; Fléche, G. Fuertes, P. *J. Catal.*, **1995**, *152*, 116.
8.   Sorokin, A.; Kachkarova-Sorokina, S.; Donzé, C.; Pinel, C.; Gallezot, P. *Top.Catal.* **2004**, *27*, 67.
9.   Donzé, C.; Pinel, C.; Gallezot, P.; Taylor, P. *Adv. Synth. Catal.* **2002**, *344* 906.
10. Bragd, P.; Besemer, A.C.; van Bekkum, H. *Carbohydr. Polym.* **2002**, *49*, 397 and references therein.
11. Sorokin, A.B.; Kachkarova-Sorokina, S.L.; Gallezot, P. WO Patent 2004/007560 A1, **2004**.

12. P. Parovuori, A. Hamunen, P. Forssell, K.Autio and K. Poutanen, *Starch/Stärke* **1995**, *47*, 19.

13. Desvergnes-Breuil, V.; Pinel, C.; Gallezot, P. *Green Chem.* **2001**, *3,* 175.

14. Donzé, C., Pinel, C.; Gallezot, P. *Catal. Commun.* **2003**, *4*, 465.

15. Garcia, R.; Besson, M.; Gallezot, P. *Appl. Catal. A* **1995**, *127*, 165

16. Fordham, P.; Besson, M.; Gallezot, P. *Appl. Catal. A* **1995**, *133,* L179.

17. Abbadi, A.; van Bekkum, H. *J. Mol. Cat. A* **1995**, *97,* 111.

18. Fordham, P.; Besson, M.; Gallezot, P. *Catal. Lett.* **1997,** *46*, 195.

19. Caley, P.N.; Everett, R.C. U.S. Patent 3,350,871, **1967**.

20. Zimmerman, D.; Isaacson, R.B. U.S. Patent 3,814,725, **1974**

21. Arntz, D.; Haas, T.; Müller, A.; Wiegand, N. *Chem. Ing. Tech.,***1991**, *63,* 733.

22. Lam, K.T.; Powell, J.P.; Wieder, P.R. WO Patent 97,16250, **1997**.

23. Che, T.M. U.S. Patent 4,642,394, **1987**.

24. Chaminand, J.; Djakovitch, L.; Gallezot, P.; Marion, P.; Rosier, C. *Green Chem.* **2004**, *6*, 359.

Chapter 6

Novel Polymeric Materials from Soybean Oils: Synthesis, Properties, and Potential Applications

Dejan D. Andjelkovic, Fengkui Li, and Richard C. Larock

Department of Chemistry, Iowa State University, Ames, IA 50011

A variety of promising new polymeric materials, ranging from soft rubbers to hard, tough and rigid plastics, have been prepared by the cationic copolymerization of readily available soybean oils and alkenes initiated by boron trifluoride diethyl etherate (BFE). The resulting polymers possess good thermal and mechanical properties, including good damping and shape memory properties. As such, these new biomaterials appear promising as replacements for a variety of petroleum-based polymers.

The growing worldwide demand for petroleum-based polymeric materials has raised environmental concerns about these non-biorenewable, indestructible materials, and also increased our dependence on crude oil. The rapid growth of the industrial world has placed increased pressures on the finite petroleum reserves and forced us to look for alternative resources. Biorenewable resources represent a promising new alternative, but they require new approaches and developments in order to be successfully utilized.

Biopolymers produced from renewable and inexpensive natural resources have drawn considerable attention over the past decade, due to their low cost, ready availability, environmental compatibility, and their inherent biodegradability (1). Application of these biomaterials has a huge potential market, because of the current emphasis on sustainable technologies. Many naturally occurring biopolymers, such as cellulose, starch, dextran, as well as those derived from proteins, lipids and polyphenols, are widely used for material applications (2,3). The exciting new area of biorenewables, which lies on the border of molecular biology and polymer chemistry, offers many opportunities to expand the range of exciting new bio-based materials. Particularly promising is the development of new biopolymers from functionalized, low molecular weight natural substances, like natural oils, utilizing polymerization methods widely used for petroleum-based polymers.

Natural oils represent one of the most promising renewable resources. These molecules possess a triglyceride structure with fatty acid side chains possessing varying degrees of unsaturation (4-6). The presence of carbon-carbon double bonds makes these biological oils ideal monomers of natural origin for the preparation of biopolymers. Soybean oil is the most abundant vegetable oil, accounting for approximately 30% of the world's vegetable oil supply (7). The bulk of the soybean oil (~80%) produced annually is used for human food. The remaining 20% is used for animal feed (~6%) and non-food uses, such as soaps, lubricants, coatings and paints (~14%). The most promising new applications for soybean oils are expected to be in non-food uses. So far, these applications have mainly involved the polymerization of soybean oil derivatives, such as fatty acids, epoxidized oils, polyols, etc. (4,8). The advantages of soybean oil as a monomer are its ready availability on a huge scale, its low cost (~$0.24/lb) (9) and high purity, the possibility of genetically engineering the number of C=C bonds present in the oil (4), its significantly higher molecular weight when compared to other conventional alkene monomers (~880 g/mol), and the existence and availability of many structurally related vegetable oils. These advantages make natural oils ideal starting materials for biopolymer synthesis.

Considerable recent effort has been directed towards the conversion of vegetable oils into solid polymeric materials. These vegetable oil-based polymers generally possess viable mechanical properties and thus show promise as structural materials in a variety of applications. For example, Wool and co-workers have prepared rigid thermosets and composites via free-radical copolymerization of soybean oil monoglyceride maleates and styrene (10-12). The new maleate monomers are obtained by glycerol transesterification of soybean oil, followed by esterification with maleic anhydride (10). It has been found that the original C=C bonds of the soybean oil side chains participate in this free-radical copolymerization. Petrovic and co-workers, on the other hand, have successfully converted the C=C bonds of soybean oil into polyols by epoxidizing the C=C bonds of the triglyceride oil, and then carrying out oxirane ring-opening of the epoxidized oil. The newly synthesized polyols are then reacted with a variety of isocyanates to produce polyurethanes (*13-17*).

More recently, a variety of exciting new polymeric materials have been prepared in our group by the cationic copolymerization of soybean and other vegetable oils with a variety of alkene comonomers (*18-35*). These biopolymers possess industrially viable thermophysical and mechanical properties and thus may find structural applications. This chemistry takes advantage of the original C=C bonds of the soybean (*18-27*), tung (*28,29*), corn (*30*) and fish oils (*31-35*) to effect crosslinking. In this chapter, we shall focus primarily on the synthesis and characterization of soybean oil-based polymers, which result from the direct copolymerization of the C=C bonds of soybean oils with other comonomers via cationic polymerization (*18-27*).

Soybean Oils as Cationic Monomers

Commercially available soybean oils have a triglyceride structure. Oleic (C18:1), linoleic (C18:2) and linolenic acid (C18:3) are the primary fatty acid components of soybean oils (Table I). Regular soybean oil (SOY) has approximately 4.5 C=C bonds per triglyceride, while low saturation soybean oil (LSS) has approximately 5.1 C=C bonds. LSS oil is a commercially available soybean oil with considerably more linoleic acid (*4*). The fatty acid side chains in these two soybean oils are non- conjugated. Conjugated LSS oil (CLS) has been prepared in our laboratories from LSS by Rh-catalyzed isomerization (*36*).

Table I. Compositions of the Various Soybean Oils

Oils	C=C		Fatty Acids				
	Number[a]	Conjugated	C16:0	C18:0	C18:1	C18:2	18:3
SOY	4.5	No	10.5	3.2	22.3	54.4	8.3
LSS	5.1	No	4.5	3.0	20.0	63.6	9.0
CLSS	5.1	Yes	4.5	3.0	20.0	63.6	9.0

[a] The average number of C=C bonds per triglyceride has been calculated by ^{1}H NMR spectral analysis.

The cationic polymerization of soybean oils is possible only when certain basic thermodynamic requirements are met. This means that a negative change in free energy is required ($\Delta H - T\Delta S < 0$). Since the entropic component decreases during cationic polymerizations, because of the loss of translational degrees of freedom caused by connecting triglyceride units together, the thermodynamic feasibility of the polymerization will depend on enthalpic factors. Thus, cationic polymerizations must be sufficiently exothermic to compensate for the loss in entropy. In the case of soybean oil, this is accomplished through conversion of the C=C bonds in the triglyceride to C-C single bonds in the polymer. The C=C bonds in the soybean oil represent sites for electrophilic attack of the reactive species generated by the initiator. Because of that, electronic effects within the triglyceride monomers will have a crucial effect on their reactivity. Soybean oil triglycerides, like all other alkene monomers, are expected to polymerize cationically by the addition of monomers to the growing carbocation. Thus, triglyceride monomers must be nucleophilic and capable of stabilizing the intermediate carbocation (37). Since nucleophilicity increases with increasing substitution on the C=C bond, due to the positive inductive effect of the alkyl substituents, the C=C bonds of the SOY and LSS are considered more nucleophilic than those of ethylene and propylene. They are therefore capable of stabilizing the positive charge to a greater extent and are prone to cationic polymerization. For the same reason, conjugated C=C bonds in the CLS are considered even more reactive toward cationic polymerization, since they will generate very stable allylic carbocations as intermediates. Thus, the SOY, LSS and CLS oils are cationically polymerizable monomers.

Unlike ethylene, propylene, isobutylene or styrene, soybean oils are considered polyfunctional monomers due to the presence of multiple C=C bonds within the triglyceride. Along with their relatively high molecular weights (~ 880 g/mol) and ability to efficiently stabilize intermediate carbocations, it is not surprising that their cationic polymerization easily affords high molecular weight polymers with crosslinked polymer networks.

Polymer Synthesis and Microstructure

Several classes of compounds are capable of initiating the cationic polymerization of vegetable oils, including protonic and Lewis acids (35). Among these, boron trifluoride diethyl etherate (BFE) has proven to be the most efficient catalyst. Its initiation and propagation mechanisms are well understood and described (37,38). Simple homopolymerization of soybean oils typically results in viscous oils or soft rubbery materials containing approximately 50 weight % of unreacted oil (27). These rubbery materials appear to be very weak and possess rather limited utility. In order to overcome this problem, soybean oils have been copolymerized with either divinylbenzene (DVB) or mixtures of DVB and styrene (ST) (Scheme 1). In general, copolymerization with the more reactive DVB comonomer yields hard, rigid polymers with very high crosslink densities and a nonuniform crosslinking structure (19). Because of that, these hard plastics are brittle, which drastically limits their applications. On the other hand, copolymerization with ST-DVB mixtures yields a variety of polymers ranging from elastomers to tough and rigid plastics with much more uniform

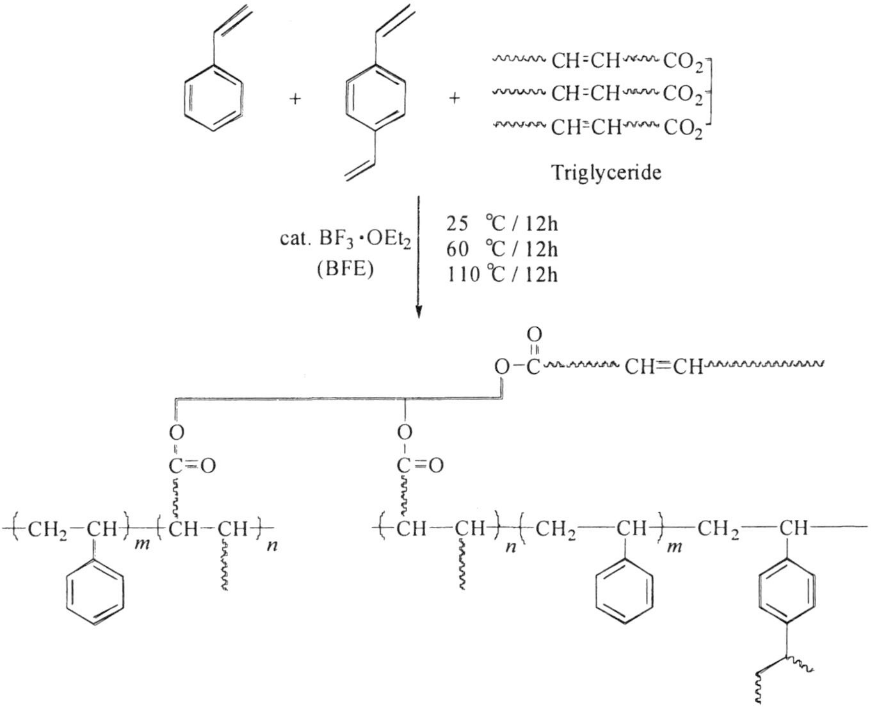

Scheme 1. Copolymerization of soybean oil triglyceride with ST and DVB.

crosslinking structures (*20*). During the copolymerization, the BFE initiator typically has to be modified by a fatty acid ester, such as soybean oil methyl esters or more commonly Norway fish oil ethyl ester (NFO), to allow homogeneous copolymerization. In the absence of the initiator modifier, the copolymerization becomes heterogeneous (*19*), presumably due to poor miscibility between the oil and the initiator. Since the modifier is completely miscible with both the soybean oil triglyceride and the alkene comonomers, it enhances their mutual solubility and provides a homogeneous reaction mixture. Upon mixing; reaction mixtures are typically cured for 12 h at room temperature, 12 h at 60 °C and finally 24 h at 110 °C. It has been found that, besides DVB, the soybean oil triglyceride also contributes directly to crosslinking (*27*). Gelation times of the reaction mixtures vary from a few minutes to several hours and depend on the type of soybean oil used in the mixture, the stoichiometry and the curing temperatures. The activation energies for the gelation process are 95-122 kJ/mol, which appear to be slightly higher than those of epoxy resins (70-90 kJ/mol) (25). Generally, 20-50 weight % of the soybean oil reactants are converted into crosslinked polymers at gelation and subsequent vitrification. Fully-cured thermosets are obtained after post-curing at elevated temperatures. It has been noted that an increase in the ST and DVB content in the starting mixture increases the percentage of the soybean oil incorporation. However, the soybean oil is not readily incorporated when the alkene content dominates. The most efficient consumption of the soybean oil occurs when 45-55 weight % of the oil is employed.

The resulting bulk polymers appear as opaque materials with a glossy dark brown color and a slight odor. These polymers can be made into various shapes by *in situ* reaction molding or cutting. Their microstructures have been determined by Soxhlet extraction with methylene chloride as a solvent, followed by subsequent ^{1}H NMR spectral analysis. Analyses have shown that bulk polymers are composed primarily of soybean oil-ST-DVB crosslinked polymers (insoluble fraction), a small amount of linear or less crosslinked soluble polymers, minimal amounts of unreacted soybean oil, and residual initiator fragments (soluble fraction). It has been determined that the ST and DVB completely participate in the copolymerization to form crosslinked polymers. The yield of the crosslinked polymer increases with the increase of DVB content in the mixture. Among the three soybean oils employed, CLS gives the smallest percent of the soluble fraction due to its higher reactivity. The maximum incorporation of the soybean oil into the crosslinked polymer occurs when the soybean oil constitutes approximately 45-50 wt % of the reactants. Careful choice of the reaction conditions results in almost complete conversion of the CLS oil into the crosslinked polymer (27). Similarly, the more reactive CLS oil yields polymers with higher crosslink densities than the SOY and LSS oils with the same stoichiometries. The crosslink densities of the soybean oil-based polymers range from 1×10^2 to 4×10^4 mol/m^3 (21).

Thermophysical and Mechanical Properties

Soybean oil-based polymers exhibit good thermal stability below 200 °C under both air and nitrogen atmospheres. These thermosetting materials typically show three-stage thermal degradation at 200-400 °C (stage I), 400-500 °C (stage II), and >500 °C (stage III), with the second stage being the fastest (19,20). The first stage degradation is attributed to evaporation and decomposition of the unreacted oil and other soluble components in the bulk material. The second step corresponds to degradation and char formation of the crosslinked polymer structure, while the third stage corresponds to gradual oxidation of the char residue. In general, these biomaterials lose 10% of their weight in the temperature range of 310-350 °C, depending on the content of the unreacted oil present in the bulk polymers. For example, Figure 1 shows the thermal degradation behavior of the polymer LSS45-ST32-DVB15-(NFO5-BFE3). The formula refers to a polymer prepared by the copolymerization of 45 weight % LSS, 32 weight % ST and 15 weight % DVB initiated by adding 5 weight % NFO mixed with 3 weight % BFE.

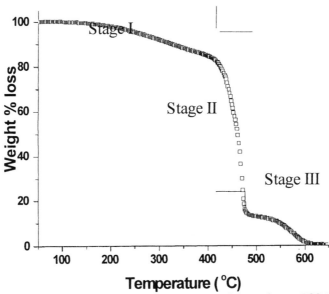

Figure 1. Three stages of the thermal degradation of the polymer LSS45-ST32-DVB15-(NFO5-BFE3) in air at 20 °C/min heating rate.

DMA analysis shows that these soybean oil-based polymers are typical thermosetting polymers (*21*). This is evidenced by the rubbery plateau in the DMA curve, which indicates the existence of a crosslinked network structure. The glass transition regions are fairly broad and are ascribed to the segmental heterogeneities upon crosslinking. Some soybean oil-DVB polymers exhibit two glass transition temperatures (T_g's) during DMA analysis (*19*). The high temperature transition (α_1) at about 80 °C represents the glass transition of the crosslinked polymer, while the low temperature transition (α_2) at 0-10 °C corresponds to the glass transition of the unreacted oil. Conversely, soybean oil-ST-DVB polymers possess a single glass transition, indicating better mixing of the small amounts of unreacted oil with the polymer main chains (*21*). Greater amounts of DVB in the starting mixture result in polymers with higher crosslink densities and consequently higher T_g's. Similarly, more reactive CLS oil affords polymers with higher T_g's and crosslink densities than the corresponding SOY and LSS oils. The glass transition temperatures of these soybean oil polymers vary from 0 °C to 105 °C.

The bulk polymers have been further characterized through tensile measurements in order to elucidate their tensile mechanical properties. Analyses have proven that these materials exhibit characteristics typical of materials ranging from elastomers to tough and rigid plastics. For example, Figure 2 shows the evolution of the polymeric material from an elastomer, through a ductile plastic with yielding behavior, to a rigid plastic, just by varying the DVB content from 0% to 47%, while keeping the total ST + DVB concentration constant (*22*).

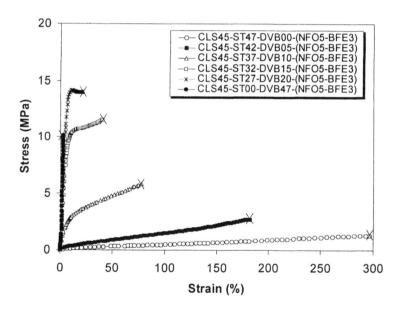

Figure 2. Tensile stress-strain behavior of soybean oil polymers. (Reproduced from reference 22. Copyright 2001 John Wiley & Sons, Inc.)

Table II summarizes some thermal and mechanical properties of several soybean oil polymers and compares them with two of the most useful industrial thermoplastics, low density polyethylene (LDPE) and polystyrene (PS). The results show that the CLS polymers possess higher mechanical properties than the SOY and LSS polymers with the same stoichiometries. Also, the highest toughness of the resulting polymers is reached when the amount of the soybean oil is equivalent to the amount of alkene comonomers. With a proper choice of the polymer composition and the reaction conditions, soybean oil polymers can be tailored to possess comparable or better properties than those of LDPE and PS.

Table II. Summary of the Properties of Soybean Oil-Based Polymers

Polymer	T_g (°C)	v_e (mol/m^3)	$T_{max}{}^a$ (°C)	E^b (MPa)	$\sigma_b{}^c$ (MPa)	$\varepsilon_b{}^d$ (%)	Toughness (MPa)
Polyethylene (LDPE)[e]	-68	-	355	370	9.6	46	5.2
Polystyrene[e]	90	-	420	1330	30.3	4	0.5
CLS45-ST47-DVB00-I[f,g]	10	1.0x10^2	448	12	1.3	300	2.0
CLS45-ST32-DVB15-I[f,h]	76	2.2x10^3	475	225	11.5	41	4.0
CLS35-ST39-DVB18-I[f,i]	82	3.4x10^3	477	500	21.0	3	0.8
SOY45-ST32-DVB15-I[f]	68	1.8x10^2	468	71	4.1	57	1.7
LSS45-ST32-DVB15-I[f]	61	5.3x10^2	470	90	6.0	64	2.9
CLS45-ST32-DVB15-I[f]	76	2.2x10^3	475	225	11.5	41	4.0

[a] The temperature at the maximum degradation rate. [b] Young's modulus. [c] Break strength. [d] Elongation at break. [e] Analyses were performed on commercially available LDPE and PS samples. [f] I = modified BFE initiator, typically (NFO5-BFE3). [g] A typical elastomer. [h] A ductile plastic. [i] A rigid plastic.

Good Damping and Shape Memory Properties

Many important engineering plastics and composite materials possess damping properties important for their applications in the aircraft, automotive, and construction industries (39). This class of polymeric materials is used for the reduction of unwanted noise, as well as for the prevention of vibration fatigue failure (39). The damping properties of materials arise from their ability to transform mechanical energy into some other forms of energy, for example, by dissipation of mechanical energy through heat, while undergoing vibrations (40). Most common thermoplastics exhibit efficient damping in the 20-30 °C range. Good damping materials, on the other hand, should exhibit a high loss factor (tan δ > 0.3) over a temperature range of at least 60-80 °C (41). Soybean oil-ST-DVB polymers with proper compositions have been shown to have good damping properties (23). With increasing DVB content, the damping profiles evolve from a narrow, extremely intense damping peak to less intense, but significantly broadened, damping peaks, as shown on Figure 3. DVB, as an efficient crosslinker, increases the degree of crosslinking with increasing content, thus reducing the damping intensity and considerably broadening the damping region of the bulk polymers. It can be seen that the temperature regions for efficient damping vary from 80-110 °C when DVB constitutes 5-15 weight % of the polymer, which indicates very good damping properties for the resulting materials. The overall damping capacity of the soybean oil polymers

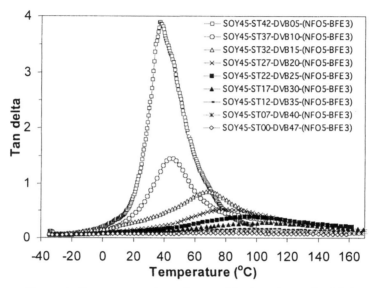

Figure 3. Temperature dependence of the loss factor for SOY45-(ST+DVB)47-(NFO5-BFE3) copolymers prepared by varying the DVB content. (Reproduced from reference 23. Copyright 2002 John Wiley & Sons, Inc.)

(TA) is represented by the area under the loss factor-temperature curve. In general, an increase in polymer crosslink density decreases its TA value and noticeably broadens the damping temperature region. It has also been shown that greater incorporation of the soybean oil into the crosslinked network results in higher TA values. Therefore, it is not surprising that CLS polymers possess higher TA values than the LSS and SOY polymers with the same crosslink densities (23). These findings suggest that one can tailor good damping soy oil-based materials by carefully designing their structures. The greater the number of fatty acid ester groups in the polymer network, the higher the damping intensities. Unfortunately, the triglycerides alone afford strong damping peaks over only a very narrow temperature range. The crosslinking thus appears necessary and beneficial to a certain degree, since it yields segmental inhomogeneities and thus effectively broadens the damping region. Overall, the two opposite effects (crosslinking density and triglyceride incorporation) need to be balanced in order to achieve optimum damping properties for the soybean oil polymers. So far, the most promising damping soy-based materials show loss factor maxima (tan $\delta)_{max}$ of 0.8-4.3, TA values of 50-124 and temperature ranges for efficient damping of ΔT = 80-110 ^{o}C (23).

Along with the good damping properties, the soybean oil-ST-DVB polymers show classical shape memory properties (24). Shape memory polymers are a group of "intelligent" materials, which have the ability to recover their permanent shape on demand after intended or accidental deformation (43). Generally, some kind of external stimuli is required to trigger the shape memory effect. Thermoresponsive shape memory polymers, such as soybean oil copolymers, require heat to revert from their temporary shape to their permanent shape. Structural requirements play a crucial role, since they determine the thermal and mechanical properties of the material and their shape memory properties. As described earlier, soybean oil copolymers are polymer networks composed of long polymer chains interconnected by chemical crosslinks. Soybean oil materials with appropriate compositions have high T_g's and appear as hard plastics at room temperature. However, upon heating they become elastomers capable of undergoing rapid deformation by an external force. Once cooled below its T_g (glassy state), the polymer is capable of retaining the deformation for an unlimited time, due to the very limited chain mobility. Raising the temperature above the T_g triggers the shape memory effect and the polymer returns to its original shape.

Several parameters, including the polymer's ability to deform at $T > T_g$, the ratio of fixed deformation to the original deformation at room temperature, and the polymer's final recovery upon being reheated, have been determined using a bending test (24). Simple variation in the composition of the soybean oil-ST-DVB polymers reveals that the higher deformability at the elastomeric state is always accompanied by a lower ratio of fixed deformation to the original deformation at room temperature, and vice versa. It has been determined that crosslinking density strongly affects deformability in the elastomeric state and that a relatively low degree of crosslinking is favorable for the deformation. The increased rigidity of the polymer backbone allows better fixation of the mechanical deformation at ambient temperature. Unfortunately, increasing rigidity and decreasing crosslinking density represent two opposite effects and do not allow simultaneous optimization of the polymer's deformability and fixation of its deformation. Substituting DVB with more rigid but less effective crosslinkers, such as dicyclopentadiene (DCP) and norbornadiene (NBD), circumvents this problem and allows the preparation of materials with an appropriate combination of chain rigidity and crosslink density. The resulting polymers are able to fix over 97% of their deformation at room temperature and completely recover their original shape upon being reheated. Additionally, soybean oil polymers also possess good reusability, with a constant shape memory effect over at least seven processing cycles (*24*).

Potential applications

The current state of the art in the area of biomaterials design and synthesis indicates that there exist a large number of materials with specific thermal and mechanical properties comparable to those of widely used industrial plastics (*1-3*). Soybean oil-based materials possess such qualities as well. These biomaterials can be made into a variety of complex structures currently made from petroleum-based plastics, by either cutting or injection molding techniques. The soybean oil-based resins can also be fabricated into biocomposites having excellent mechanical properties (*11*). These biocomposites may serve as replacements in a range of applications where engineering plastics are currently being used.

Due to their good damping properties, soybean oil polymers may also find applications in places where the reduction of noise and prevention of vibration fatigue failure is required (*39*). Several conventional polymers, such as polyacrylates, polyurethanes, polyvinyl acetate, as well as natural, silicone and SBR rubbers, which are currently used as damping materials, might be replaced by soybean oil polymers. Unlike the above-mentioned thermoplastics, soybean oil polymers are capable of efficient damping over wider temperature and frequency ranges (*23*).

Soybean oil polymers are the first thermoresponsive shape memory materials prepared from renewable natural resources (*24*). Their good shape memory properties extend their application range even further. They have great potential as materials for construction (heat-shrinkable rivets, gaskets and tube joints), electronics (electromagnetic shielded materials and cable joints), printing and packaging (shrinkable films, laminate covers, trademarks), mechanical devices (automatic valves, lining material, joint devices), medical materials (bandages, splints, orthopedic apparatus), as well as a variety of household, sport and recreational applications (*42*).

Conclusions

A series of new polymeric materials ranging from elastomers to tough and rigid plastics have been prepared by the cationic copolymerization of different soybean oils and alkenes in the presence of a modified BFE initiator. Polymerization reactions have been conducted at relatively low temperatures and pressures, employing at least 50% of natural and renewable starting materials. Manipulations in polymer composition and reaction conditions allow efficient control over the polymer structure and properties. Soybean oil polymers possess a good combination of thermal and mechanical properties, including excellent damping and shape memory properties. Because of this unique combination of valuable properties, these novel biomaterials show promise as replacements for conventional plastics.

The commercialization of this new technology greatly depends on the mutual efforts of polymer chemists, material scientists and chemical engineers, since it involves not only technical problems, but also economic and political issues that need to be resolved. The time is coming when we must seriously look into alternative sources of energy and raw materials. It is our belief that the use of natural and renewable materials represents the future of the polymer industry in the years to come.

Acknowledgements

The authors gratefully acknowledge the Iowa Soybean Promotion Board, the Iowa Energy Center, the Consortium for Plant Biotechnology Research, Archer Daniels Midland, and the USDA for funding our own research in this area.

References

1. *Biopolymers from Renewable Resources*; Kaplan, D. L., Ed.; Macromolecular Systems – Materials Approach; Springer: New York, 1998.
2. *Biorelated Polymers – Sustainable Polymer Science and Technology*; Chiellini, E.; Gil, H.; Braunegg, G.; Buchert, J.; Gatenholm, P.; van der Zee, M.; Kluwer Academic-Plenum Publishers: New York, 2001.
3. Kaplan, D. L.; Wiley, B. J.; Mayer, J. M.; Arcidiacono, S.; Keith, J.; Lombardi, S. J.; Ball, D.; Allen, A. L. In *Biomedical Polymers: Designed to Degrade Systems;* Shalaby, W. S., Ed.; Carl Hanser: New York, 1994; Chapter 8.
4. O'Brien, R. D.; *Fats and Oils – Formulating and Processing for Applications*; 2nd Ed., CRC Press LLC: Boca Raton, FL, 2004.
5. Eckey, E. W.; *Vegetable Fats and Oils*; The ACS Monograph Series; New York: Reinhold Publishing Co., 1954; Chapters 1-4.
6. Salunkhe, D. K.; Chavan, J. K.; Adsule, R. N.; Kadam, S. S. *World Oilseeds: Chemistry, Technology and Utilization*; New York: Van Nostrand Reinhold, 1991.

7. Fehr, W. R. In *Oil Crops of the World: Their Breeding and Utilization*; Robbelen, G., Downey, R. K. and Ashri, A., Eds.; McGraw-Hill: 1989, Chapter 13.

8. Gunstone, F. D. *Industrial Uses of Soybean Oil for Tomorrow*, Special Report '96, Iowa State University and the Iowa Soybean Promotion Board, 1995.

9. *Chemical Market Reporter*, October 11, 2004, pp 24-26.

10. Can, E.; Kusefoglu, S.; Wool, R. P. *J. Appl. Polym. Sci.* **2001**, *81*, 69.

11. Khot, S. N.; Lascala, J. J.; Can, E.; Morye, S. S.; Williams, G. I., Palmese, G. R.; Kusefoglu S. H.; Wool, R. P. *J. Appl. Polym. Sci.* **2001**, *82*, 703.

12. Can, E.; Kusefoglu, S.; Wool, R. P. *J. Appl. Polym. Sci.* **2002**, *83*, 972.

13. Petrovic, Z. S.; Guo, A.; Zhang, W. *J. Polym. Sci. Part A Polym. Chem.* **2000**, *38*, 4062.

14. Javni, I.; Petrovic, Z. S.; Guo, A.; Fuller, R. *J. Appl. Polym. Sci.* **2000**, *77*, 1723.

15. Guo, A.; Javni, I.; Petrovic, Z. S. *J. Appl. Polym. Sci.* **2000**, *77*, 467.

16. Guo. A.; Demydov, D.; Zhang, W.; Petrovic, Z. S. *J. Polym. Environ.* **2002**, *10*, 49.

17. Petrovic, Z. S; Zhang, W.; Zlatanic, A.; Lava, C. C.; Ilavsky, M. *J. Polym. Environ.* **2002**, *10*, 5.

18. Larock, R. C.; Hanson, M. W. U.S. Patent 6,211,315, 2001.

19. Li, F.; Hanson, M. W.; Larock, R. C. *Polymer* **2001**, *42*, 1567.

20. Li, F.; Larock, R. C. *J. Appl. Polym. Sci.* **2001**, *80*, 658.

21. Li, F.; Larock, R. C. *J. Polym. Sci. Part B Polym. Phys.* **2000**, *38*, 2721.

22. Li, F.; Larock, R. C. *J. Polym. Sci. Part B Polym. Phys.* **2001**, *39*, 60.

23. Li, F.; Larock, R. C. *Polym. Adv. Tech.* **2002**, *13*, 436.

24. Li, F.; Larock, R. C. *J. Appl. Polym. Sci.* **2002**, *84*, 1533.

25. Li, F.; Larock, R. C. *Polym. Int.* **2002**, *52*, 126.

26. Li, F.; Larock, R. C. *Polym. Mater. Sci. Eng.* **2002**, *86*, 379.

27. Li, F.; Larock, R. C. *J. Polym. Environ.* **2002**, *10*, 59.

28. Li, F.; Larock, R. C. *J. Appl. Polym. Sci.* **2000**, *78*, 1044.

29. Li, F.; Larock, R. C. *Biomacromolecules* **2003**, *4*, 1018.

30. Li, F.; Hasjim, J.; Larock, R. C. *J. Appl. Polym. Sci.* **2003**, *90*, 1830.

31. Li, F.; Marks, D.; Larock, R. C.; Otaigbe, J. U. *SPE ANTEC Technical Papers* **1999**, *3*, 3821.

32. Li, F.; Marks, D.; Larock, R. C.; Otaigbe, J. U. *Polymer* **2000**, *41*, 7925.

33. Li, F.; Larock, R. C.; Otaigbe, J. U. *Polymer* **2000**, *41*, 4849.

34. Li, F.; Perrenoud, A.; Larock, R. C. *Polymer* **2001**, *42*, 10133.

35. Marks, D.; Li, F.; Larock, R. C. *J. Appl. Polym. Sci.* **2001**, *81*, 2001.

36. Dong, X.; Chung, S.; Reddy, C. K.; Ehlers, L. E.; Larock, R. C. *J. Am. Oil Chem. Soc.* **2001**, *78*, 447.

37. *Cationic Polymerizations – Mechanisms, Synthesis, and Applications*; Matyjaszewski, K., Ed.; Marcel Dekker, Inc.: New York, 1996.

38. Odian, G.; *Principles of Polymerization*, 3rd Ed.; John Wiley & Sons, Inc.: New York, 1991, Chapter 5.
39. *Sound and Vibration Damping with Polymers*; Corsaro, R. D., Sperling, L. H., Eds.; ACS Symposium Series No. 424; American Chemical Society: Washington, DC, 1990.
40. Aklonis, J. J.; MacKnight, W. J. *Introduction to Polymer Viscoelasticity*; 2nd Ed.; Wiley-Interscience: New York, 1983.
41. Yao, S. In *Advances in Interpenetrating Polymer Networks*; Klempner, D.; Frisch, K. C., Eds.; Technomic Publishing Co.: Lancaster, PA, 1994, Vol. IV.
42. Lendlein, A.; Kelch, S. *Angew. Chem. Int. Ed. Engl.* **2002**, *41*, 2034.

Chapter 7

Cyclic Voltammetry as a Potential Predictive Method for Supported Nanocrystalline Gold Catalysts for Oxidation in Aqueous Media

Graham J. Hutchings[1], Silvio Carrettin[1], Paul McMorn[1],
Patrick Jenkins[1], Gary A. Attard[1], Peter Johnston[2], Ken Griffin[2],
and Christopher J. Kiely[3]

[1]School of Chemistry, Cardiff University, P.O. Box 912,
Cardiff CF10–3TB, United Kingdom
[2]Johnson Matthey, Orchard Road, Royston,
Herts. SG8 5HE, United Kingdom
[3]Department of Materials Science and Engineering, Lehigh University,
5 East Packer Avenue, Bethlehem, PA 18015–3195

Nanocrystalline supported Au is a potent catalyst for oxidation of alcohols and polyols in aqueous media using oxygen in which exclusive formation of the mono-acid can be observed. However, relatively non-selective catalysts for the same reaction can also be prepared. Herein we present a detailed characterization study of wholly selective and non-selective Au/graphite catalysts used in the oxidation of glycerol to glyceric acid using a combination of cyclic voltammetry (CV) and transmission electron microscopy (TEM). Analysis using TEM, a technique that is often incisive with supported Au catalysts, in this particular case shows that the non-selective catalysts comprise relatively larger Au crystallites as compared to those of the more selective catalysts. In addition, a set of three Au/C with different selectivities for glyceric acid were characterized using TEM but no discernable differences were observed which could be correlated with the selectivity differences. A detailed study using CV shows distinct differences between the four catalysts. In particular, differences in the activity and selectivity of these Au/graphite catalysts towards glycerol oxidation can be correlated to differences in the relative rates of formation of the selective oxygen species and surface poisons as determined using cyclic voltammetry. It is considered that these cyclic voltammetry correlations can have predictive potential for supported Au catalysts.

Introduction

In recent years, there has been an explosion of interest in the use of gold catalysts for a wide range of reactions (1,2), including selective oxidation (3-7) and hydrogenation (8-10). Particular interest has focused on the nanocrystals of Au supported on a range of oxides or carbon. These supported Au catalysts are often effective at low reaction temperatures; for example Au /α-Fe$_2$O$_3$ is active at 0 °C for CO oxidation (11) and Au-Pd/Al$_2$O$_3$ is effective as a catalyst at 2 °C for the direct oxidation of H$_2$ to give H$_2$O$_2$ (12,13). Most recently, building on the earlier pioneering studies by Rossi and co-workers (3-5) we have shown that Au/C is an effective catalyst for the oxidation of glycerol in which a 100% specificity to glyceric acid can be achieved using molecular oxygen as the oxidant under mild reaction conditions but only in the presence of base (6,7). In the absence of base, the Au/C catalyst is inactive and no glycerol oxidation is observed. This feature appears to be unique to Au catalysts since Pd and Pt catalysts can oxidise glycerol, albeit non-selectively, under acidic conditions (14).

The oxidation of glycerol is of immense importance as glycerol is widely available as a bio-renewable feedstock and can be the source of a wide range of chemical intermediates. However, the electro-oxidation of biorenewable fuels has been extensively studied, given the longstanding interest in the design of effective fuel cells. There are numerous studies concerning gold electrodes and their electrocatalytic activity in fuel cells based on biofuels, such as glycerol (15-21). Using extremely acidic conditions, gold is found to be completely inactive towards glycerol electrooxidation, in contrast to platinum. However, under alkaline conditions gold is actually more active than platinum.

In this paper we aim to combine the two approaches used in the selective oxidation using supported Au nanocrystals and the non-selective electro-oxidation studies using gold electrodes in order to gain new insights into the mechanism by which Au can be a selective oxidation catalyst. In particular, we have now used cyclic voltammetry (CV) to demonstrate the origin of this effect for Au catalysts and, in this paper, we present the first results using this technique for supported gold catalysts which show that the active surface oxygen species formed under oxidising conditions on Au is observed only in the presence of base. Furthermore, based on this approach, we identify two parameters that correlate with the activity and selectivity of Au/C catalysts for the oxidation of glycerol.

Experimental

Catalyst preparation

1 Wt% gold catalysts supported on graphite were prepared as follows. The carbon support (graphite, Johnson Matthey, 113.2 g) was stirred in deionised water (1 l) for 15 min. An aqueous solution of chloroauric acid (41.94% Au, Johnson Matthey, 2.38 g in 70 ml H_2O) was added slowly drop-wise over a period of 30 min. After addition, the slurry was then refluxed for 30 min and following cooling, formaldehyde was added as a reducing agent. The catalyst was recovered by filtration and washed until the washings contained no chloride. The catalyst was then dried for 16 h at 106 °C. This method was also used to prepare 0.25 wt% Au/C and 0.5 wt% Au/C catalyst using proportionately smaller amounts of chloroauric acid. In addition, a further sample of 1% Au/C catalyst was prepared in the same way but using hydrazine as the reducing agent. Furthermore, the carbon support utilised was of somewhat higher surface area. This resulted in a much larger double layer capacitance for this 1% Au on carbon relative to all of the other catalysts as expected on the basis of their relative surface areas.

Oxidation of glycerol

Glycerol oxidation reactions were carried out using a 50 ml Parr autoclave. The catalyst was suspended in an aqueous solution of glycerol (0.6 mol/l, 20 ml). The autoclave was pressurised to the required pressure with oxygen and heated to 60 °C. The reaction mixture was stirred (1500 rpm) for 3 h, following which the reaction mixture was analysed. Analysis was carried out using HPLC with ultraviolet and refractive index detectors. Reactant and products were separated on an ion exclusion column (Alltech QA-1000) heated at 70 °C. The eluent was a solution of H_2SO_4 (4×10^{-4} mol/l). Reaction mixture samples (10 µl) were diluted with a solution of an internal standard (100 µl, 0.2 mol/l isobutanol) and 20 µl of this solution was analysed. It is essential that a standard is added so that the carbon mass balance can be determined. Attempts to find a suitable internal standard that could be added prior to reaction were unsuccessful as all such standards were readily oxidised under the reaction conditions. Hence, it was deemed necessary to add the standard immediately following the reaction as described.

Catalyst characterisation

Samples were structurally characterised in a JEOL 2000 EX high resolution electron microscope operating at 200 kV. The catalyst powders were made suitable for TEM examination by grinding them in high purity ethanol using an agate pestle and mortar. A drop of the suspension was then deposited onto, and allowed to evaporate, on a holey carbon grid.

Cyclic Voltammetry

The electrochemical cell and the electronic apparatus used for voltammetric characterisation have been described in detail elsewhere (22). Briefly, the electrochemical cell consisted of a two-compartment glass cell containing a Pt mesh counter electrode together with a Pd/H reference electrode. The working electrode was a fine gold wire mesh (0.5 cm^2) into which 5 mg of the supported gold catalyst was pressed. All electrolytes (Merck Suprapure/ 18.2 MΩ cm $^{-1}$ Milli-Q ultra-pure water) were thoroughly degassed prior to any measurement in order to preclude contributions to the voltammetric signal from any dissolved oxygen. The potential sweep rate utilised was 10 mV/s. This choice of sweep rate was found to be optimal in order to minimise Ohmic drop associated with the large electric currents generated at the high surface area catalysts whilst maintaining resolution of all redox features.

Results

Oxidation of glycerol using Au/C catalysts

The three catalysts (0.25, 0.5 and 1.0 wt% Au/graphite) prepared using formaldehyde as reducing agent were evaluated for the oxidation of glycerol in the presence of NaOH (60 °C, 3 h, glycerol (12 mmol), NaOH, water (20 ml), catalyst (220 mg), 1500 rpm stirring, oxygen (3 bar)) and gave conversions (selectivity to glyceric acid) of 18% (54%), 26% (61%) and 56% (100%) respectively. The 1% Au/graphite prepared using hydrazine as reducing agent was also examined and this was completely inactive even when using these basic reaction conditions. In the absence of base all the catalysts were found to be inactive for glycerol oxidation.

Transmission Electron Microscopy Analysis

Transmission electron microscopy characterisation of the catalysts showed they all comprised Au nanocrystals with a broad size range (*ca.* 5-50 nm) with a mean particle diameter of *ca.* 25 nm of multiply twinned nanocrystals supported on micron scale graphite flakes. In the 1 wt% Au/graphite specimen, Au particles as small as 5 nm and as large as 50 nm in diameter were detected. The majority, however, were about 25 nm in size and were multiply twinned in character. Decreasing the loading to 0.5 wt% or 0.25 wt% did not appreciably change the particle size distribution; the particle number density per unit area was observed to decrease proportionately however, which may be correlated to the decrease in glycerol conversion observed with these catalysts. However, these TEM studies do not give any real insight into the decrease in selectivity to glyceric acid observed with the lower Au loadings since the particle size distributions are unchanged. At first sight it is interesting to note that the Au particles were considerably larger than the 2-4 nm size that is considered to be optimal for low temperature CO oxidation, yet they were still highly active for the glycerol to glyceric acid reaction. However, they do contain a significant fraction of 5 nm scale particles. The 1% Au/graphite sample prepared using hydrazine was also examined by TEM and this material showed mainly very large Au crystallites (*ca.* 50nm in diameter) and no small gold particles were observed. Hence it is feasible that not all the gold is active in the three Au/C catalysts prepared using formaldehyde as reducing agent and that the activity may be related to the minority population of small Au nanocrystals. However it will be suggested later that a key parameter that may also rationalise these findings is the ratio of the number of "rim" sites (see Scheme 1 Site I, located at the boundary between the particle and the carbon support) and the number of surface adsorption sites associated with the surface gold atoms not adjacent to the support (see Scheme 1 Site II).

Characterisation using Cyclic Voltammetry

A detailed cyclic voltammetric study was carried out on these catalysts. Initial blank experiments were conducted using a gold sample holder and the graphite support in both aqueous H_2SO_4 and NaOH to determine the signals associated with metallic Au and the uncoated support. Figure 1 shows the CV traces obtained from the gold mesh support with no catalyst present in 0.5M NaOH. The usual voltammetric features reported previously for gold (*15*) are

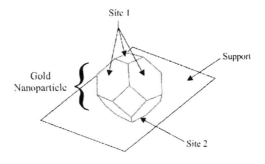

Scheme 1

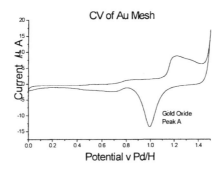

Figure 1. Cyclic voltammogram in 0.5M NaOH of gold mesh electrode used as sample holder for all gold catalysts. Sweep rate = 10 mV/s.

observed including the electrosorption of oxide species at potentials > 1.1V and the corresponding oxide stripping peak at 1.05V (peak A). In addition a broad "double layer" charge between 0.3 and 0.8V ascribed previously to the "incipient oxidation" of gold:

$$Au + OH^- = AuOH + e$$

is also observed (*21*). This AuOH species has been shown to be responsible for the enhanced electrocatalytic properties of gold electrodes under alkaline conditions (*21*).

In contrast to CVs from supported Pt catalysts (*23*), it is also seen that there is no hydrogen underpotential deposition region (H UPD) in the potential range 0 – 0.3V since gold cannot chemisorb hydrogen at room temperature. The large difference in surface area between gold in the supported catalyst and the gold sample holder ensured that contributions from the sample holder itself to the CV signal were negligible. This is seen by comparing the magnitudes of the electric currents in Figure 1 with those in Figure 2. Figure 2 shows CVs obtained in 0.5M NaOH for the 0.25%, 0.5% and 1% Au/C catalysts. Two gold reduction

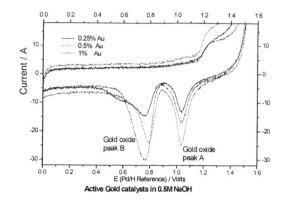

Figure 2. Cyclic voltammetry of Au/graphite catalysts in aqueous NaOH (0.5 mol/ℓ). Sweep rate = 10mV/s

peaks at 1.04V and 0.78 – 0.8V labelled A and B respectively are observed. A is the well characterised "gold bulk oxide" peak seen in Figure 1 and corresponds to place exchange processes occurring between surface gold atoms and electrosorbed oxide (*24*). In contrast, peak B at more negative potentials is not observed in acidic electrolytes (Figure 3) and therefore may be ascribed to the formation of the catalytically active species previously observed on single crystalline and polycrystalline massive gold electrodes (Figure 4 of reference 21). The magnitude of peak B relative to that in reference 21 is far larger, attesting to the unusual adsorption properties of the gold nanoparticles supported on graphite in relation to massive gold electrodes. In fact, more recently (*25*), it has been noted that the reduction of peroxo-species (formed during oxygen evolution) may also occur at these potentials. Future studies in our group will attempt to determine the true nature of peak B.

The sequence of intensities of peak A in Figure 2 is 0.25% < 1% < 0.5% Au. However signal B at 0.78 – 0.8V which is characteristic only of the active Au/graphite catalysts in aqueous NaOH shows some structure with the signal for the 1 wt% Au/graphite catalyst being significantly broader and more intense than for the other two catalysts (0.25% < 0.5% < 1% Au). This may be of importance with respect to the nature of the selective oxidant. For the inactive 1% Au/C catalyst, peak A was hardly observed at all (Figure 4). This would be consistent with relatively large gold particles of low surface area in agreement with the TEM. However, it must also be remembered that the higher surface area of the carbon support gives rise to a much larger contribution to the CV signal than in the previous three samples. This would have the effect of making the gold CV peaks relatively small, even if they originated from gold particles of similar size to those of the active catalysts. A further set of CV experiments was

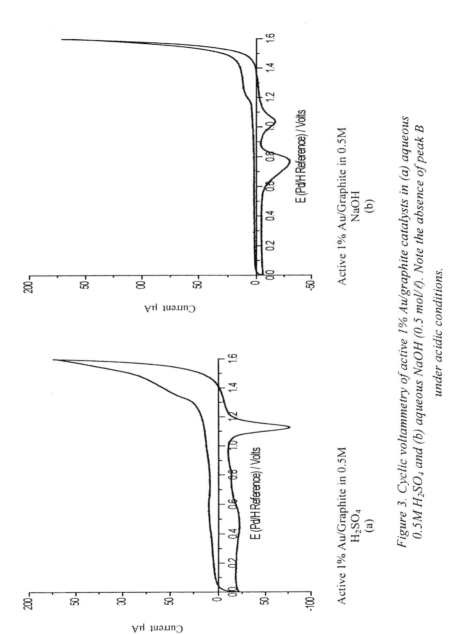

Active 1% Au/Graphite in 0.5M
H₂SO₄
(a)

Active 1% Au/Graphite in 0.5M
NaOH
(b)

*Figure 3. Cyclic voltammetry of active 1% Au/graphite catalysts in (a) aqueous
0.5M H₂SO₄ and (b) aqueous NaOH (0.5 mol/ℓ). Note the absence of peak B
under acidic conditions.*

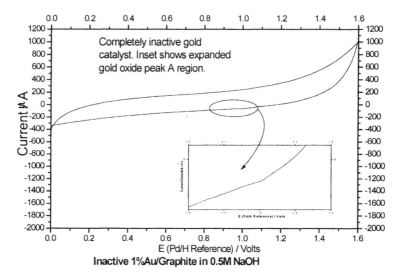

Completely inactive gold catalyst. Inset shows expanded gold oxide peak A region.

Inactive 1%Au/Graphite in 0.5M NaOH

Figure 4. Cyclic voltammogram of completely inactive 1% Au/graphite catalyst. Expanded potential range shows small peak A and negligible peak B

carried out with the Au/graphite catalysts in the presence of glycerol and NaOH/H_2SO_4, thereby studying the behaviour *in situ* under reaction conditions (Figures 5 - 8). Only the first potential cycle is reported since multiple potential excursions to 1.6V would lead to disruption of the surface by oxide formation/stripping. Hence comparison of a "clean" multiply potentially cycled electrode surface with the real catalyst would not be straightforward. Therefore, potential cycling would produce data that are unrelated to glycerol oxidation in the catalysis experiments in the autoclave.

Under acidic conditions, perturbation of the voltammetric signal from the 1% Au/graphite catalyst in the absence of glycerol is found to be negligible for potentials < 1.2V (Figure 5). However, at more positive potentials (strongly oxidising conditions) electrooxidation currents associated with strongly adsorbed intermediates commences in the presence of glycerol. It is recalled at this point that gold catalysts are reported to be inactive towards selective oxidation of glycerol in acidic media and this is consistent with the complete absence of any electrooxidation currents in the range 0 – 1.2V. It is predicted nonetheless, based on the voltammetric results that at much higher pressures of oxygen (associated with a surface electrochemical potential > 1.2V) that even under acidic conditions, some activity with gold should be observed. In contrast to the results obtained under acidic conditions, in 0.5M NaOH all of the gold catalysts gave rise to significant glycerol electrooxidation currents in the potential range 0 – 1.2V (Figures 6 – 8). In the forward potential sweep, all catalysts showed a broad signal associated with the electrooxidation of glycerol at *ca.* 0.9 – 1.3V (labelled C) and a narrower feature on the reverse sweep (labelled D). By comparing Figures 3 and 8, for the oxidation of glycerol on active 1% Au/graphite in 0.5M H_2SO_4 and 0.5M NaOH is without doubt the result of glycerol with peak B previously formed (in alkaline solution), which is not present in acidic solutions. Peak D is the consequence of oxidation of glycerol on the AuOH coverage formed after elimination from the electrode surface of the gold oxides, which takes place during the sweep towards more negative potentials by electrochemical reactions and also by chemical interaction of the glycerol in solution and the gold oxides. This behaviour also emphasises the poisoning effect on the reaction of bulk gold oxides which quench the reaction at potentials > 1.3V on the forward sweep and also down to 1.1V on the negative sweep (due to hysteresis in the "irreversible" formation/desorption of the bulk oxide phase (*24*)). However both peak C and D should also be moderated by the formation of site-blocking, strongly adsorbed molecular fragments derived from glycerol decomposition. Since peak D corresponds to the removal of bulk oxide from the gold surface leaving behind only the Au-OH species (peak B and double layer AuOH) with a minimal amount of molecular fragments being adsorbed (since these have been oxidised during the previous positive potential sweep), this leads to peak D being the most intense and corresponds to the catalyst being in its most active state. Peak

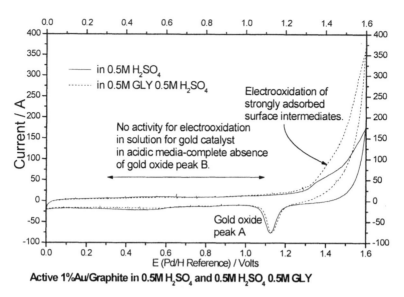

Figure 5. Cyclic voltammetry of 1% Au/graphite catalysts in H₂SO₄ (0.5 mol/ℓ) in the presence and absence of glycerol (0.5 mol/ℓ)

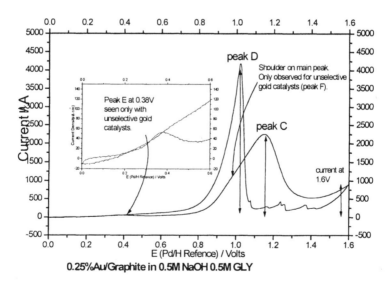

Figure 6. Cyclic voltammogram of 0.25% Au/graphite catalyst in aqueous NaOH (0.5 mol/ℓ) and glycerol (0.5 mol/ℓ). Sweep rate = 10 mV/s

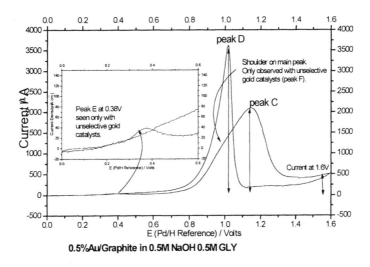

Figure 7. Cyclic voltammetry of 0.5% Au/graphite catalyst in aqueous NaOH (0.5 mol/ℓ) and glycerol (0.5 mol/ℓ)

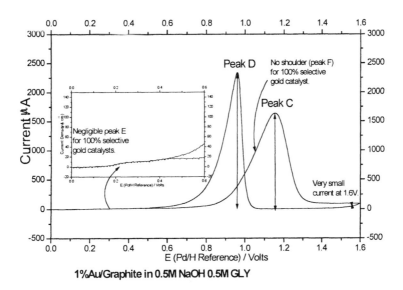

Figure 8. Cyclic voltammetry of 1% Au/graphite catalyst in aqueous NaOH (0.5 mol/ℓ) and glycerol (0.5 mol/ℓ).

C reflects a similar situation although the relative amount of strongly adsorbed molecular fragments is increased (since these have not yet been oxidised at most positive potentials) and hence corresponds to a smaller concentration of Au-OH species due to site-blocking via glycerol decomposition. Both of these factors lead to peak C being less intense than peak D. This suggests that there should be a strong correlation between activity and the relative intensities of peaks C and D. This proposition is explored subsequently. In addition, it should be noted that the 0.25% and 0.5% Au/C catalysts gave rise to two less intense peaks labelled E (0.38V) and F (1.0V). For the active catalyst displaying total specificity to glycerate (1 wt% Au/graphite) peaks E and F are both absent and we consider this to be a key finding. Furthermore, current density at potentials >1.3V associated with the electrooxidation of strongly adsorbed glycerol fragments increases in the order:

1% Au/C < 0.5% Au/C < 0.25% Au/C < 1% Au/C (inactive)

In order to highlight these trends, Figure 9 shows the glycerol electrooxidation CVs of the 1% Au/C (active) together with the 1% Au/C (inactive) catalysts. It is apparent that the ratio of peak intensities for C/D and the amplitude of the current density positive of 1.3V are quite different in each case.

Discussion

The four catalysts evaluated in detail in this study show very different activities for glycerol oxidation and selectivities to glyceric acid. The TEM analysis suggests that the active catalysts prepared using formaldehyde as reducing agent all contain some small Au nanocrystals (<10nm in diameter). In contrast the Au/C material prepared in an analogous manner using hydrazine contained no such small particles and was inactive. The CV study, however, revealed differences between all four samples. In particular, two parameters have been identified that exhibit clear differences: (a) the relative intensities of specific peaks observed in the CV and (b) the amplitude of the current density at >1.3V. Therefore, in Figure 10, these two parameters are plotted versus catalyst activity, namely i) the ratio of currents (j) of peak C/peak D and ii) the ratio of current density at 1.6V/the current density of peak C. Both these parameters express the rates of surface blocking (poisoning) relative to oxidation by adsorbed Au-OH species. Inspection of Figure 10 demonstrates a smooth correlation between glycerol conversion and both of these parameters. Therefore the following mechanistic insights into the selective oxidation of glycerol under basic conditions can be suggested. Activity is a reflection of the rate of formation/oxidation of surface poisons (strongly adsorbed molecular fragments) relative to the formation of an Au-OH surface intermediate formed *only* at free gold sites associated with gold nanoparticles under alkaline conditions. The

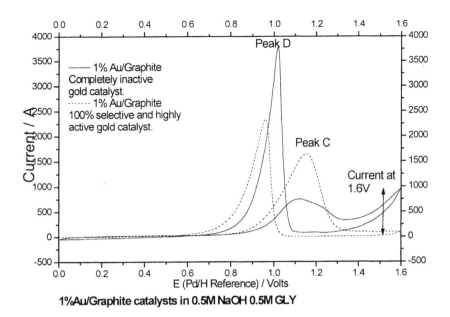

Figure 9. Cyclic voltammetry of 1% Au/graphite catalysts in aqueous NaOH (0.5 mol/ℓ) and glycerol (0.5 mol/ℓ). (——) 1 wt% Au/graphite (active); (b) (----) 1 wt% Au/graphite (inactive). Sweep rate = 10mV/s.

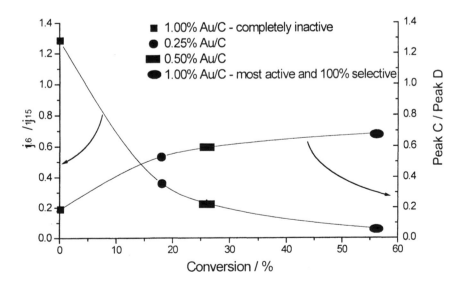

Figure 10. Plot of current density at 1.6 V ($j_{1.6}$)/current at 1.15V ($j_{1.15}$) together with plot of ratio of peak C/peak D versus percentage conversion for various supported Au/C catalysts used for glycerol oxidation.

reason why the active 1% Au/C catalysts are resistant to molecular fragmentation is speculated to be a matter of surface structure. Weaver *et al* (*16*) have already shown that there is a strong structural effect for the electrooxidation of glycerol at single crystal gold electrodes and that defect sites appear to be sites of preferential poison formation. Hence, although TEM analysis reveals only that the inactive catalyst contains somewhat larger gold particles than the other three (whereas the three active gold catalysts display a similar particle size distribution to each other with a very broad range of Au nanocrystal sizes), it does not reveal directly the nature of the adsorption sites themselves. The source of the non-selective oxidation with these catalysts is clearly the strongly adsorbed intermediates together with the extra products formed at more negative potentials (peaks E and F) since, when these are absent, the catalyst is 100% selective. Hence, we propose, based on the combination of the characterisation and the catalytic results, that the peaks C and D in the 100% selective catalyst must correspond exclusively to glycerate formation since this is the only product formed with this catalyst. As for the other peaks, E in particular, at relatively negative potentials, may be attributed to the initial glycerol deprotonation step often observed for organic molecules such as glucose (*26*) and methanol (*27*) on platinum prior to surface electrooxidation. As yet we have no clear proposals concerning peak F but, since it occurs at a more negative potential than peak C, it could logically correspond to a less oxidised intermediate than glycerate such as glyceraldehyde. The electrooxidation features obtained exclusively with the 0.25, 0.5 and 1wt%(inactive) Au/graphite samples are clearly the source of the non-selective oxidation, as has been proposed for Pd and Pt (*17*). These observations together support the notion that the fewer the number of redox processes observed in CV, the more selective will be the Au catalyst towards glycerol oxidation.

Our study shows that the combination of catalytic experiments, together with cyclic voltammetry, can provide significant insights into the origin of the selective catalyst activity of Au nanocrystals. In particular, in the absence of base, it is apparent that glycerol is hardly adsorbed. We have shown that the adsorption of glycerol can only occur when Au-OH is formed as has already been demonstrated for Au{100}type surfaces (*21*). We have commented previously that the base permits the initial deprotonation of glycerol and this step must be essential for adsorption on the surface of the Au nanocrystal. However, the cyclic voltammetry measurements reveal a far more important feature that, in the absence of base, the oxygen-containing species responsible for the activation of glycerol and selective formation of glycerate is not formed (species B in Figure 2). The low intensity of peak B in Figure 1 for the gold holder is intriguing and suggests that the gold nanoparticles possess unique properties. It is doubtful that these unique properties could be ascribed simply to finite size effects associated with changes in electronic structure since, according to TEM, even the smallest catalysts particles correspond to diameters

> 4 - 5nm. Therefore we suggest that it is the interface between the particle rim and the carbon support that is the key structural feature that bestows enhanced oxidative capability relative to massive gold. This is similar to the proposals of Haruta (*11*), Kung (*28*) and Bond and Thompson (*1*) for the oxidation of CO. It would also be consistent with the very poor activity exhibited for the 1%Au on carbon high surface area support in that the electronic structure of the particle rim is heavily influenced by the electronic structure of the adjoining support. More specifically the relative conductivity of both types of carbon support may be crucial. If this suggestion is correct, these sites enable preferential adsorption of an active Au-OH or Au-peroxo intermediate, essential for the selective oxidation of glycerol. Consequently the ratio of peak B (rim adsorption - Site I) to peak A (adsorption of oxide at Site II) should be maximised in order to generate the most active and selective gold catalysts. Catalysts are being prepared presently to test this hypothesis. Hence, the base is required for two essential features of selective oxidation, i.e. substrate adsorption and formation of the selective oxidant in agreement with previous electrochemical findings (*15-21*). Cyclic voltammetry clearly permits a greater insight into oxidation processes using Au catalysts than has been possible using the alternative techniques used to date. Future work using underpotential deposition (UPD) of Pb as a surface probe (*29,30*) will try to elucidate the nature of the active site responsible for surface poison formation in relation to terrace and defect sites. In addition, the predictive aspects of this technique, in terms of the number/position. relative intensity and presence/absence of redox peaks corresponding to catalytic activity/selectivity variations, is expected to be quite general. So much so that the substantial amount of existing electrochemical data may prove invaluable in assessing activity for, as yet, untried catalytic reactions using supported gold.

Acknowledgement

We thank Johnson Matthey and the EPSRC for financial support.

References

1.   Bond, G.C.; Thompson, D.T., *Cat. Rev.-Sci. Eng.* **1999**, *41*, 319.
2.   Prati, L. *Gold Bull.* **1999**, *32*. 96.
3.   Prati, L.; Rossi, M. *J. Catal.* **1998**, *176*, 552.
4.   Porta, F.; Prati, L.; Rossi, M.; Colluccia, S.; Martra G. *Catal. Today* **2000**, *61*, 165.
5.   Bianchi, C.; Porta, F.; Prati L.; Rossi M. *Top. Catal.* **2000**, *13*, 231.

6. Carrettin, S.; McMorn, P.; Johnston, P.; Griffin, K.; Hutchings, G.J. *Chem. Commun.* **2002**, 696.
7. Carrettin, S.; McMorn, P.; Johnston, P.; Griffin, K.; Kiely, C.J.; Hutchings, G.J. *Phys. Chem. Chem. Phys.* **2003**, *5*, 1329.
8. Bailie, J.; Hutchings, G.J. *Chem. Commun.* **1999**, 2151.
9. Bailie, J.E.; Abdullah, H.A.; Anderson, J.A.; C.H. Rochester,; Richardson, N.V.; Hodge, N.; Zhang, J.G.; Burrows, A.; Kiely, C.J.; Hutchings, G.J. *Phys. Chem. Chem. Phys.* **2001**, *3*, 4113.
10. Claus, P.; Bruckner, A.; Mohr, C.; Hoffmeister, H. *J. Am. Chem. Soc.* **2000**, *122*, 11430.
11. Haruta, M.; Yamada, N.; Kobayashi, T.; IJima, S. *J. Catal.* **1989**, *115*, 301.
12. Landon, P.; Collier, P.J.; Papworth, A.J.; Kiely, C.J.; Hutchings, G.J. *Chem. Commun.* **2002**, 2058.
13. Landon, P.; Collier, P.J.; Chadwick, D.; Papworth, A.J.; Burrows, A.; Kiely, C.J.; Hutchings, G.J. *Phys. Chem. Chem. Phys.* **2003**, *5*, 1917.
14. Fordham, P.; Garcia, R.; Besson, M.; Gallezot, P. *Stud. Surf. Sci. Catal.* **1996**, *101*, 161.
15. Beden, B.; Cefin, I.; Kakyaoglu, A.; Takky, D; Lamy, C. *J. Catal.* **1987**, *104*, 37.
16. Hamelin, A.; Ho, Y.; Chang, S.-C.; Gao, X.; Weaver, M.J. *Langmuir*, **1992**, *8*, 975.
17. Vitt, J.E.; Larew, L.A.; Johnson, D.C. *Electroanal.* **1990**, *2*, 21.
18. Kahaoglu, A.; Beden, B.; Lamy, C. *Electrochim. Acta*, **1984**, *29*, 1489.
19. Enea, O.; Ango, J.P. *Electrochim. Acta*, **1989**, *34*, 391.
20. Alonso, C.; Gonzalez-Velsasco, J. *Z. Phys. Chem.* **1990**, *271*, 799.
21. Avramov-Ivic, M.L.; Leger, J.M.; Lamy, C.; Jovic, V.D.; Petrovic, S.D. *J. Electroanal. Chem.* **1991**, *308*, 309.
22. Evans, R.W.; Attard, G.A. *J. Electroanal. Chem.* **1993**, *345*, 337.
23. Price, Matthew, M Phil., Cardiff (2000).
24. Schneeweiss, M.A.; Kolb, D.M.; Liu, D.; Mandler, D. *Can. J. Chem.* **1997**, *75*, 1703.
25. El-Daeb, M.S.; Ohsaka, T. *J. Electroanal. Chem.* **2003**, *553*, 107.
26. Popovic, K.D.; Tripkovic, A.V.; Adzic, R.R. *J. Electroanal. Chem.* **1992**, *339*, 227.
27. Jarvi, T.; Stuve, E. *Electrocatal.* **1998**, 75.
28. Costello, C.K.; Lang, J.H.; Law, H.Y.; Lin, J.N.; Marks, L.D.; Kung, M.C.; Kung, H.H. *Applied Catalysis A*, **2003**, *243*, 15.
29. Hamelin, A. *J. Electroanal. Chem.* **1984**, *165*, 167.
30. Hamelin, A.; Lipkowski, J. *J. Electroanal. Chem.* **1984**, *171*, 317.

Chapter 8

How Biobased Products Contribute to the Establishment of Sustainable, Phthalate Free, Plasticizers and Coatings

J. van Haveren[1], E. A. Oostveen[1], F. Micciché[1], and J. G. J. Weijnen[2]

[1]Agrotechnology and Food Innovations B.V., Bornsesteeg 59, P.O. Box 17, 6700 AA Wageningen, The Netherlands
[2]Sigma Coatings B.V., Oceanenweg 2, 1040 HB, Amsterdam, The Netherlands

Biobased components for the development of environmentally friendly, durable products are being described. The potential and versatility of isosorbide diesters as substitutes for the currently phthalate based plasticisers for PVC and other resins, is shown. Also high solid alkyd resins for decorative paints, completely based on commercially available renewable resources are being described. Paints comprising alkyd resins based on inulin or sucrose and unsaturated fatty acids or oils, showed very good gloss, excellent levelling and a low intrinsic viscosity making them suitable to be used in high solid alkyd paints. A new concept was used to prepare effective, cobalt free, drying catalysts for alkyd paint systems. Iron in combination with ascorbic acid (vitamin C), and optionally other ligands, was found to be a very effective drying catalyst.

Introduction

During the past few decades a substantial amount of effort has been devoted to the development of renewable products that are readily biodegradable, such as bioplastics for solid disposable articles, packaging applications, biobased surfactants, co-builders and bleaching activators for detergency applications and starch and sugar based products for the cosmetics industry.[1]

All these applications have in common that the "operational lifetime" of these products is quite limited and that in case these products are based on renewable materials they are usually, in a post-consumer phase, readily biodegradable under composting or aquatic conditions.

For many products in our daily life, however, a much longer operational lifetime is required and for numerous products such as construction materials or coatings the predominant aim is to make them as durable as possible aiming at a long service life.

In order to come to an economy that is sustainable in all respects it is of utmost importance that biobased materials are able to make an enlarged contribution to products where a long service lifetime is required. The aim of the work described in this paper is to make a contribution to biobased products requiring a long service time. In this paper research and development activities towards phthalate free, isosorbide based plasticisers, high solid alkyd resins and cobalt free drying catalysts are described.

Results and discussion

Plasticisers

Plasticisers are used to lower the glass transition temperature of numerous materials such as sealants, inks and nail polish. As a result materials get softer and more flexible. However, their main application is to plasticise polyvinylchloride (PVC).[2] In 2004 the world consumption of plasticisers was approximately 4 million tonnes, with 85-90 % of these plasticisers being used to plasticise PVC. Plasticised PVC is used in the production of a large variety of different materials including packaging, medical devices, cable and wire, roofing, tents, shower curtains, floor coverings, wallpaper and toys. Plasticisers based on phthalate chemistry currently dominate the plasticiser market to a very large extent; almost 90 % of all plasticisers are based on phthalates.

For over 20 years there has been a continued debate on the potentially negative environmental and health effects of phthalate based plasticisers. Phthalates are diesters of phthalic acid anhydride and alcohols. The most commonly used phthalate for decades was DEHP (diethylhexylphthalate). In recent years numerous studies indicate estrogenic effects especially for DEHP and DBP (dibutylphthalate) and as a result very recently the market share of DEHP (and DBP) is decreasing.[3] DINP (diisononylphthalate) and DIDP (diisodecylphthalate) are now increasingly used instead of DEHP, predominantly because they have a performance profile which is very close to that of DEHP and fewer studies have dealt with DINP and DIDP than with DEHP. Non-phthalate based plasticisers such as epoxidised soy bean oil, alkyl adipates or citrates have a minor market share because they are not generally applicable, because they are not primary plasticisers or because they have other negative aspects such as high volatility or unpleasant odor.

Switching from DEHP to DEHP analogues nonetheless is unlikely to end the controversy and therefore there remains a need for a new, undisputed, safe and general applicable type of plasticiser. In close co-operation with the PVC industry, a new family of plasticisers based on isosorbide esters has been developed.

Isosorbide diesters

Isosorbide is derived from the natural sugar polyol sorbitol (a common food ingredient). Sorbitol is commercially produced by hydrolysis of starch into glucose and hydrogenation of glucose to sorbitol.[4] In Western Europe sorbitol is produced in a volume of approximately 300,000 tonnes per annum at approximately 0.70 €/kg. Cargill, Roquette and ADM are among world's largest sorbitol manufacturers. Isosorbide, which can be obtained from sorbitol by two consecutive dehydration steps, has been identified recently by the chemical industry as one of the main future renewable platform chemicals (Figure 1).

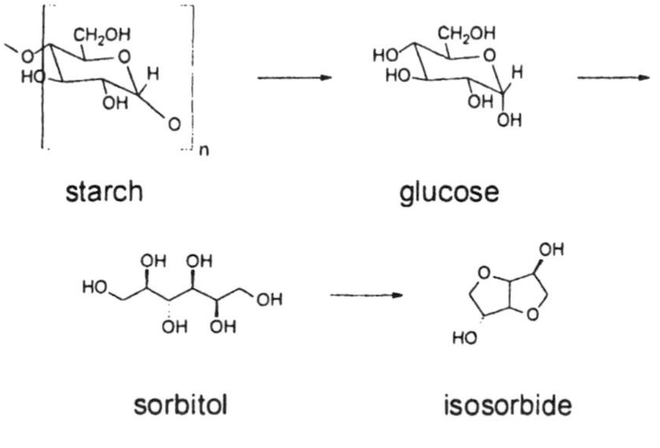

starch   glucose

sorbitol   isosorbide

Figure 1. preparation of isosorbide from natural raw materials.

Phthalates, such as DEHP, are (di)esters of phthalic acid anhydride. In Figure 2 the isosorbide analogue of DEHP is shown.

DEHP

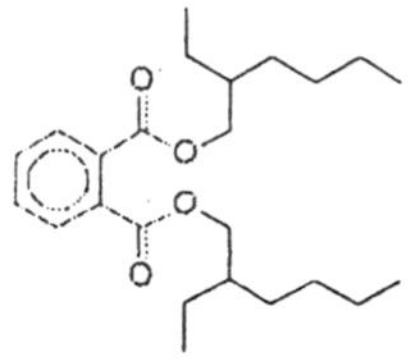

IsDEH

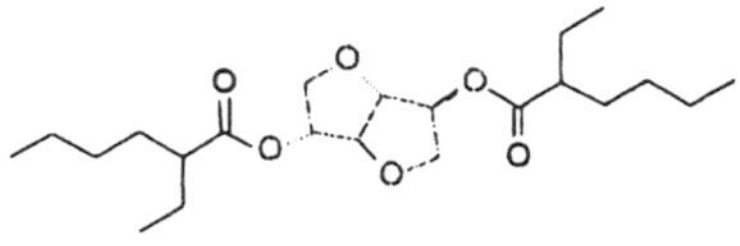

Figure 2. DEHP vs. IsDEH.

Isosorbide diesters can be prepared in high yield, high purity and excellent colour using new technology.[5] Among the critical conditions is the use of an appropriate ion-exchange resin. Both isosorbide and sorbitol can be used as starting materials. The new technology allows for the direct conversion of sorbitol to isosorbide diesters in a single one-pot procedure. The synthetic procedure has been proven to be highly reproducible up to 50-L scale.

Synthetic variation is limited only by the availability of the alkanoic acids. Depending on whether the alkanoic acid is linear or branched and the chain length, the alkanoic acid can be derived from either petrochemical or renewable resources. Over 15 different types of isosorbide diesters have been synthesised.

Chemical conversion of carbohydrates–including esterification – often leads to colour formation. A new work-up procedure is capable of removing the color and byproducts until a near waterwhite product of 45 hazen is obtained.

Performance of isosorbide diesters as plasticisers

It was found that isosorbide diesters are functionally comparable with the equivalent phthalates. Required properties, such as volatility, can be "tuned" by variation of the ester groups. Extensive testing in PVC formulations has shown that isosorbide diesters are primary plasticisers, are highly compatible with PVC, have good thermal stability, have excellent optical properties and have good migration/extraction characteristics.

Plasticiser efficiency is reflected in the Shore hardness values. Figure 3 shows that isosorbide C7/C8 esters are highly comparable to the industry standard DEHP (35 phr indicates 35 weight% of plasticiser compared to hundred parts of PVC).

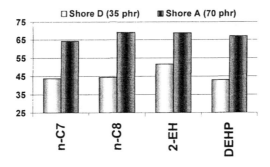

Figure 3. Shore hardness values for isosorbide esters vs. DEHP in PVC.

Environment and Health

Tests have shown that isosorbide esters are fully biodegradable under aquatic aerobic conditions. Although they are readily biodegradable under aquatic conditions, there are no indications that PVC plasticised with isosorbide based plasticisers is more vulnerable to microbial attack than PVC plasticised by phthalates.

Toxicology tests have shown that isosorbide esters are not acutely toxic, cause no sensitisation, and are not mutagenic (Ames test). Recent toxicology results also indicate that isosorbide esters are not estrogenic.

Isosorbide esters have also been successfully tested in several non-PVC materials like cellulose acetate, sealants and rubbers and appeared to be excellent plasticisers compared with the currently used plasticisers.

Biobased alkyd resins

Alkyd resin based paints in the decorative paint industry; current situation

Many items in our everyday life are coated, with the coatings having either a protective, signal or a decorative function. Coatings obviously aim (besides their esthetical function) at enhancing the *durability* of a product. Moreover, nowadays true *sustainability* of coatings is required as well, requiring coatings with properties such as being solvent free, easy to apply, recyclable and producing less waste. Such coatings, when based on *undepletable, renewable* resources will make an important contribution to enhanced *sustainability*.

Alkyd resins and alkyd resin based paints are of utmost importance for the decorative coatings industry. Currently alkyd resins are being produced in quantities of approximately 400,000 tonnes/yr in Western Europe. Traditionally alkyd paints are organic solvent based, using up to 50 % of volatile organic components (VOC) in a paint formulation.[6] These types of coatings are especially popular because of their ease of appliance and high gloss characteristics. New existing and anticipated legislation, aiming at reducing VOC levels, is the current driving force for the coatings industry, especially for decorative paints. Manufacturers of solvent-based paints have to respond to these forces, leading to basically three main options:

- Switch to nonalkyd water based systems
- Switch to water based alkyd systems
- Switch to (high) solid systems containing less/no VOC.

Another environmental issue connected with the use of current alkyd resin based paints, is the usage of cobalt based drying catalysts. Studies on mice and rats have shown that exposure to cobalt sulphate in aerosol form is related to the formation of cancer.[7] Consequently there is, especially in Europe, a driving force to move away from cobalt based drying catalysts towards effective and more environmentally acceptable drying catalysts.

High-solid alkyd resins based on renewable resources

Conventional air drying paints usually comprise alkyd oligomers as binders. These alkyds used in coating formulations are suitable for application at ambient temperature and are synthesised by a polycondensation reaction of polycarboxylic acids, polyhydric alcohols and unsaturated fatty acids or oils (Figure 4). Highly suitable polycarboxylic acids include phthalic acid anhydride, terephthalic acid anhydride, trimellitic acid, adipic acid and maleic acid.

The polyalcohols normally used in the alkyd synthesis include: (di)pentaerythritol, glycerol, ethylene glycol, trimethylolpropane and neopentylglycol. Whereas the fatty acid or oil component is derived from renewable resources, the majority of the aforementioned polyacids and polyalcohols are alkyd building blocks derived from petrochemical resources.

Conventional synthesis of alkyd resins leads to polycondensates with a rather high polydispersity, resulting in the high molecular weight component contributing to a high intrinsic viscosity and the low molecular component leading to a negative impact on drying properties. In order to obtain high solid paints, it is important to reduce the intrinsic viscosity of the alkyd resin. This might be achieved by developing an alkyd oligomer with a suitable average molecular mass and a narrow molecular weight distribution (low polydispersity).

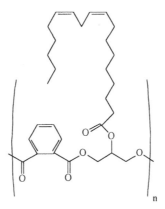

Figure 4. Structure of a typical alkyd resin

Alkyd resins completely based on renewable resources have been the subject of a few studies. EP 0741 175 A2 to Hoechst[8] describes waterborne alkyd emulsions based on renewable resources. The patent describes the use of sorbitol (easily derived from starch) as the polyhydric alcohol part and succinic acid anhydride as the polycarboxylic acid part. Hájek[9] described the use of sorbitol and xylitol in alkyd resin synthesis, whereas Bagchi et al.[10] described the partial replacement of conventional polyols, i.e., glycerol and pentaerythritol, by sorbitol.

US 3,870,664[11] and US 3,870,664 to Research Corporation New York[12] disclose oil modified sucrose resins obtainable by reaction of a partial esterified sucrose ester with a cyclic dicarboxylic anhydride and a diepoxide. Although this work mentions alkyd resins based on renewables, it is not specifically aimed at deriving high solid alkyd paints based on resins with a low intrinsic viscosity.

Within the framework of a research project together with Dutch based paint producer SigmaKalon, the development of alkyd resins with low intrinsic viscosity was aimed for. In order to achieve this, we explored the potential of two types of renewable, carbohydrate derived resources as the polyhydric alcohol part of alkyd resins, i.e. sucrose and inulin.[13,14,15] These carbohydrate based resources were chosen because of various reasons:

1) Inulin is an oligomer of (2>1)-β-D-fructofuranan with a terminal α-D-glucopyranosyl group. It is commercially produced from chicory or Jerusalem artichoke. Depending on the type of plant and harvesting time, inulin has different molecular weight distributions and different average chain lengths between about 8 and 25. Because of its oligomeric nature, with regard to alkyd resins inulin could substitute both the petrochemical based polyhydric alcohol part as well as the polycarboxylic acid part. Inulins with a polydispersity lower

than 1.5 are commercially available. Potentially alkyd resins based on inulin could therefore have a narrow molecular weight distribution.

2) Sucrose is a cheap, easily accessible disaccharide commercially produced in millions of tonnes quantities/annum from sources such as sugar beet and sugar cane.

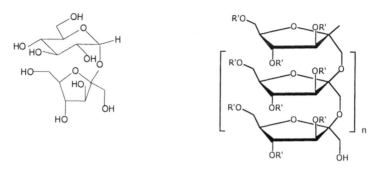

Figure 5. Structures of sucrose and inulin, respectively.

Synthesis of alkyd resins

A synthesis procedure for the sucrose and inulin based alkyd resins was developed by dissolving non-modified or partially acetylated sucrose or inulin in DMAA (N,N-dimethylacetamide), adding sunflower oil or methyl linoleate and reacting the mixture at 140 °C. Catalysts such as potassium carbonate or lithium hydroxide were found to be the most effective.

Derivatisation of inulin with unsaturated fatty acid methyl esters or oils results directly in products with alkyd type characteristics. In order to synthesise oligomers from sucrose, the addition of renewable dicarboxylic acids (or polycarboxylic acids) is also required along with sucrose and fatty acid methyl esters. Examples of such renewable dicarboxylic acids are dimethyl succinate and dimethyl sebacate.

In case of sucrose this results in structures of alkyd resins as depicted in Figure 6. This type of alkyd resin was synthesised by using either sucrose or sucrose (octa)acetate as the starting material. After synthesis and work-up, characteristics of the alkyd resins were determined by techniques such as NMR and GPC.

Characteristics of paint formulations based on renewable alkyds

The most promising binder systems were synthesised at 500 -2000 gram scale and used to prepare paint formulations. Characteristics of paint formulations based on inulin type alkyd resins are shown in Table I.

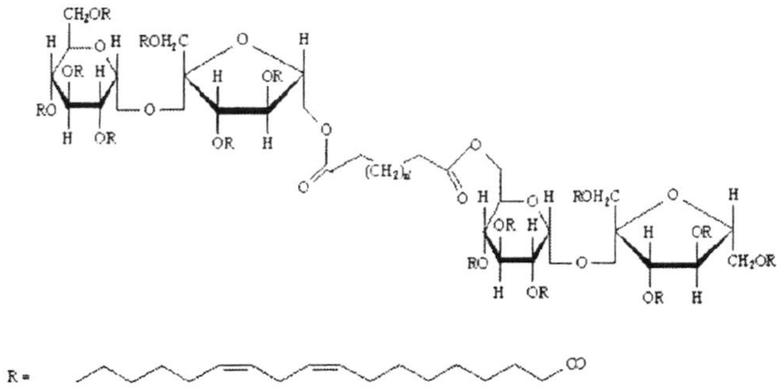

Figure 6. Alkyd resins from sucrose

The inulin esters exhibit a narrow molecular weight distribution. A low polydispersity of a polyester is favourable for a low intrinsic viscosity and consequently only a limited amount of cutting solvent is required to obtain a practical processing viscosity.

From the data it is clear that the drying times of the paints based on the renewable alkyd resins are comparable to commercial reference products and that also wrinkling, elasticity, hardness and gloss is comparable to commercial products. In an accelerated yellowing test (6 hours in NH_4OH), paints containing alkyd resins based on inulin tend to yellow more quickly than conventional alkyd resins. This, however, probably can be related to the presence of (small amounts) of reducing sugars in the (un)reacted inulin. In the presence of NH_4OH this will lead to Maillard reactions and result in yellowing/browning. A more optimised synthesis procedure potentially can overcome this.

Water sensitivity, adhesion and gloss retention upon QUV-A weathering appeared to be strongly dependent on the hydroxyl value of the inulin esters; lower OH-values improve these properties. [13]

Characteristics of high solid alkyd paint formulations containing sucrose based alkyd resins are shown in Table II.[14]

Table I. Characteristics of 3 paint formulations containing inulin based alkyd resins; compared to 3 commercial alkyd based paints.

Test	Paint 1	Paint 2	Paint 3	A*	B*	C*
VOC (g/l)	210	260	134			
Solid content	83.9	80.1	82.2			
Viscosity (dPa.S)	12.8	9.4	8.3			
BK drying recorder: 23 °/50% RH						
Run back	1 h, 25'	1 h, 45'	1 h, 40'	1 h, 45'	1 h, 30'	1 h, 25'
start of gel tear	1 h, 25'	1 h, 45'	1 h, 55'	1 h, 45'	1 h, 30'	1 h, 25'
End of gel tear	4 h	6h, 30'	>12 h	8 h, 15'	9 h, 30'	6h, 35'
End of track	>12 h	>12 h	>12 h	>12 h	>12 h	>12 h
Wrinkling: 300 μm wet film thickness	0	0	0	0	0	0
Elasticity (depth of indentation in mm)	6.9	8.1	7.2	8.5	8.5	9.2
Hardness (1 day)	3	3	3	3	3	3
Hardness (14 days)	1	1	1	1	1	1
Gloss at ∠ 20°/60°	76/87	73/86	74/86	60/82	77/87	76/88
Levelling	1	1	1	1	1	1
Yellowing (6h in NH$_4$OH atmosphere)	- 14.9	- 13.4	- 18.3	- 10.4	- 11.5	-12.7

* A and B are commercial high solid alkyd based formulations where the alkyd contains an aromatic polybasic acid . A is an air-drying high-solid paint for outside and inside use based on modified alkyd resins with a VOC of 257 g.l. C is a rapid through drying high solid paint based on modified urethane- alkyd resins with a VOC of 263 g/l.

Table II. Paint formulations containing a sucrose based alkyd resin (OVN -
227 and OVN 228); methyllinoleate was used as the fatty acid component.
Low shear viscosity was measured with a "Haake VT 500" viscometer
using a cylindrically shaped E30 spindle at a rotation speed of 179 rpm and
at 23 °C . High shear viscosity was measured in accordance with the ICI
cone and plate method (ASTM D 4287) at a shear rate of 20.000 s^{-1}
0 = excellent performance, 5= very poor performance

Test	OVN-227	OVN-228	OVN-231	Comm. product
Low shear viscosity (dPa.s)	14.0	14.0	13.8	
High shear viscosity (dPa.s)	9.8	> 10	9.6	4.6
Solids (%)	89.6	93.7	85.7	80.5
VOC (g / L)	134	83	180	260
Whiteness	77.1	75.4	74.8	76.5
Drying (RT; 50 % RV)	0	0	0	0
Drying (5°C; 90 % RV)	0	0	0	0
Gloss	98.8	88.9	83.5	85.9
Water sensitivity (4 days)	0	0	0	0
Levelling	0 - 1	0 - 1	0 - 1	1 - 2
Hiding power	1	1	1	1

Paint formulations containing sucrose based alkyd resins showed very attractive properties. The formulations exhibited a fast (through) drying, very good levelling, good waterfastness and adhesion to wood, and good hiding power. A minor negative aspect was the tendency to have a relatively high viscosity at high shear, which needs adaptation. Furthermore, results of QUV-A tests showed that the gloss retention of paint formulations containing the sucrose based alkyd resins was just as good as commercially produced products. This is a strong indication that alkyd paints containing sucrose based alkyd resins will be as durable as conventional alkyd paints, partly based on renewables and partly based on petrochemical resources. On top of that for the sucrose based products there are interesting options for a price effective production.

Cobalt free drying catalysts

A quick drying of alkyd paints is of enormous commercial importance. Common solvent borne alkyd paints contain small amounts of cobalt based driers (e.g. cobalt ethylhexanoate) besides the main constituents of alkyd resins (binders), pigments and solvents. The cobalt salts increase the oxidative cross-linking rate of the unsaturated fatty acids which are present as constituents of alkyd resins.

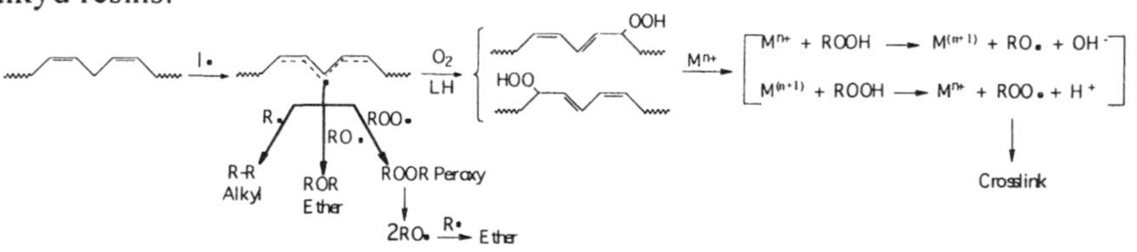

Figure 7. Unsaturated fatty acid oxidation by radical pathway and crosslinking reactions.

As a result of the fatty acid oxidation during air drying, hydroperoxides are formed. Their decomposition into free radicals, catalysed by metal ions, together with the evaporation of solvent (VOC emissions), contributes to the hardening of the film. In spite of their important role in the drying process of alkyd paints, cobalt driers, however, are facing environmental pressure because studies indicate that specific cobalt salts might be carcinogenic in aerosols.[7] Therefore finding substitutes for cobalt based catalysts is of high importance to the alkyd producing and using industry.

Several substitutes for cobalt based drying agents are under development. Most of these are based on either manganese[17] or vanadium salts.[18] Manganese - bipyridine complexes have been especially promoted as alternatives.[17]

Although these alternatives can be applied in specific systems, their overall performance does not match up to that of cobalt based driers; the paint films usually remain too soft, whereas manganese also has a negative impact on the film colour. At A&F a biomimetic approach is being followed to develop alternative cobalt free drying catalysts.

Mechanisms by which nature *in vivo* oxidises (Figure 8) unsaturated fatty acids by the action of oxygen, iron and ascorbic acid (reducing agent) are being translated and adopted for usage in both water borne and solvent based alkyd systems.

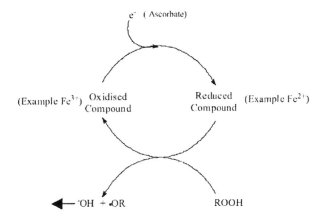

Figure 8: Redox cycle enabling the oxidation of unsaturated fatty acids. Ascorbic acid (AsA) reduces Fe^{3+} back to Fe^{2+}.

The principle was verified by studying diluted aqueous emulsions of methyl linoleate. It was shown that in this system the combination of Fe(II)/AsA induces hydroperoxide formation, in case a neutral (Brij 700) or negatively charged surfactant (SDS) is used to emulsify the methyl linoleate.

In order to study more real water borne and solvent based paint systems, besides ascorbic acid (AsA), 6-O-palmitoyl-L-ascorbic acid (AsAp) has also been used as a reducing agent. This latter derivative is, in contrast to L-ascorbic acid, easily soluble in solvent based systems.

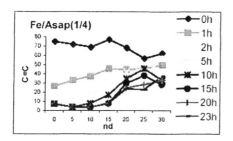

Figure 9. Structures of L-ascorbic acid and 6-O-palmityol-L-ascorbic acid

Iron based catalysts, at different Fe/AsA and Fe/AsAp ratios, were tested for their ability to serve as drying catalysts for both waterborne and solvent borne systems, including varnishes and complete paint formulations (Table III).[19]

Results of Table III demonstrate that addition of imidazole (Im) as a ligand for the iron further enhances the drying capacity of the iron/ascorbic acid palmitate based system. Hardness development is improved compared to manganese based systems. Colour formation is a little bit less than in case of manganese. This, however, might be solved by choosing the appropriate ligand or reducing the amount of iron in the system.

In Figure 10 the first results of a confocal Raman spectroscopy study on short oil-based alkyd films are presented. Depicted is the change in intensity of C=C double bonds as function of time throughout the layer of films. Striking is the enhanced activity of the iron/ascorbyl palmitate driers in the deeper layers of the film (improved through drying), with respect to the conventional cobalt catalyst. An improved through drying in deeper layers might result in a decreased necessity to use auxiliary driers based on Ca or Zr, as is common in the case of cobalt based driers.

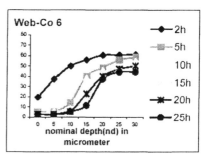

Figure 10. Intensity of C=C peaks versus thickness of short oil-based alkyd films determined by confocal Raman spectroscopy at 1 h time intervals

Table III: Fe/AsAp systems as drying catalyst for a high gloss white paint based on Setal 16 LV WS – 70. As Co-based catalyst Nuodex WEB Co 6 was used

Drier	% of active drier	Incubation Time (days)	Dust free drying (h)	Tack-free drying (h)	Total drying time (h)	Hardness (s/K) 5 days	Skin formation	Whiteness Index
Fe/Im/AsA6p 1/4/1	0.07 Fe	12	5.06	3.8	10.8	56	N.O.	80.6
Fe/Im/AsA6p 1/4/2	0.07 Fe	12	2.0	4.9	11.3	56.4	O.	80.8
Fe/Im/AsA6p 1/4/3	0.07 Fe	12	8.6	4.2	17.3	68.6	O.	81.5
Fe/Im/AsA6p 1/4/4	0.07 Fe	12	--------	--------	> 20	59.7	N.O.	81.2
Co + Exkin 2	0.07 Co	12	3.3	4.2	14.1	73	N.O.	82.8
Mn	0.04 Mn	12	0.66	6	12.5	52	O.	83
Mn	0.07 Mn	12	0.66	4.76	9.9	51.3	O.	82.1

Drier	% of active drier	Incubation Time (days)	Dust free drying (h)	Tack-free drying (h)	Total drying time (h)	Hardness (s/K) 5 days	Skin formation	Whiteness Index
Fe/Im/AsA6p 1/4/1	0.07 Fe	18	0.7	2.66	4.5	73.6	N.O.	80.6
Fe/Im/AsA6p 1/4/2	0.07 Fe	18	0.63	0.43	2.2	80.6	O.	80.8
Fe/Im/AsA6p 1/4/3	0.07 Fe	18	0.6	1.53	5.8	85	O.	81.5
Fe/Im/AsA6p 1/4/4	0.07 Fe	18	1.16	1.53	6.6	80	N.O.	81.2
Co + Exkin 2	0.07 Co	12	3.3	4.23	14.1	73	N.O.	82.8
Mn	0.04 Mn	12	0.66	6	12.5	49	O.	83
Mn	-0.07 Mn	12	0.66	4.76	9.9	51.3	O.	82.1

*Incubation time is defined as the time between adding the catalyst system to the paint formulation and applying the paint as a film on a glass plate.

Conclusions and outlook

Research projects carried out at A&F clearly indicate that effective plasticisers for PVC and other resins and binders and additives for alkyd based decorative paints can be based on those raw materials which are the least undisputed from an environmental point of view i.e. renewable resources. These products are not only renewable based but also durable. Provided a positive life cycle analysis, after further R&D work plasticisers and high-solid or water borne alkyd paints based on renewables therefore can make an important contribution to a sustainable chemical industry.

Acknowledgements

Part of the work on the plasticiser development was sponsored by several Dutch Ministries: the Ministry of Economic Affairs, the Ministry for Housing, Spatial Planning and the Environment, the Ministry for Education, Culture and Science and the Ministry of Agriculture, Nature Management and Fisheries. Research and development work on the high solid alkyd resins and reactive diluents was done in a joined cooperation project between A&F and SigmaKalon.

Research on the iron based drying catalyst has been financially supported within the framework of the Dutch IOP MT/ZM (Environmental technology/Heavy metals) programme. O. Oyman (Technical University of Eindhoven) is acknowledged for supplying Figure 10. DSM Coatings Resins is thanked for supplying URADIL alkyd emulsion and Sasol Servo Delden B.V. for supplying a manganese based catalyst.

References

1. H. Röper, *Starch* **2002**, *54*, 89-99.
2. Plastic Additives Handbook 5th Ed., Carl Hanser Verlag, Munich 2001, 515.

3. D. Cadogan, Phthalates: EU Risk Assessment, Progress & Impact, *Plasticiser Symposium 2004*, 28-29 September, Brussels.
4. H.R. Kricheldorf *J. Macromol. Sci. Rev. Macromol. Chem. Phys.* **1997**, *C37*, 566-633.
5. D. S. van Es, A. E. Frissen, H. Luitjes, *Improved Synthesis of Anhydroglycitol Esters of Improved Colour*, WO 01/83488 A1 to A&F.
6. D. Stove, Paints, Coatings and Solvents, VCH Weinheim, 1993, 41-52.
7. J. R. Bucher et al., *Inhalation and Carcinogenicity Studies of Cobalt Sulphate, Toxicological Studies,* 49 (1999), 56-67.
8. J. Zöller, G. Merten, E. Urbano, M. Gobec, *Wässrige Alkydharzemulsionen aus Nachwachsenden Rohstoffe*, EP 0741 175 A2 to Hoechst AG.
9. K. Hájek *Farb. Lacken* **1977**, *83*, 798- 804.
10. D. Bagchi and R. K. Malakar, *J. Coat. Technol.***1986**, *58*, 51-57.
11. R. N. Faulkner, *Resinous reaction product of a sucrose partial ester, a cyclic acid anhydride and a diepoxide.* US 3,870,664.
12. US 3,870,664 to Research Corporation New York.
13. E. A. Oostveen, J. W. Weijnen, J. van Haveren, M. Gillard, *Polysaccharide esters and their use as binders in coatings,* WO 03064477 to SigmaKalon.
14. E. A. Oostveen, J. W. Weijnen, J. van Haveren, M. Gillard, *Air drying paint compositions comprising carbohydrate based polyesters,* WO 03064498 to SigmaKalon.
15. E. A. Oostveen, J. W. Weijnen, J. van Haveren, M. Gillard, *Reactive diluents and coatings comprising them* WO 03064548 to SigmaKalon.
16. J. H. Bieleman, Lackadditive, Wiley-VCH Verlag GmbH, 1998, 212-240.
17. J. H. Bieleman *Macromol. Symp.* **2002**, *187*, 811-821.
18. http://www.borchers.com/index.cfm?PAGE_ID=1062
19. E.A. Oostveen, J. van Haveren, F. Miccichè, R. V. D. Linde, *Drier for Air Drying Coatings,* WO 03093384.

Chapter 9

Carbon Dioxide as a Renewable C1 Feedstock: Synthesis and Characterization of Polycarbonates from the Alternating Copolymerization of Epoxides and CO$_2$

Scott D. Allen, Christopher M. Byrne, and Geoffrey W. Coates[*]

Department of Chemistry and Chemical Biology, Cornell University, Ithaca, NY 14853

The alternating copolymerization of epoxides and CO$_2$ using well-defined β-diiminate zinc complexes is described. Zinc complex **4** is a highly active catalyst for the alternating copolymerization of propylene oxide and CO$_2$. We show that this catalyst is also active for the alternating copolymerization of several other aliphatic and alicyclic epoxides. The ability to copolymerize both aliphatic and alicyclic epoxides is a unique feature of this catalyst system.

Introduction

Because CO$_2$ is a nontoxic, nonflammable, and inexpensive substance, there is continued interest in its activation with transition metal complexes and its subsequent use as a C1 feedstock ($1,2$). Even though CO$_2$ is used to make commodity chemicals such as urea, salicylic acid and metal carbonates, efficient catalyst systems that exploit this feedstock as a comonomer in polymerization reactions have been elusive ($3,4$). One reaction that has been considerably successful is that of CO$_2$ with epoxides to yield aliphatic polycarbonates (Scheme 1) (5). Of particular significance is the synthesis of poly(propylene carbonate) (PPC), because the starting materials—propylene oxide (PO) and CO$_2$—are inexpensive.

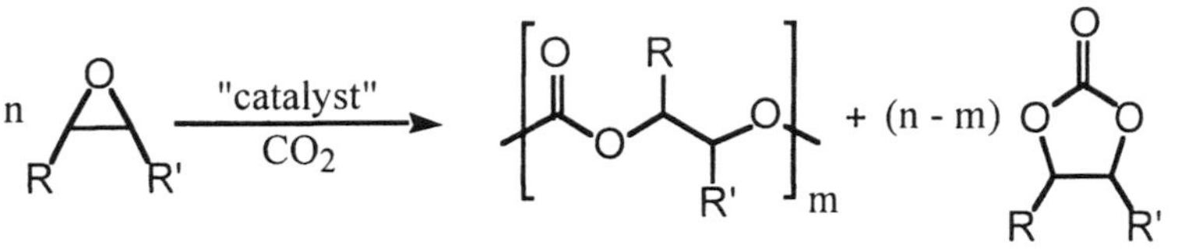

Scheme 1. Synthesis of polycarbonates from epoxides and CO_2.

The pioneering work of Inoue in the late 1960s began a field of research that has gained increasing attention over the last four decades (*6,7*). Despite the considerable number of catalyst systems reported, there have been relatively few accounts describing the controlled synthesis of polycarbonates from monomers other than PO and 1,2-cyclohexene oxide (CHO).

Copolymerization of Alternative Epoxides

Because polymer properties are governed by the constitution and orientation of the side chains, a great deal of research has been aimed at controlling and modifying these features in aliphatic polycarbonates. Initial research focused on catalyst discovery; thus PO and CHO were the model epoxides. A modest amount of research, however, has been directed at the synthesis of other polycarbonates, and more recently, the properties of such materials have been examined.

In efforts to elucidate the mechanism of the original $ZnEt_2/H_2O$ catalyst system, Inoue described the synthesis of polycarbonates from several optically active epoxides including styrene oxide (SO) (*8,9*), 3-phenyl-1,2-epoxypropane (*10*), cyclohexylepoxyethane (*11*) and 1,2-epoxybutane (EB) (*9*). The properties of the polycarbonates, however, are not discussed in these accounts. Other epoxides copolymerized by Inoue using the $ZnEt_2/H_2O$ catalyst system include 1,2-, and 2,3-EB and isobutylene oxide (*12*) and glycidol ethers and carbonates (*13*).

After discovering an active aluminum porphyrin catalyst system, Inoue pursued the synthesis of hydroxyl-functionalized polycarbonates (*14*). The desired glycidol/CO_2 copolymer could only be generated by first protecting the active hydroxyl group of glycidol with a trimethylsilyl group and then performing a post-polymerization deprotection step. Endo, however, recently reported that glycidol can be polymerized without the protection of the active hydroxyl group by using simple alkali metal halides (*15*).

Functional aliphatic polycarbonates were synthesized by Lukaszczyk through the copolymerization of allyl glycidyl ether (AGE) and CO_2 (*16*). Wang also synthesized AGE/CO_2 copolymers in addition to other glycidyl ether/CO_2–based polycarbonates and reported the mechanical and thermal properties of

118

several of these copolymers (*17*). Additionally, Tan reported the synthesis of several functionalized polycarbonates to include: 1) AGE/CO_2 copolymers that were used as precursors for silica nanocomposites (*18*), and 2) block copolymers of CHO/CO_2 and 4-vinyl-1,2-cyclohexene oxide (VCHO)/CO_2 (*19*). Darensbourg synthesized a silyl ether–functionalized CHO derivative that, when copolymerized with CO_2, forms a liquid CO_2–soluble polymer that can be further modified through cross-linking (*20*).

Although the synthesis of these polycarbonates has been accomplished, few reports have described the thermal and mechanical properties of these materials (*21*). To gain a better understanding of the potential applications for these materials, a variety of aliphatic and alicyclic polycarbonates must be synthesized and studied.

Synthesis of Polycarbonates Using Zinc Complexes

High-Activity Catalysts for Epoxide/CO_2 Copolymerization

We previously showed that well-defined, structurally characterized β-diiminate (BDI) zinc complexes (Figure 1) catalyze the ring-opening polymerization of cyclic esters (*22,23*) and the alternating copolymerization of

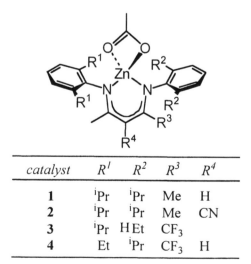

catalyst	R^1	R^2	R^3	R^4
1	^{i}Pr	^{i}Pr	Me	H
2	^{i}Pr	^{i}Pr	Me	CN
3	^{i}Pr	HEt	CF$_3$	
4	Et	^{i}Pr	CF$_3$	H

Figure 1. β-Diiminate zinc acetate complexes.

epoxides and CO_2 (*24-30*). The alternating copolymerization of CHO and CO_2 occurs rapidly under mild reaction conditions using catalyst **1** (Scheme 2) (*24,26*). Under the same conditions, however, the alternating copolymerization of other epoxides such as PO does not occur. Instead, the cycloaddition of PO and CO_2 takes place, and propylene carbonate (PC) is isolated in low yields (Scheme 2) (*28*).

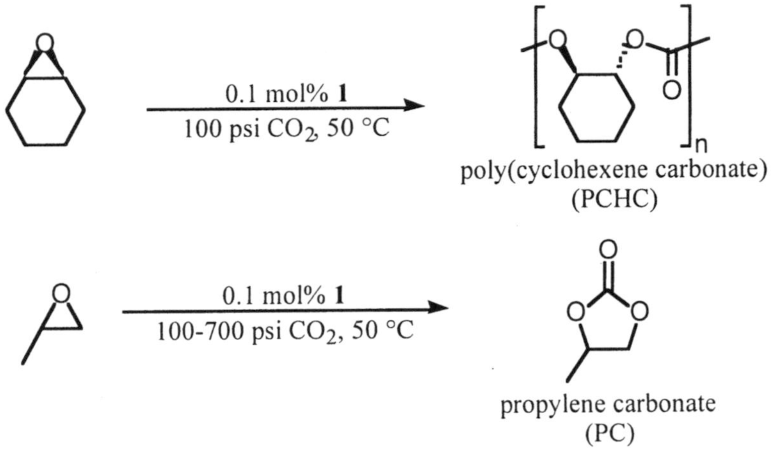

poly(cyclohexene carbonate)
(PCHC)

propylene carbonate
(PC)

Scheme 2. Reactivity of cyclohexene oxide and propylene oxide with CO_2 in the presence of zinc complex 1.

More recently, we have shown that zinc complexes bearing BDI ligands substituted with electron-withdrawing groups are extremely active for CHO/CO_2 copolymerizations (*27,29*) and the alternating copolymerization of PO and CO_2 (*28*). As seen in Scheme 3, catalyst **4** is an unusually active and selective catalyst for PO/CO_2 copolymerization. Under optimized conditions (300 psi CO_2 and 25 °C in neat PO), complex **4** converts PO to PPC with 87% selectivity. The isolated material has a narrow molecular weight distribution and consists of >99% carbonate linkages (Figure 2).

The regiochemistry of PPC can be determined using ^{13}C NMR spectroscopy (*31,32*) and by comparing the relative intensities of the resonances corresponding to the head-to-head, head-to-tail, and tail-to-tail regiosequences (Figure 3a). The polycarbonate isolated using catalyst **4** is regioirregular, with 54% head-to-tail regiosequences (Figure 3b) (*28*).

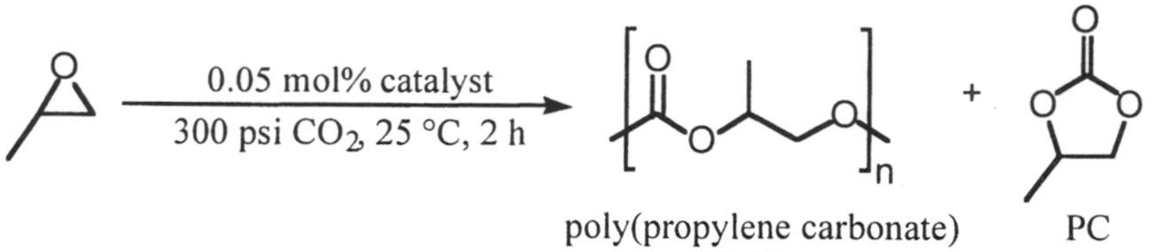

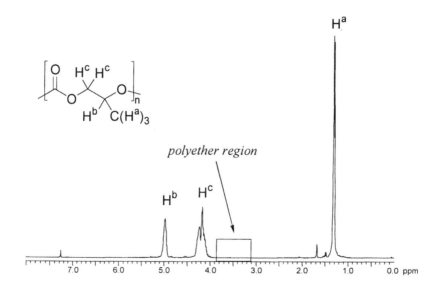

catalyst	activity (TOF[a])	selectivity (PPC:PC)
2	47	85:15
3	26	72:28
4	219	87:13

[a] Turnover frequency (TOF) = (mol PO / (mol Zn •h)).

Scheme 3. Propylene oxide (PO)/CO₂ alternating copolymerization activity of electron-deficient β-diiminate zinc complexes.

Hᵃ

polyether region

Hᵇ Hᶜ

7.0   6.0   5.0   4.0   3.0   2.0   1.0   0.0 ppm

Figure 2. Representative ¹H NMR spectrum (400 MHz, CDCl₃) of poly(propylene carbonate) showing >99% carbonate linkages

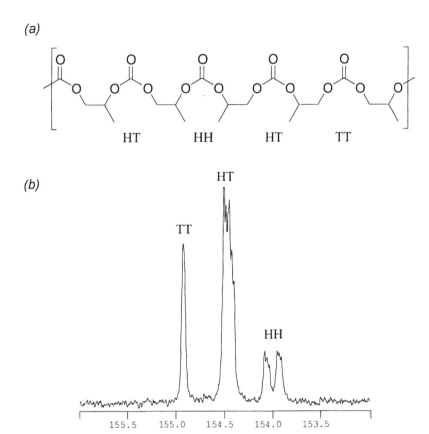

Figure 3. (a) Possible diad level regiosequences in poly(propylene carbonate) (PPC); HH = head-to-head, HT = head-to-tail, TT = tail-to-tail. (b) Representative ^{13}C NMR spectrum (125 MHz, CDCl$_3$) of the carbonate region of PPC synthesized using catalyst 4 (300 psi CO$_2$, 25 °C).

Copolymerization of Other Aliphatic Epoxides

Given the high activity and selectivity of catalyst **4** and the variety of commercially available epoxides, we explored the versatility of **4** as a catalyst for the alternating copolymerization of other aliphatic epoxides. The copolymerizations of aliphatic epoxides with saturated and unsaturated hydrocarbon chains were examined first, and the results are shown in Table I.

Table I. Copolymerization of Aliphatic Epoxides and CO_2 using Catalyst 4.

entry	epoxide	time (h)	4 (mol %)	activity (TOF[a])	selectivity[b] (polymer:cyclic)
1		2	0.05	220	87:13
2		4	0.05	87	85:15
3		4	0.1	80	50:50
4		24	0.1	20	< 1:100
5		6	0.1	87	35:65

[a] Turnover frequency (TOF) = (mol PO / (mol Zn • h)).

[b] Determined by ^1H NMR spectroscopy.

Even though catalyst **4** copolymerizes the epoxides with longer alkyl chains, the activity and selectivity suffers, as seen with EB and 1,2-epoxy-hex-5-ene (EH). The activity drops off dramatically, from 220 h^{-1} for PO to only 87 h^{-1} for EB. Under the same reaction conditions, EH is very slowly polymerized, so to attain an appreciable amount of polymer, we used longer reaction times and higher catalyst loadings. With both EB and EH, as the polymerization activity decreases, the selectivity to polymer also decreases as more of the cycloaddition product is formed. In each case, however, polymers with narrow molecular weight distributions are still attained and each polycarbonate has >99% carbonate linkages.

Table I gives the results of the polymerization of the unsaturated epoxides butadiene monoepoxide (BDME) and SO (entries 4 and 5, respectively). Surprisingly, BDME does not form any polymer but instead slowly converts exclusively to the cycloaddition by-product. SO, however, undergoes copolymerization but with low selectivity; only 35% of the epoxide is converted to polycarbonate. Increasing the CO_2 pressure in both systems does not improve the selectivity to polymer.

Next, we probed the functional group tolerance of catalyst **4** using heteroatom-substituted epoxides such as epichlorohydrin and methyl glycidyl ether. These epoxides do not convert to the polycarbonate but instead slowly convert to the cycloaddition by-products under all temperatures and pressures studied.

Copolymerization of Alicyclic Epoxides

As expected, the copolymerization of CHO and CO_2 using catalyst **4** occurs rapidly under the previously reported optimized conditions (*26*) and has an activity similar to that of the most active catalysts reported to date (Table II, entry 1) (*27*). The isolated polycarbonate has >95% carbonate linkages and a narrow molecular weight distribution. Unlike the copolymerization of aliphatic epoxides, the copolymerization of alicyclic epoxides does not generate any of the cycloaddition product.

We recently expanded the range of polymerizable alicyclic epoxides to include the renewable epoxide limonene oxide (LO) (*30*) and the related VCHO, as shown in Table II (entries 2 and 3, respectively). The polymerization of a cis/trans mixture of (*R*)-LO is considerably slower, presumably owing to the trisubstituted epoxide ring and the preference for the polymerization of the trans isomer (*30*). Conversely, VCHO undergoes rapid polymerization, with an activity comparable to that of CHO. This result supports the claim that the low activity of LO is due to the methyl group on the epoxide, not the vinyl substituent.

Table 2. Copolymerization of Alicyclic Epoxides and CO_2 using Catalyst 4.

entry	epoxide	temp. (°C)	time (min)	4 (mol %)	activity (TOF[a])
1		50	10	0.1	1890
2		25	120	0.4	37
3		50	10	0.1	1490

[a] Turnover frequency (TOF) = (mol PO / (mol Zn •h)).

Thermal Properties of Polycarbonates

Using differential scanning calorimetry (DSC) and thermogravimetric analysis (TGA), we examined the thermal properties of polycarbonates.

Aliphatic Polycarbonates

Figure 4 shows the TGA and DSC thermograms for the aliphatic polycarbonates synthesized using catalyst **4**. The data for all the thermal analyses are summarized in Table III. Since the thermal properties of the

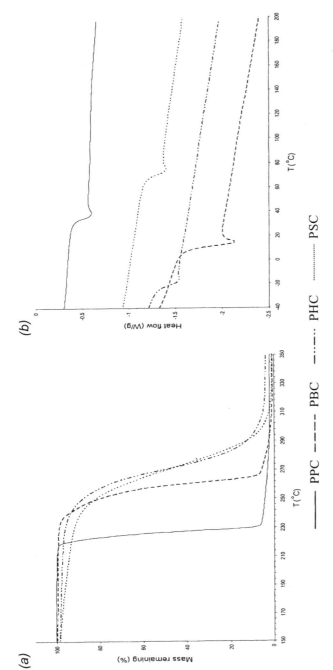

Figure 4. Thermal properties of aliphatic polycarbonates: poly(propylene carbonate) (PPC); poly(1,2-butylene carbonate) (PBC); poly(1,2-hex-5-ene carbonate) (PHC); poly(styrene carbonate) (PSC). (a) Thermogravimetric analysis curves (samples were run under an N_2 atmosphere with a heating rate of 20 °C/min). (b) Differential scanning calorimetry curves (samples were run under an N_2 atmosphere with a heating and cooling rate of 10 °C/min; data shown are from the second heat).

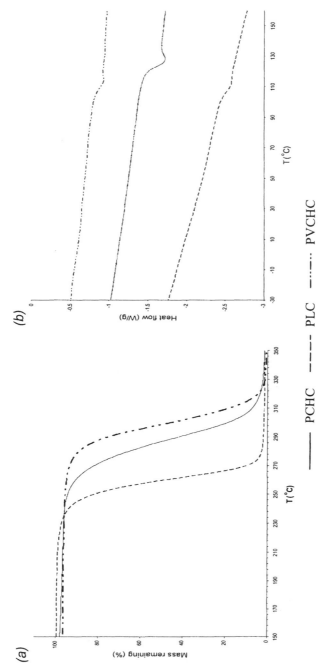

Figure 5. Thermal properties of alicyclic polycarbonates: poly(1,2-cyclohexene carbonate) (PCHC), poly(limonene carbonate) (PLC), poly(4-vinyl-1,2-cyclohexene carbonate) (PVCHC). (a) TGA thermograms (samples were run under an N_2 atmosphere with a heating rate of 20 °C/min). (b) DSC thermograms (samples were run under an N_2 atmosphere with a heating and cooling rate of 10 °C/min; data shown are from the second heat).

Table III. Thermal Properties of Epoxide/CO_2 Copolymers.

polycarbonate	M_n (kg/mol)	PDI (M_n/M_w)	T_g (°C)	T_d (°C)
PPC	28.3	1.17	35	220
PBC	17.1	1.24	12	240
PHC	14.6	1.14	-22	243
PSC	20.8	1.14	71	246
PCHC	26.6	1.23	122	253
PLC	11.0	1.18	106	244
PVCHC	38.2	1.20	110	282

polymers depend on their molecular weight and molecular weight distribution, these values are also included in Table III. The decomposition temperature (T_d = onset decomposition temperature) is dependent on the side chain as determined by TGA. The T_d for PPC is 220 °C, whereas all of the other aliphatic polycarbonates exhibit slightly higher T_d values ranging from 240 to 248 °C. The glass transition temperature (T_g) also depends on the side chain substituent (Figure 4b). PPC exhibits a T_g of 35 °C, whereas poly(1,2-butylene carbonate) and poly(1,2-hex-5-ene carbonate) exhibit sub-ambient T_gs of 12 and –22 °C, respectively. The low T_g values of the polycarbonates with long alkyl side chains have been attributed to internal plasticization *(21)*. PSC, however, has a relatively high T_g of 71 °C, which is comparable to that of polystyrene (T_g = ~100 °C).

Alicyclic Polycarbonates

The thermal properties of the alicyclic epoxides are significantly different from those of the aliphatic polycarbonates, as shown in Figure 5 and Table III. First, the T_d values are generally higher; PCHC and poly(4-vinyl-1,2-cyclohexene carbonate) have T_d values of 253 and 282 °C, respectively. Conversely, PLC has a T_d of 244 °C, similar to those of the aliphatic polycarbonates described earlier. Also of noticeable difference are the T_g values for the alicyclic polycarbonates. The rigidity of the cyclohexane ring limits the flexibility of the polymer chains, resulting in the higher T_gs.

Conclusion

We have shown that catalyst **4** is a highly active catalyst not only for the alternating copolymerization of PO and CO_2 but also for a variety of aliphatic and alicyclic epoxides. The copolymerizations generate polycarbonates with >95% carbonate linkages and narrow molecular weight distributions. In the case of the aliphatic polycarbonates, the slower copolymerizations generate significant amounts of the cycloaddition by-product, but this by-product is not observed in the case of the alicyclic epoxides. Given the versatility of this complex, future efforts will be devoted to the synthesis of polycarbonate copolymers and other polycarbonates with controlled architectures.

References

1. Arakawa, H.; Aresta, M.; Armor, J. N.; Barteau, M. A.; Beckman, E. J.; Bell, A. T.; Bercaw, J. E.; Creutz, C.; Dinjus, E.; Dixon, D. A.; Domen, K.; DuBois, D. L.; Eckert, J.; Fujita, E.; Gibson, D. H.; Goddard, W. A.; Goodman, D. W.; Keller, J.; Kubas, G. J.; Kung, H. H.; Lyons, J. E.; Manzer, L. E.; Marks, T. J.; Morokuma, K.; Nicholas, K. M.; Periana, R.; Que, L.; Rostrup-Nielson, J.; Sachtler, W. M. H.; Schmidt, L. D.; Sen, A.; Somorjai, G. A.; Stair, P. C.; Stults, B. R.; Tumas, W. *Chem. Rev.* **2001**, *101*, 953-996.

2. Leitner, W. *Coord. Chem. Rev.* **1996**, *153*, 257-284.

3. Aresta, M.; Dibenedetto, A.; Tommasi, I. *Energy Fuels* **2001**, *15*, 269-273.

4. Aresta, M.; Tommasi, I. *Energy Conserv. Mgmt.* **1997**, *38*, S373-S378.

5. For reviews on epoxide/CO_2 polymerizations, see: (a) Kuran, W. *Prog. Polym. Sci.* **1998**, *23*, 919-992. (b) Super, M.; Beckman, E. *Trends Polym. Sci.* **1997**, *5*, 236-240. (c) Darensbourg, D. J.; Holtcamp, M. W. *Coord. Chem. Rev.* **1996**, *153*, 155-174. (d) Coates, G. W.; Moore, D. R. *Angew. Chem., Int. Ed.* **2004**, *43*, 2-23.

6. Inoue, S.; Koinuma, H.; Tsurata, T. *Makromol. Chem.* **1969**, *130*, 210-220.

7. Inoue, S.; Koinuma, H.; Tsurata, T. *Polym. Lett.* **1969**, *7*, 287-292.

8. Hirano, T.; Inoue, S.; Tsuruta, T. *Makromol. Chem.* **1975**, *176*, 1913-1917.

9. Inoue, S.; Hirano, T.; Tsuruta, T. *Polym. J.* **1977**, *9*, 101-106.

10. Hirano, T.; Inoue, S.; Tsuruta, T. *Makromol. Chem.-Macromol. Chem. Phys.* **1976**, *177*, 3245-3253.

11. Hirano, T.; Inoue, S.; Tsuruta, T. *Makromol. Chem.-Macromol. Chem. Phys.* **1976**, *177*, 3237-3243.

12. Inoue, S.; Matsumoto, K.; Yoshida, Y. *Makromol. Chem.-Macromol. Chem. Phys.* **1980**, *181*, 2287-2292.

13. Takanashi, M.; Nomura, Y.; Yoshida, Y.; Inoue, S. *Makromol. Chem.-Macromol. Chem. Phys.* **1982**, *183*, 2085-2092.

14. Inoue, S. *J. Macromol. Sci.-Chem.* **1979**, *A13*, 651-664.

15. Motokucho, S.; Sudo, A.; Sanda, F.; Endo, T. *J. Polym. Sci. Pol. Chem.* **2004**, *42*, 2506-2511.
16. Lukaszczyk, J.; Jaszcz, K.; Kuran, W.; Listos, T. *Macromol. Rapid Commun.* **2000**, *21*, 754-757.
17. Guo, J. T.; Wang, X. Y.; Xu, Y. S.; Sun, J. W. *J. Appl. Polym. Sci.* **2003**, *87*, 2356-2359.
18. Tan, C. S.; Juan, C. C.; Kuo, T. W. *Polymer* **2004**, *45*, 1805-1814.
19. Hsu, T. J.; Tan, C. S. *J. Chin. Inst. Chem. Eng.* **2003**, *34*, 335-344.
20. Darensbourg, D. J.; Rodgers, J. L.; Fang, C. C. *Inorg. Chem.* **2003**, *42*, 4498-4500.
21. Thorat, S. D.; Phillips, P. J.; Semenov, V.; Gakh, A. *J. Appl. Polym. Sci.* **2003**, *89*, 1163-1176.
22. (a) Cheng, M.; Attygale, A. B.; Lobkovsky, E. B.; Coates, G. W. *J. Am. Chem. Soc.* **1999**, *121*, 11583-11584. (b) Chamberlin, B. M.; Cheng, M.; Moore, D. R.; Ovitt, T. M.; Lobkovsky, E. B.; Coates, G. W. *J. Am. Chem. Soc.* **2001**, *123*, 3229-3238.
23. Rieth, L. R.; Moore, D. R.; Lobkovsky, E. B.; Coates, G. W. *J. Am. Chem. Soc.* **2002**, *124*, 15239-15248.
24. Cheng, M.; Lobkovsky, E. B.; Coates, G. W. *J. Am. Chem. Soc.* **1998**, *120*, 11018-11019.
25. Cheng, M.; Darling, N. A.; Lobkovsky, E. B.; Coates, G. W. *J. Am. Chem. Soc.* **2000**, 2007-2008.
26. Cheng, M.; Moore, D. R.; Reczek, J. J.; Chamberlain, B. M.; Lobkovsky, E. B.; Coates, G. W. *J. Am. Chem. Soc.* **2001**, *123*, 8738-8749.
27. Moore, D. R.; Cheng, M.; Lobkovsky, E. B.; Coates, G. W. *Angew. Chem.-Int. Ed.* **2002**, *41*, 2599-2602.
28. Allen, S. D.; Moore, D. R.; Lobkovsky, E. B.; Coates, G. W. *J. Am. Chem. Soc.* **2002**, *124*, 14284-14285.
29. Moore, D. R.; Cheng, M.; Lobkovsky, E. B.; Coates, G. W. *J. Am. Chem. Soc.* **2003**, *125*, 11911-11924.
30. Byrne, C. M.; Allen, S. D.; Lobkovsky, E. B.; Coates, G. W. *J. Am. Chem. Soc.* **2004**, *126*, 11404-11405.
31. Lednor, P. W.; Rol, N. C. *J. Chem. Soc.-Chem. Commun.* **1985**, 598-599.
32. Chisholm, M. H.; Navarro-Llobet, D.; Zhou, Z. P. *Macromolecules* **2002**, *35*, 6494-6504.

Chapter 10

Electrocatalytic Oxidation of Methanol

Corey R. Anthony and Lisa McElwee-White

Department of Chemistry and Center for Catalysis, University of Florida,
Gainesville, FL 32611–7200

The electrochemical oxidation of methanol to dimethoxymethane (DMM) and methyl formate (MF) is catalyzed by several RuPt, RuPd and RuSn heterobimetallic complexes. The product ratios and current efficiencies depend on the catalyst, solvent, potential and presence of water. Complete selectivity for DMM and high current efficiency can be obtained by electrolysis of CpRu(TPPMS)$_2$(SnCl$_3$) in methanol.

Introduction

Fuel cells have been postulated as the power generation system of the immediate future, possibly replacing not only fossil fuel based engines (1,2) but also advanced alkali batteries (2,3). Due to their simplicity, high energy efficiency and low pollution, direct methanol fuel cells (DMFCs) are especially suited for portable electronic devices (2,3). The electrooxidation of methanol to CO$_2$ is a mechanistically complex reaction (Scheme 1) with stable intermediates causing a large overpotential. Through the use of catalysts the overpotential associated with the overall process can be lowered.

Although there has been moderate success with platinum anodes for DMFC, the presence of a second metal can decrease the overpotential and increase the lifetime of the anode. The beneficial effects of an additional metal have been

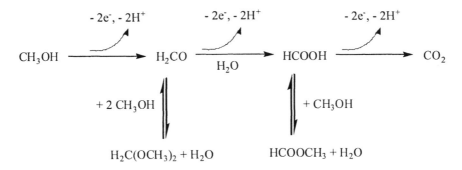

Scheme 1. Electrooxidation of methanol

observed with several combinations, including PtSn (*4*), PtRe (*5*), PtNi (*6*), PtRu (*7-11*), PtRuOs (*12*), PtRuMoW (*13*), and PtRuOsIr (*14,15*). The role of the non-Pt metal in these anodes has been a topic of interest. Although electronic effects have been invoked (*6,16,17*), the major effect of an additional metal has generally been ascribed to a bifunctional oxidation mechanism (*18,19*). For the PtRu anodes (the most effective combination), the postulate is that Pt sites engage in methanol binding and dehydrogenation, while the Ru serves as a source of "active oxygen" for the formation of CO_2 (*20*).

Our work on the electrooxidation of alcohols with homogeneous heterobimetallic catalysts was initially motivated by the advances in DMFC anodes (*8,21-23*). The strategy initially employed for this project was to utilize the bifunctional oxidation mechanism observed in DMFC anodes to design discrete bimetallic complexes as catalysts for the electrooxidation of renewable fuels. These complexes can be viewed as utilizing all of the metal as opposed to bulk metal anodes, where the reaction occurs only at active surface sites. Although bimetallic complexes are not accurate models of the proposed surface binding site on bulk metal anodes, our investigations address the question of whether it is possible to reproduce the essential functions of the proposed electrooxidation mechanism in discrete heterobinuclear complexes (*24-26*).

Experimental Section

Electrochemistry. Electrochemistry was performed at ambient temperature in a glove box under N_2 using an EG&G PAR model 263A potentiostat/galvanostat. Cyclic voltammetry (CV) was performed with a three-electrode configuration consisting of a glassy carbon working electrode (3 mm diameter), a Pt flag counter electrode and a Ag/Ag$^+$ reference electrode. For experiments in 1,2-dichloroethane (DCE) or methanol, the reference electrode consisted of an acetonitrile or methanol solution respectively of freshly prepared 0.01 M AgNO$_3$ and 0.1 M tetrabutylamonium triflate (TBAT) along with a silver wire. The Ag$^+$ solution and wire were contained in a 75 mm glass tube fitted at the bottom with a Vycor tip. Constant potential electrolysis utilized similar equipment except the working electrode was a vitreous carbon electrode. All potentials are reported vs. NHE and are not corrected for the junction potential of the Ag/Ag$^+$ reference electrode. The E^0 values for the ferrocenium/ferrocene couple in 0.7 M TBAT/DCE and 0.1 M TBAT/MeOH were +0.50 V and +0.795 V.

Electrolysis products were analyzed by gas chromatography on a Shimadzu GC-17A chromatograph containing a 15 m x 0.32 mm column of AT-WAX (Alltech, 0.5 μm film) on fused silica. The column was attached to the injection port with a neutral 5 m x 0.32 mm AT-WAX deactivated guard column. The electrolysis products were quantitatively determined with the use of n-heptane as an internal standard. Products were identified by comparison to authentic samples.

Ligands: triphenylphosphine (PPh$_3$), bis(diphenylphosphino)-methane (dppm), (m-sulfonatophenyl)diphenylphosphine (TPPMS) and 1,5-cyclooctadiene (COD).

Synthesis and Electrochemistry of Heterobimetallic Catalysts

Heterobimetallic catalysts have been of long-standing interest, due to the possibility of exploiting the different reactivities of the two metals in chemical transformations (27-30). It has long been recognized that two metals in close proximity may exhibit new reactivities which are different from their parent mononuclear compounds (31-33). In designing our initial catalysts two structural motifs were utilized in tandem: bridging bidentate phosphines for stability and μ-halides for electronic communication between the metal centers (Figure 1).

The synthesis of our first electrocatalyst was accomplished by reacting CpRu(Cl)(PPh$_3$)(η^1-dppm) with Pt(COD)Cl$_2$ to afford the Ru/Pt complex CpRu(PPh$_3$)(μ-Cl)(μ-dppm)PtCl$_2$ (**1**) (34). Cyclic voltammetry of complex **1** exhibits three redox couples (35) (Figure 2, Table I which have been assigned to the reversible Ru(II/III), irreversible Pt(II/IV) and Ru(III/IV) oxidation processes, respectively.

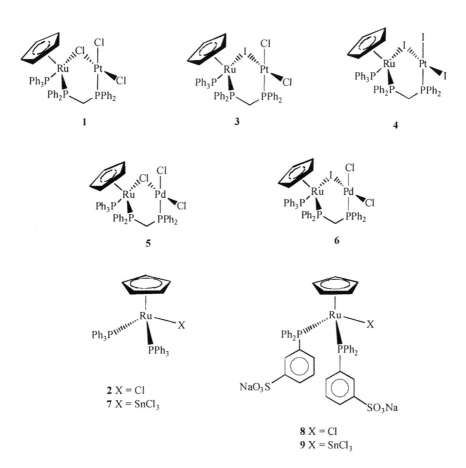

Figure 1. Structures of heterobimetallic catalysts and mononuclear model compounds.

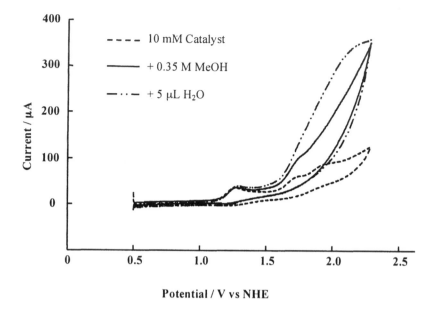

Figure 2. Cyclic voltammograms of 1 under nitrogen in 3.5 mL of DCE/0.7 M
TBAT; glassy carbon working electrode; Ag/Ag⁺ reference electrode; 50 mV/s,
solutions as specified.

Electrochemical oxidation of complex **1** in the presence of methanol leads
to considerable enhancement of the oxidative currents (Figure 2), consistent
with an electrocatalytic oxidation process. The onset of this catalytic current
coincides with the irreversible Pt(II/IV) oxidative wave at 1.70 V. The bulk
electrolysis of **1** and dry methanol were performed at 1.70 V (onset of catalytic
current) in 0.7 M TBAT/DCE. Gas chromatographic analysis of the solution
indicated that dimethoxymethane (DMM, formaldehyde dimethyl acetal) and
methyl formate (MF) are formed (Scheme 1). This result is consistent with the
electrooxidation of dry methanol on PtRu anodes, which yields DMM after acid-
catalyzed condensation of the formaldehyde product with excess methanol *(36)*.
Bulk electrolysis of methanol in the presence of heterobimetallic complex **1**
resulted in higher current efficiencies than those obtained from the mononuclear
model compound $CpRu(PPh_3)_2Cl$ (**2**) (Table II). No oxidation products were
found when the electrolysis was performed at 1.70 V in the absence of a Ru
complex or in the presence of the Pt model compound $(\eta^2\text{-dppm})PtCl_2$. These
results suggest that Pt enhances the catalytic activity of the Ru metal center.

Table I. Formal Potentials for Complexes 1-9[a]

Complex	Couple	E_{pa} (V)	$E_{1/2}$ (V)[b]	Couple	E_{pa} (V)	$E_{1/2}$ (V)[b]	Couple	E_{pa} (V)
1	Ru(II/III)	1.25	1.21	Pt(II/IV)	1.69		Ru(III/IV)	1.91
2	Ru(II/III)	0.93	0.87					
3	Ru(II/III)	1.29	1.25	Pt(II/IV)	1.54	1.47	Ru(III/IV)	1.90
4	Ru(II/III)	1.10		Pt(II/IV)	1.49	1.43	Ru(III/IV)	1.98
5	Ru(II/III)	1.30		Pd(II/IV)	1.49		Ru(III/IV)	1.92
6	Ru(II/III)	1.29		Pd(II/IV)	1.55	1.50	Ru(III/IV)	1.98
7	Ru(II/III)	1.48	1.44					
8[c]	Ru(II/III)	0.79	0.73					
9[c]	Ru(II/III)	1.29						

[a]All potentials obtained in 0.7 M TBAT/DCE unless otherwise specified and reported vs. NHE. [b]$E_{1/2}$ reported for reversible waves. [c]Potential obtained in 0.1 M TBAT/MeOH and reported vs. NHE.

Based on the electrocatalytic activity of the RuPt complex 1, we decided to prepare additional Ru-containing heterobimetallic complexes (Figure 1). Modifications were made in the identity of the second metal, the ligands on the second metal and the bridging moiety. These changes were designed to provide variation in 1) the oxidation potential of the metal centers; 2) the degree of electronic communication between the metals and 3) the distance between the metal centers. Figure 1 summarizes the additional catalysts that were prepared by reacting $CpRu(PPh_3)(\eta^1-Ph_2P(CH_2)PPh_2)X$ [X = Cl, or I] with the appropriate Pd or Pt complex (19).

Cyclic voltammograms of the heterobimetallic complexes 3-6 generally exhibit three redox waves (Table I) within the solvent window. In the presence of methanol, the Ru/Pd and Ru/Pt complexes all display a current increase at the Pd(II/IV) and Pt(II/IV) waves, indicating a catalytic methanol oxidation process similar to what was observed for complex 1 (Figure 2). Bulk electrolyses of methanol with complexes 3-6 were carried out for product identification and quantification. For comparison purposes, bulk electrolyses of methanol were performed at the same potential (1.70 V vs. NHE) as compound 1 (25). Therefore, the electrocatalysis with these complexes was performed at potentials positive of both the Ru(II/III) couple and the first oxidative wave of the second metal but before the Ru(III/IV) wave.

Table II. Product Ratios and Current Efficiencies for Dry Methanol Oxidation[a]

	Product ratios (DMM/MF)						
Charge (C)	Ru/Pt (1)	Ru/Pt (3)	Ru/Pt (4)	Ru/Pd (5)	Ru/Pd (6)	CpRu(PPh$_3$)$_2$Cl + Pd(COD)Cl$_2$	CpRu(PPh$_3$)$_2$Cl (2)
25	2.45	2.27	2.23	3.18	1.85	3.85	4.20
50	2.35	1.68	1.66	2.41	1.56	2.95	4.00
75	1.51	1.24	1.40	1.54	1.21	2.58	3.73
100	1.23	0.98	1.26	0.94	0.91	1.97	2.87
Current Efficiency (%)[c]	18.6	43	39	24.6	42	18	12

[a] Electrolyses were performed at 1.70 V vs. NHE. A catalyst concentration of 10 mM was used. Methanol concentration was 0.35 M. [b] Determined by GC with respect to *n*-heptane as an internal standard. Each ratio is reported as an average of 2-5 experiments. [c] Average current efficiencies after 75-100 C of charge passed.

The oxidation products formed during the bulk electrolysis of methanol with complexes 3-6 are DMM and MF, just as were observed for complex 1. Product distribution and current efficiencies for the formation of DMM and MF are shown in Table II. Initially, all the Ru complexes yield higher proportions of DMM. However, as the reaction progresses, this ratio shifts toward the more highly oxidized product, MF. Differences in the catalytic activity of the heterobimetallic and monometallic complexes can also be observed in Table II. All the heterobimetallic complexes favor the formation of MF and have higher current efficiencies than the Ru model compound CpRu(PPh$_3$)$_2$Cl. The Cl-bridged complexes 1 and 5 gave lower current efficiencies (24.6 and 18.6% respectively) (*24*) compared to the I-bridged complexes 3, 4 and 6 (approximately 43%). This significant increase in current efficiency has been attributed to the higher stability of the I-bridges that maintain the close proximity of the metal centers. The stability is also reflected in the Pd(II/IV) and Pt(II/IV) oxidation waves (Table I), which are irreversible in 1 and 5, but reversible in 3 and 6, implying greater stabilities for the oxidized I-bridged species (*35*).

In order to probe whether the reaction involves both metal centers in close proximity, a mixture of CpRu(PPh$_3$)$_2$Cl and Pd(COD)Cl$_2$ was used as a model for the Ru/Pd catalysts (Table II). The product distribution and current efficiency for the mixture was similar to that of complex 5 and the Ru model compound CpRu(PPh$_3$)$_2$Cl. A higher current efficiency was afforded by complex 6 (stable halide bridge). The differences in product distribution and current efficiency between the mixture and complex 6 indicate that the bimetallic structure is significant to the catalytic activity of the complexes.

The condensation of H_2CO and HCOOH with CH_3OH (Scheme 1) produces 1 mole of water for each mole of DMM and MF formed. The effect of additional water on the electrocatalytic reaction was then probed by introducing 5 μL of water before starting the electrolysis. The CVs of complexes 1, 3 – 6 all display a slight increase in current when water is introduced. The presence of water, as predicted in Scheme 1 favors the formation of MF (Table III). A similar effect was previously described for heterogeneous catalysts (37). The effect of water on the ratio of DMM to MF can also be observed during the electrolysis of dry methanol since as the reaction progresses the amount of water formed increases while the ratio of DMM to MF decreases.

Table III. Product Ratios and Current Efficiencies for Wet Methanol Oxidation with Complexes 1, 4, 5 and 6[a]

	Product ratios (DMM/MF)			
Charge	Ru/Pt (1)	Ru/Pt (4)	Ru/Pd (5)	Ru/Pt (6)
25	1.68	2.23	1.38	1.26
50	1.34	1.66	0.98	1.06
75	1.17	1.40	0.84	0.72
100	0.67	1.26	0.70	0.58
Current Efficiency (%)[c]	19.5	39	20.6	41

[a] Electrolyses were performed at 1.70 V vs. NHE. A catalyst concentration of 10 mM was used. Methanol concentration was 0.35 M. [b] Determined by GC with respect to *n*-heptane as an internal standard. Each ratio is reported as an average of 2-5 experiments. [c] Average current efficiencies after 75-100 C of charge passed.

Catalysts with a Lewis Acidic Site

One possible role of the non-Ru metal center in the RuPt and RuPd catalysts is that of a Lewis acid. A Lewis acidic site could aid the coordination of methanol before oxidation by the Ru metal center. To investigate this possibility we prepared $CpRu(PPh_3)_2(SnCl_3)$ (7), which had previously been reported to selectively oxidize methanol to methyl acetate at elevated temperatures (38,39).

In the CV of complex 7 a single reversible oxidation wave is observed at 1.44 V. A similar oxidation process has previously been assigned for $CpRu(PPh_3)_2Cl$ (40) (Table 1) as the one electron oxidation of the Ru metal center. Based on this assignment and the fact that Sn^{2+} is not redox active within the solvent window, the oxidation wave observed for complex 7 was assigned to the Ru(II/III) couple. In the presence of methanol, there is a significant increase in the current indicative of an electrocatalytic oxidation process (Figure 3). For

7, the onset of catalytic current coincides with the Ru(II/III) oxidation process. This contrasts significantly with the behavior of the RuPt and RuPd complexes where the onset coincided with the oxidation of the non-Ru metal center (Figure 2). The electrolysis potential was varied to investigate what effect this would have on the reaction. Hence for complexes **2** and **7** the electrooxidation was performed at 1.55 V (onset of catalytic current for complex **7**) and at 1.70 V. DMM and MF are the only products formed during the electrolysis of methanol with $CpRu(PPh_3)_2Cl$ and $CpRu(PPh_3)_2(SnCl_3)$. This contrasts with the thermal reaction of methanol with $CpRu(PPh_3)_2(SnCl_3)$ (*38,39*) where methyl acetate was the only product observed. Varying the oxidation potential has a significant effect on the product ratio (selectivity) of the methanol electrooxidation reaction (Table IV). Although $CpRu(PPh_3)_2(SnCl_3)$ and $CpRu(PPh_3)_2Cl$ form more DMM than MF at lower potentials, this effect is more pronounced for the RuSn complex.

Table IV. Bulk Electrolysis Data for Complexes 2, 7, 8 and 9[a]

Complex	Oxidation Potential (V)	Current Efficiency[b] (%)	DMM (%)	MF (%)
$CpRu(PPh_3)_2Cl$ (**2**)	1.55	5.6	76.8	23.2
$CpRu(PPh_3)_2Cl$ (**2**)	1.70	7.3	75.2	24.8
$CpRu(PPh_3)_2(SnCl_3)$ (**7**)	1.55	18.2	95.0	5.0
$CpRu(PPh_3)_2(SnCl_3)$ (**7**)	1.70	13.1	67.4	32.6
$CpRu(TPPMS)_2Cl$[c] (**8**)	1.25	63.2	100	0
$CpRu(TPPMS)_2Cl$[c] (**8**)	1.40	76.9	98.5	1.5
$CpRu(TPPMS)_2(SnCl_3)$[c] (**9**)	1.25	89.4	100	0
$CpRu(TPPMS)_2(SnCl_3)$[c] (**9**)	1.40	90.1	90.5	9.5

[a] All electrolyses performed for five hours in 0.7 M TBAT/DCE with 1.0 mmoles methanol unless otherwise specified. [b] Moles of product formed per mole of charge passed. [c] Electrolyses performed in 0.1 M TBAT/MeOH

The effect of methanol concentration was initially probed using complexes **3** and **6** (*26*). In the presence of 1.41 M methanol, current efficiencies of approximately 74% were obtained for both catalysts. This value was significantly higher than the 43% obtained at a methanol concentration of 0.35 M. This increase in current efficiency was attributed to improved electron

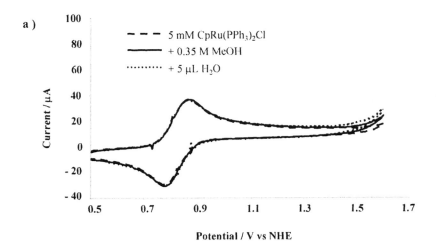

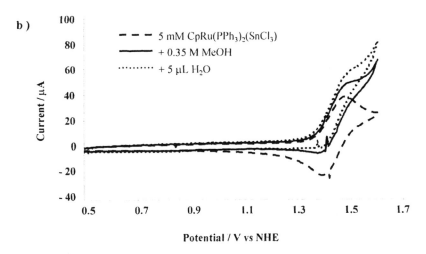

Figure 3. Cyclic voltammograms of (a) CpRu(PPh₃)₂Cl and (b) CpRu(PPh₃)₂(SnCl₃) under N₂ in 3.5 mL of DCE/0.7 M TBAT; glassy carbon working electrode; Ag/Ag+ reference electrode; 50 mV/s. (Reproduced from reference 41. Copyright 2004 Elsevier.)

transfer kinetics and to the higher concentration of substrate when methanol is present in greater quantities. To further investigate this effect, the electrolyses of the water soluble complexes $CpRu(TPPMS)_2(Cl)$ (**8**) and $CpRu(TPPMS)_2(SnCl_3)$ (**9**) were performed in 0.1 M TBAT/MeOH (*41*).

Cyclic voltammograms of complexes **8** and **9** both exhibit a single oxidation wave assigned to the Ru(II/III) oxidation process (Table 1). The electrooxidation of neat methanol with complexes **8** and **9** was performed at 1.25 V and at 1.40 V. Higher current efficiencies for the electrooxidation of methanol are obtained with the Ru/TPPMS complexes **8** and **9** than are obtained with complexes **2** and **7** (Table IV). Complex **7** has been shown to form significantly more DMM than MF at lower potentials; this effect is also observed during the electrolysis with complex **9**. When the potential was decreased from 1.40 V to 1.25 V the amount of DMM in the product mixture increased from 90.5% to 100%. To our knowledge, no other examples of selective electrooxidation of methanol to DMM have been reported.

Conclusion

The electrochemical oxidation of methanol to DMM and MF was performed with the homogeneous catalysts **1–9**. The heterobimetallic complexes afforded higher current efficiencies than the corresponding mononuclear model compounds. The presence of water favors formation of the more highly oxidized product, MF. Iodide-bridged complexes **3**, **4** and **6** produced significantly higher current efficiencies than **1** and **5** because of their increased stability during the electrolysis.

The Ru/Sn complexes **7** and **9** favor the formation of DMM over MF. The current efficiency and selectivity for DMM are affected by the oxidation potential and the electrolysis solvent, with both maximized when the electrolysis is performed at lower potential in neat methanol. Complete selectivity for DMM was achieved when the electrocatalyst was $CpRu(TPPMS)_2(SnCl_3)$.

References

1. Yoshida, P. G. *Fuel cell vehicles race to a new automotive future*. Office of Technology Policy, U.S. Dept. of Commerce, 2003.
2. Stone, C.; Morrison, A. E. *Solid State Ionics* **2002**, *152-153*, 1-13.
3. Kim, D.; Cho, E. A.; Hong, S.-A.; Oh, I.-H.; Ha, H. Y. *J. Power Sources* **2004**, *130*, 172-177.

4. Wang, K.; Gasteiger, H. A.; Markovic, N. M.; Ross, P. N. *Electrochim. Acta* **1996**, *41*, 2587-2593.
5. Cathro, K. J. *J. Electrochem. Soc.* **1969**, *116*, 1608-1611.
6. Park, K. W.; Choi, J. H.; Kwon, B. K.; Lee, S. A.; Sung, Y. E.; Ha, H. Y.; Hong, S. A.; Kim, H.; Wieckowski, A. *J. Phys. Chem. B* **2002**, *106*, 1869-1877.
7. Gasteiger, H. A.; Markovic, N.; Ross, P. N.; Cairns, E. J. *J. Phys. Chem.* **1993**, *97*, 12020-12029.
8. Wasmus, S.; Kuver, A. *J. Electroanal. Chem.* **1999**, *461*, 14-31.
9. Leger, J. M.; Lamy, C. *Ber. Bunsen-Ges.* **1990**, *94*, 1021-1025.
10. Swathirajan, S.; Mikhail, Y. M. *J. Electrochem. Soc.* **1991**, *138*, 1321-1326.
11. Reddington, E.; Sapienza, A.; Gurau, B.; Viswanathan, R.; Sarangapani, S.; Smotkin, E. S.; Mallouk, T. E. *Science* **1998**, *280*, 1735-1737.
12. Ley, K. L.; Liu, R. X.; Pu, C.; Fan, Q. B.; Leyarovska, N.; Segre, C.; Smotkin, E. S. *J. Electrochem. Soc.* **1997**, *144*, 1543-1548.
13. Choi, W. C.; Kim, J. D.; Woo, S. I. *Catal. Today* **2002**, *74*, 235-240.
14. Gurau, B.; Viswanathan, R.; Liu, R. X.; Lafrenz, T. J.; Ley, K. L.; Smotkin, E. S.; Reddington, E.; Sapienza, A.; Chan, B. C.; Mallouk, T. E.; Sarangapani, S. *J. Phys. Chem. B* **1998**, *102*, 9997-10003.
15. Lei, H. W.; Suh, S.; Gurau, B.; Workie, B.; Liu, R. X.; Smotkin, E. S. *Electrochim. Acta* **2002**, *47*, 2913-2919.
16. Lin, W. F.; Zei, M. S.; Eiswirth, M.; Ertl, G.; Iwasita, T.; Vielstich, W. *J. Phys. Chem. B* **1999**, *103*, 6968-6977.
17. Goodenough, J. B.; Manoharan, R.; Shukla, A. K.; Ramesh, K. V. *Chem. Mater.* **1989**, *1*, 391-398.
18. Goodenough, J. B.; Hamnett, A.; Kennedy, B. J.; Manoharan, R.; Weeks, S. A. *J. Electroanal. Chem. Interfacial Electrochem.* **1988**, *240*, 133-145.
19. Hamnett, A.; Kennedy, B. J.; Wagner, F. E. *J. Catal.* **1990**, *124*, 30-40.
20. Kua, J.; Goddard, W. A. *J. Am. Chem. Soc.* **1999**, *121*, 10928-10941.
21. Iwasita, T. *Electrochim. Acta* **2002**, *47*, 3663-3674.
22. Jacoby, M. In *Chemical and Engineering News*, 1999, pp 31-37.
23. Hogarth, M. P.; Hards, G. A. *Platinum Met. Rev.* **1996**, *40*, 150-159.
24. Matare, G.; Tess, M. E.; Abboud, K. A.; Yang, Y.; McElwee-White, L. *Organometallics* **2002**, *21*, 711-716.
25. Tess, M. E.; Hill, P. L.; Torraca, K. E.; Kerr, M. E.; Abboud, K. A.; McElwee-White, L. *Inorg. Chem.* **2000**, *39*, 3942-3944.
26. Yang, Y.; McElwee-White, L. *Dalton Trans.* **2004**, 2352-2356.
27. Shibasaki, M.; Yoshikawa, N. *Chem. Rev.* **2002**, *102*, 2187-2209.
28. Severin, K. *Chem. Eur. J.* **2002**, *8*, 1514-1518.
29. Rida, M. A.; Smith, A. K. *Mod. Coord. Chem.* **2002**, 154-162.
30. Wheatley, N.; Kalck, P. *Chem. Rev* **1999**, *99*, 3379-3419.

31. Baranger, A. M.; Bergman, R. G. *J. Am. Chem. Soc* **1994**, *116*, 3822-3835.
32. Adams, R. D.; Barnard, T. S. *Organometallics* **1998**, *17*, 2885-2890.
33. Xiao, J.; Puddephatt, R. J. *Coord. Chem. Rev.* **1995**, *143*, 457-500.
34. Orth, S. D.; Terry, M. R.; Abboud, K. A.; Dodson, B.; McElwee-White, L. *Inorg. Chem.* **1996**, *35*, 916-922.
35. Yang, Y.; Abboud, K. A.; McElwee-White, L. *Dalton Trans.* **2003**, 4288-4296.
36. Wasmus, S.; Wang, J. T.; Savinell, R. F. *J. Electrochem. Soc.* **1995**, *142*, 3825-3833.
37. Yuan, Y.; Tsai, K.; Liu, H.; Iwasawa, Y. *Top. Catal.* **2003**, *22*, 9-15.
38. Einaga, H.; Yamakawa, T.; Shinoda, S. *J. Coord. Chem.* **1994**, *32*, 117-119.
39. Robles-Dutenhefner, P. A.; Moura, E. M.; Gama, G. J.; Siebald, H. G. L.; Gusevskaya, E. V. *J. Mol. Catal. A-Chem.* **2000**, *164*, 39-47.
40. Mandal, S. K.; Chakravarty, A. R. *Indian J Chem A* **1990**, *29A*, 18-21.
41. Anthony, C. R.; McElwee-White, L. *J. Mol. Catal. A-Chem.* **2004**, *227*, 113-117.

Chapter 11

Improved Catalytic Deoxygenation of Vicinal Diols and Application to Alditols

Kevin P. Gable and Brian Ross

Department of Chemistry, Oregon State University, Corvallis, OR 97331

Judicious modification of the ligand environment in $LReO_3$ complexes allows successful catalytic deoxygenation of vicinal diols using PPh_3 as terminal reductant. There is a steric bias toward diol units at the end of a linear chain (versus internal diols) and for erythro internal diols over threo. The potential for synthetic applications is discussed.

Introduction

Generation of organic compounds from feedstocks available in Nature usually requires manipulation of oxygen content. Most commodity chemicals (and consequently the vast majority of fine chemicals derived from them) are currently prepared by oxidation of petrochemical feedstocks (*1*). The reactions needed to generate the desired functionality require chemo- and regioselective introduction of oxygen atoms; while stereoselective oxygenation has not been important in commodity chemical production, it often becomes an issue in fine chemical production as in the pharmaceutical industry. Control over these selectivity issues coupled with the need to avoid overoxidation has created interest in understanding oxygen atom transfer reactions over the past decade (*2*).

Consideration of renewable feedstocks as petrochemical substitutes also leads one immediately to the problem of manipulating oxidation levels, but in a different sense. Carbohydrates are the most common primary class of compounds available from renewable biomass (*3*), yet these compounds are much more highly oxygenated than desired for many uses. The challenge thus becomes removal of oxygenated functionality; again, regio-, chemo- and stereoselectivity are desired control elements. Although there are reactivities that can be used to accomplish deoxygenation (e.g., dehydration/hydrogenation (4)), these often require extreme conditions and lack the degree of control that would allow retention of any of the molecular complexity provided in the original feedstock. Specific atom-transfer reactions can thus open new avenues for manipulation of these feedstocks.

Scheme 1 illustrates a comparison of possible routes from feedstock chemicals to useful commodities based on both petroleum and biomass feedstocks. In these examples, a simple 3-carbon feedstock is chosen as the ultimate precursor; manipulation of the oxidation level illustrates either selective introduction of oxygen into propane or selective removal of oxygen from glycerol.

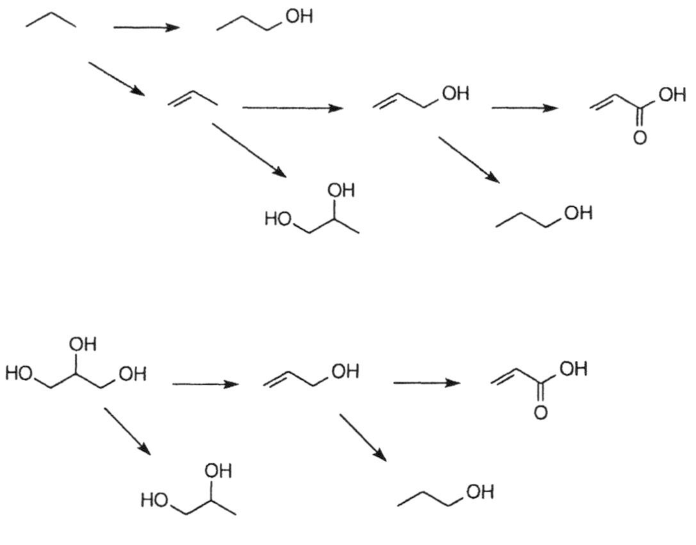

Scheme 1. Comparison of oxidative and reductive routes to common commodity chemicals containing three carbons.

One may anticipate similar schemes based on commonly available compounds with specific carbon numbers: tartrates leading to C_4 compounds; ribose or xylose leading to C_5 compounds; glucose, fructose, mannose or galactose leading to C_6 compounds. (While alditols are not, strictly speaking, carbohydrates, their production from true carbohydrates justifies their inclusion in this discussion. Glycerol itself is a true biomass-derived feedstock, being a hydrolysis byproduct in the production of biodiesel from renewable triglycerides (5).)

Rhenium-mediated deoxygenations

Work in our laboratory has for some time focused on rhenium-mediated O-atom transfer reactions (6). The initial discoveries by Davison and by Herrmann that some Re(V) diolate complexes undergo thermal cycloreversion (7) (Eq. 1) led us to perform a series of mechanistic investigations. Our observations revealed substantial asynchronous character to the bond-cleavage process, but in the end the mechanism appeared to be kinetically concerted. Several other valuable observations accompanied these studies. The impact of the ancillary ligand appeared to be negligible (8), in that the activation enthalpy for Tp'ReO(OCH$_2$CH$_2$O) (Tp' = hydrido-*tris*-(3,5-dimethylpyrazolyl)borate) was only about 4 kcal/mol higher than that of the corresponding Cp* compound; Davison also noted the similarity between our results and those where L = {CpCo{(RO)$_2$P=O)$_3$} (9). The cycloreversion was studied predominantly in nonpolar solvents (e.g., benzene), but investigation of solvent effects noted only a doubling of rate on going to acetone, clearly discounting formation of a polar intermediate.

(1)

During these studies, Cook and Andrews published the first observation of catalytic rhenium-mediated diol deoxygenation (10) (Eq. 2). Their studies were specifically directed toward alditol manipulation. These results showed a turnover frequency for 1,2-phenylethanediol that was consistent with the stoichiometric rate of diolate cycloreversion, indicating that reduction of Cp*ReO$_3$ and subsequent cyclocondensation of the diol on the rhenium were both rapid reactions. While successful catalytic turnover of a series of alditols was observed, in each case the total turnover number was limited by precipitation of a purple solid the investigators suggested might be an overreduced Re(III) species.

(2)

For a variety of reasons, we began to investigate epoxide deoxygenation. Use of the reduced Re(V) dimer $\{(Cp*ReO)_2(\mu\text{-}O)_2\}$ led to high yields of alkene (11). However, when the reaction was performed using less than stoichiometric amounts of rhenium, poor turnover was observed; as in the case of Cook & Andrews, we observed a purple solid forming. Closer investigation revealed that this solid was not a Re(III) compound, but that it formed on conproportionation of $\{(Cp*ReO)_2(\mu\text{-}O)_2\}$ with $Cp*ReO_3$ (Scheme 2). Crystallographic characterization (12) revealed the compound was an ionic tetranuclear cluster (with perrhenate as the counterion). The structure is shown in Fig. 1.

Accompanying this tetranuclear species was also a green, dicationic cluster that had first been observed by Herrmann during aerobic reactions between $Cp*ReO_3$ and PPh_3 (13); we demonstrated that a variety of oxidants could convert the purple tetranuclear cluster into the green trinuclear species. A critical key to generating a long-lived catalyst was clearly the development of a ligand system that could inhibit clustering while maintaining the substrate's access to the metal center.

The prior observation that the pyrazolylborate ligand was so similar to Cp* in diolate cycloreversions made it a prime candidate. The methylated Tp' ligand is unusual in that its cone angle is 255° (14), but the interstices of the three pyrazole rings still allow access to the metal. $Tp'ReO_3$ successfully catalyzes transfer of oxygen from epoxides to triphenylphosphine (Eq. 3) (15). The system is largely insensitive to other functional groups (16), although unprotected alcohols condense to form an inactive Re(V) bis-alkoxide, and good O-atom donors such as nitro groups also interfere. The system imposes several tradeoffs compared to the Cp* system: the trioxo precatalyst is more air- and thermally-stable than $Cp*ReO_3$; the stoichiometric reaction rates are, in general, slower; and the system is much more sensitive to steric properties of the epoxide for obvious reasons.

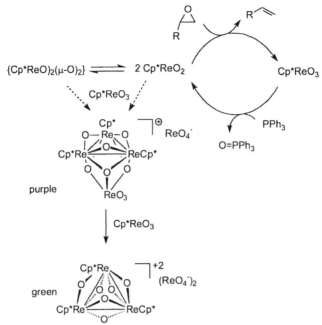

Scheme 2. Catalytic deoxygenation of epoxides with Cp*ReO₃/PPh₃

Figure 1. Cationic rhenium cluster formed by conproportionation of Cp*ReO₃
with {(Cp*ReO)₂(μ-O)₂}.

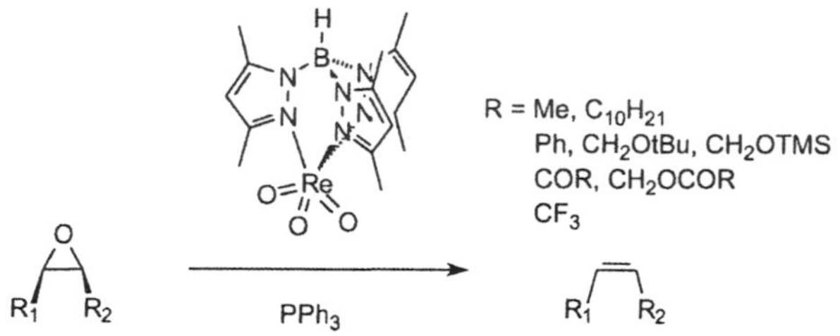

R = Me, C10H21
Ph, CH2OtBu, CH2OTMS
COR, CH2OCOR
CF3
PPh3

(3)

We next returned to catalytic deoxygenation of vicinal diols. Phenylethanediol was the first substrate chosen (in all stoichiometric reactions, the phenyl group has a significant accelerating effect). At 5 mol-percent rhenium, successful reduction was complete after 4 days at 100° C. The solvent used was toluene-d_8. One concern (Scheme 3) was that the equivalent of water generated by cyclocondensation of the diol with rhenium could have an inhibitory effect as it accumulated; this appeared not to be the case over the first 90% reaction. Independent experiments examining hydrolysis of related diolates had suggested the equilibrium constant for cyclocondensation is at least 100.

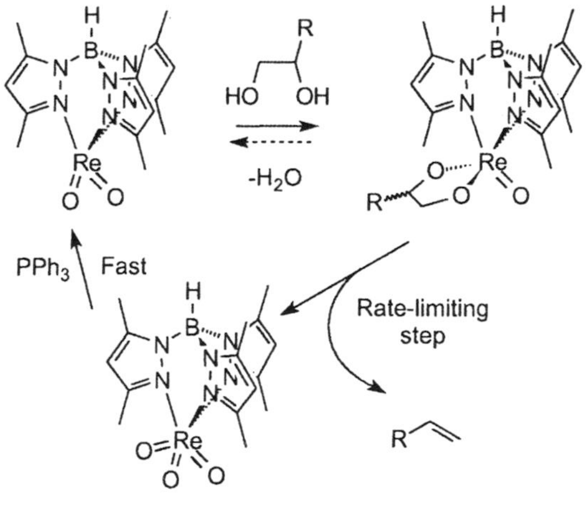

-H2O
PPh3
Fast
Rate-limiting
step

Scheme 3. Catalytic deoxygenation of vicinal diols with Tp'ReO$_3$.

One may predict the turnover frequency based on our earlier measurement of diolate cycloreversion. The observed rate constant for stoichiometric extrusion of styrene from diolate is 8.61×10^{-5} s^{-1}, predicting a maximal turnover frequency at the observed [Re] of 0.009 M of 7.77×10^{-7} M-s^{-1}. An intermediate measurement at 1002 minutes showed 33% conversion of 0.18 M diol, for a turnover frequency of approximately 9.9×10^{-7} M-s^{-1}, within experimental error of the predicted value.

It is of note that Arterburn and coworkers have developed a heterogenized rhenium catalyst that is also effective for diol deoxygenation (*17*).

Alditol substrates

Glycerol was the next target substrate. Here we had the added potential challenge that the substrate has a low solubility in the toluene solvent. However, this did not inhibit formation or reaction of the diolate, indicating rapid mass transport relative to the (slow) cycloreversion of diolate. This substrate reacts more slowly than phenylethanediol; at 121° C, a reaction mixture containing comparable concentrations requires 5 days to reach completion. This is consistent with the electronic effect of switching substituents from an aromatic to an aliphatic group; there may be some impact from product inhibition (though condensation of the vicinal diol is expected to be preferred over formation of a bis-alkoxide until late in the reaction). Again, the presence and generation of water does not appear to have any significant impact at low concentration.

Use of erythritol begins to explore the factors that control regiochemistry of the catalytic reaction. Here, cyclocondensation can lead to either a 1,2-diolate (Scheme 4; "terminal") or a 2,3-diolate ("internal"). The former will produce 3,4-dihydroxy-1-butene, while the latter will produce 1,4-dihydroxy-2-butene (presumably as the cis isomer). Secondary reduction of the terminal product will also produce 1,3-butadiene.

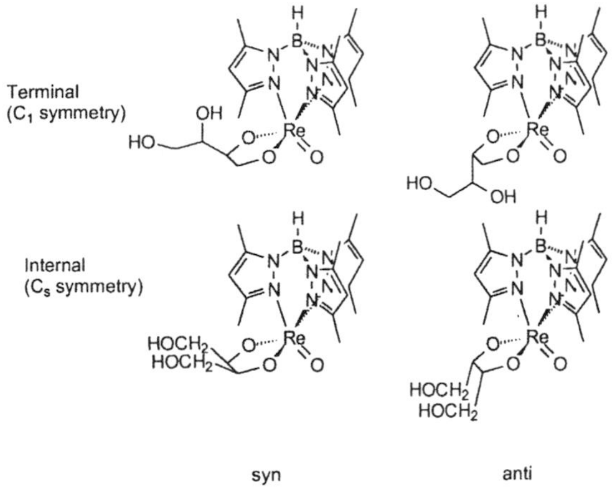

Terminal (C₁ symmetry)

Internal (Cₛ symmetry)

syn   anti

Scheme 4: Regio- and stereoisomerism in erythritolate complexes

All three products are observed (Eq. 4); the initial selectivity for terminal deoxygenation is reasonably high: the ratio of (1-ene + butadiene)/(2-ene) is 8.8:1 at 40% conversion (24 h at 121° C; 5 mol-per cent rhenium). The 1-ene:butadiene ratio is still high at this stage at 7.2:1.6. As further reaction occurs, the butadiene proportion rises, and a slight drop in the total terminal:internal reaction product ratio is seen.

$$\text{(4)}$$

Use of threitol leads to a very different outcome (Eq. 5). The reaction is slower (21% conversion after 24 h at 121° C), and there is no detectable internal alkene produced. The proportion of butadiene is also higher; 1-ene/butadiene = 1.4:1 at this point.

$$\text{(5)}$$

Consideration of structure for the diolates highlights the likely origin of the differences in selectivity. Stoichiometric reductive cyclocondensation of glycerol with Tp'ReO$_3$/PPh$_3$ gives two diolate isomers in roughly equivalent amounts. 1-D and 2-D NMR experiments allow almost total assignment of the spectroscopic signals. The CH$_2$OH signals for the two isomers are at 4.03 and 4.42 ppm. Sequential nOe experiments reveal that irradiating a methyl signal at 2.33 ppm leads to enhancement of the signal at 4.42 ppm (as well as the other syn diolate proton further downfield). Irradiation of a nearby methyl signal at 2.30 ppm results only in enhancement of diolate signals at 5.53 ppm, and not at the hydroxymethylene position. The consequences of this full assignment are that the two stereoisomers are formed in approximately equal amounts. They react at different rates, though; after heating a sample to 100° C for 15 h, the remaining diolate was enriched in the exo isomer. Further experiments are needed to establish relative rates and to what degree interconversion can occur under stoichiometric and catalytic conditions.

A similar analysis of structure for the erythritolates is more complex in that four different stereoisomers are anticipated as noted above. One aid to structural analysis lies in the symmetry properties of different isomers. The 2,3-diolates are C$_s$-symmetry, in which two of the ligand pyrazole rings are equivalent. The

1,2-diolates are nonsymmetric, and the pyrazoles are all unique. A count of the methyl signals for the ligand thus provides some sense of which isomers are present; a total of 20 signals are possible for the four isomers and 19 are readily evident. More significantly, one may discern 10 major signals, indicating that 1,2- and 2,3-isomers are present in roughly equivalent amounts: the C_1 isomers each have six unique signals, and the C_s isomers have 4 in a 2:2:1:1 integral ratio.

Given the difference between the selectivity outcome for cyclocondensation and that for production of terminal vs. internal alkenes, we can conclude that different diolates will have different reactivities. This is self-evident from our prior kinetic studies in that each alkyl substituent has a slight decelerating effect; one expects a preference for terminal deoxygenation over internal unless the internal diol has an electron-withdrawing substituent. Less important will be the difference in reactivity between syn and anti diolates, although if one or the other is particularly less reactive, it risks developing a thermodynamic sink that effectively removes active catalyst from the system. Further work is needed to quantitatively establish rates of reaction under stoichiometric conditions in order to derive a predictive model for selectivity.

Preliminary work on higher alditols confirms that catalytic deoxygenation occurs, although full characterization is not complete. Xylitol shows a mixture of mono- and bis-alkene reduction products (with terminal selectivity). A protected mannitol, the 3,4-O-acetonide, also shows a degree of double reduction.

Prospects and applications

We have successfully demonstrated the principle of selective catalytic reductive deoxygenation of vicinal diols. The challenges inherent to this transformation have hindered development of carbohydrate-based organic feedstock chemistry. While there remain aspects of our chemistry that keep it a laboratory methodology, there is now a clear route to development.

The two obvious drawbacks to these rhenium-mediated reactions are the expense of the catalyst (rhenium being one of the rarer transition metals) and the disadvantage of using PPh3 as the terminal reductant. Preliminary experiments with Ph3SiH suggest that silanes are promising reductants that could more easily be recycled; one would ideally like to develop a scheme wherein H_2 or CO was the terminal reductant. Achieving this will strongly depend on the thermodynamic properties of the M=O bonds produced in diolate cycloreversion; published data suggest the current ligand/metal system may not allow direct reduction by either.

A less significant drawback is that the temperatures required for diolate cycloreversion are relatively high. Given that switching from Cp* to Tp' induced an increase in $\Delta G^{\ddagger}$ of 2-4 kcal/mol, it appears probable that judicious ligand design would allow both a decrease in the activation barrier while maintaining the necessary steric bulk to inhibit the clustering reactions that occurred with the Cp* complexes.

More work is clearly needed to establish the basis for internal/terminal reactivity and to document the capacity for diolate isomerization under the reaction conditions. Again, ligand design would appear to allow enhancement of the selectivity for formation of a terminal diolate, and a preliminary conclusion we can draw from the reactivity of erythritol is that the terminal diolate appears to be the more reactive intermediate.

Several important applications can be immediately anticipated if these issues can be resolved. Scheme 5 illustrates potential conversions of mannitol (or, indeed, sorbitol) derivatives into linear and cyclic C_6 compounds.

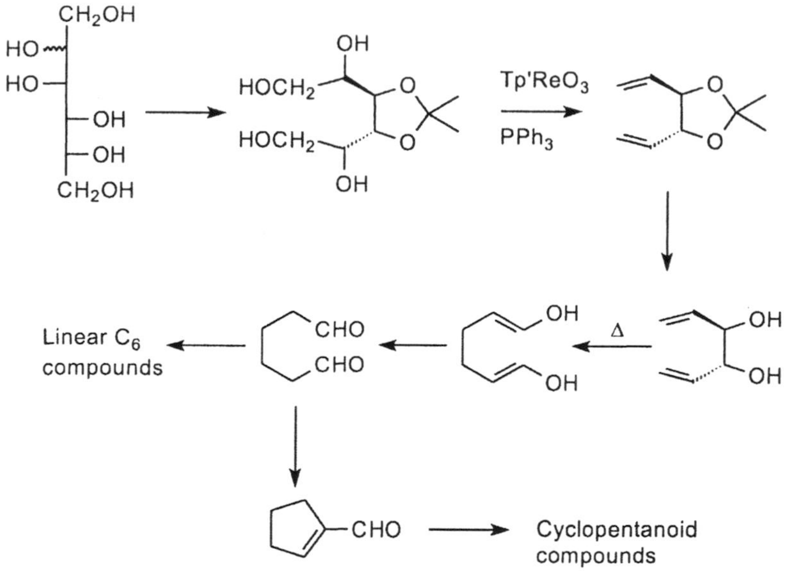

Scheme 5. Compounds available from bisdeoxygenation of mannitol or sorbitol.

Another immediate application would be direct conversion of glucose and other hexoses to glycals (Scheme 6) without protection/deprotection (18). Again, establishing the selectivity for cyclocondensation across cis vs. trans vicinal diol units must be established; the trans-diolates presumably will not cyclorevert, and dynamic interconversion of isomeric diolates may become an advantage. Solubility and choice of solvent becomes a significant issue here; development of a water-soluble catalyst would open an important means of optimizing these as practical syntheses.

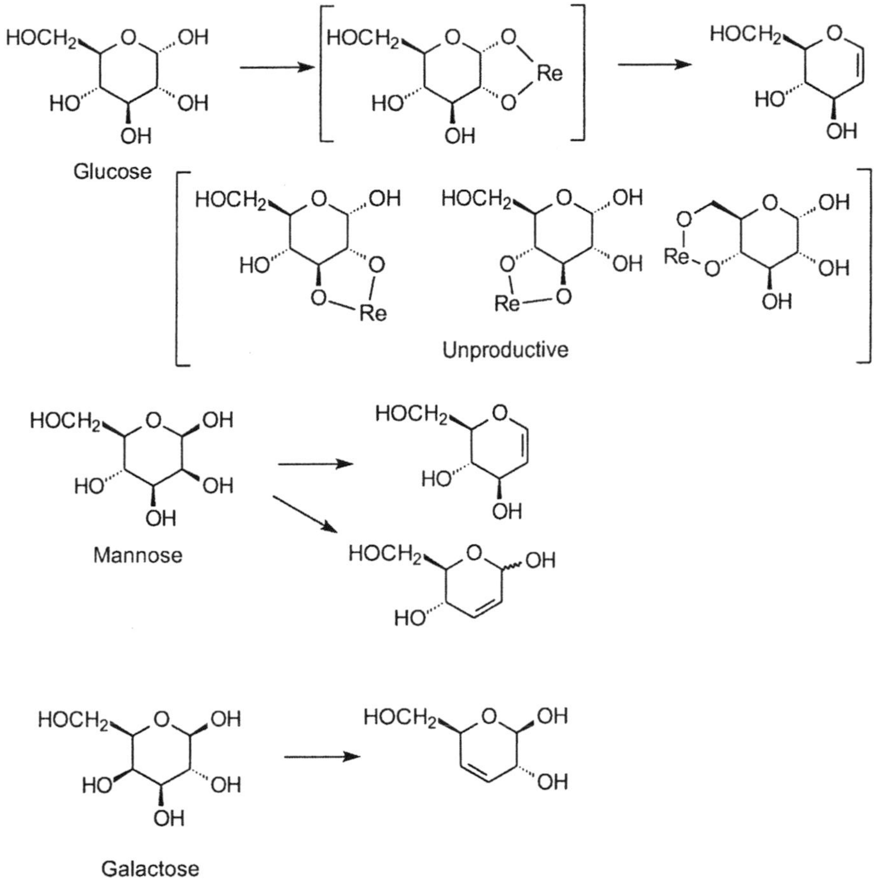

Scheme 6. Possible generation of glycals from hexoses.

Finally, it must be recognized that recent developments in organic synthetic strategy open other applications. Alkene metathesis has become a standard strategy for synthesis of cyclic compounds (*19*). Given that our methodology gives rise to new alkene-containing synthons derived from the natural chiral pool, direct entry into chiral oxygen-containing rings becomes feasible (*20*), and use in constructing chiral carbocycles is possible with minor elaboration (Scheme 7).

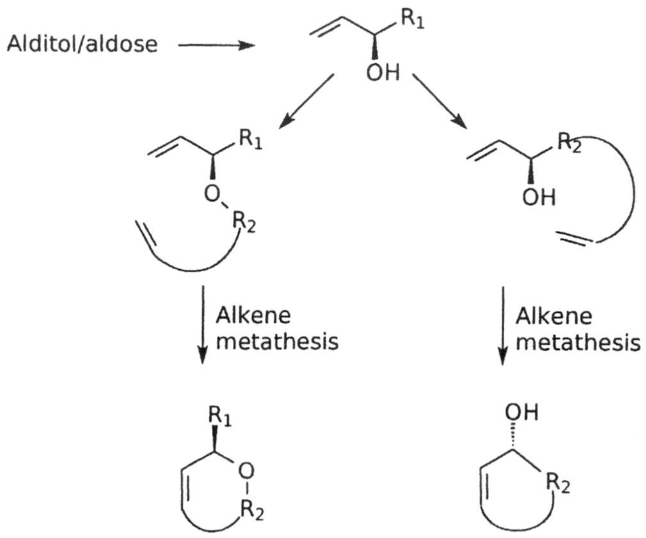

Scheme 7. Strategy for using deoxygenated products in metathesis-driven ring construction.

Conclusion

In contrast to petroleum-based chemistry, a common challenge in utilizing biomass feedstocks for fine chemical production is to reduce molecular complexity by specifically removing oxygenated functionality. The work described here advances this goal, demonstrating that a robust catalyst can achieve net reductive deoxygenation of the vicinal diol group. While important challenges remain, we have begun to understand the principles that govern selectivity.

Acknowledgment

We thank the National Science Foundation (CHE-0078505) for partial support of this work.

References

1. *Organic syntheses by oxidation with metal compounds* Mijs, W. J.; de Jonghe, C. R. H. I., Eds. Plenum: New York, NY, 1986.
2. a. Espenson, J. H.; *Adv. Inorg. Chem.* **2003**, *54*, 157-202. b. Romão, C. C.; Kühn, F. E.; Herrmann, W. A. *Chem. Rev.* **1997**, *97*, 3197-3246.

3. *Carbohydrates as Organic Raw Materials II*, Descotes, G., ed. VCH: New York, 1993.
4. Mehdi, H.; Horvath, I. T.; Bodor, A. *Abstracts of Papers, 227th ACS National Meeting,* CELL095. American Chemical Society: Washington, 2004.
5. Ma, F.; Hanna, M. A. *Bioresource Tech.* **1999**, *70*, 1-15.
6. Gable, K. P. *Adv. Organomet. Chem.* **1997**, *41*, 127-161.
7. a. Pearlstein, R. H.; Davison, A. *Polyhedron* **1988**, 7, 1981-1989. b. Thomas, J. A.; Davison, A.; *Inorg. Chim. Acta* **1991**, *190*, 231-235. c. Herrmann, W. A.; Marz, D.; Herdtweck, E.; Schäfer, A.; Wagner, W.; Kneuper, H. J. *Angew. Chem., Int. Ed. Engl.* **1987**, *26*, 462-464.
8. Gable, K. P.; AbuBaker, A.; Zientara, K.; Wainwright, A. M. *Organometallics* **1999**, *18*, 173-179.
9. Cook, J. A.; Davis, W. M.; Davison, A.; Jones, A. G.; Nicholson, T. L.; Simpson, R. D. *Abstracts of the 207th National Meeting of the American Chemical Society*, INOR066. American Chemical Society: Washington, 1994.
10. Cook, G. K.; Andrews, M. A. *J. Am. Chem. Soc.* **1996**, *118*, 9448-9449.
11. Gable, K. P.; Juliette, J. J. J., Gartman, M. A. *Organometallics* **1995**, *14*, 3138-3140.
12. Gable, K. P.; Zhuravlev, F. A.; Yokochi, A. F. T. *J. Chem. Soc., Chem. Commun.* **1998**, 799-800.
13. Herrmann, W. A.; Serrano, R.; Ziegler, M. L.; Pfisterer, H.; Nuber, B. *Angew. Chem., Int. Ed. Engl.* **1985**, *24*, 50-51.
14. Bergman, R. G.; Cundari, T. R.; Gillespie, A. M.; Gunnoe, T. B.; Harman, W. D.; Klinckman, T. R.; Temple, M. D.; White, D. P. *Organometallics* **2003**, *22*, 2331-2337.
15. Gable, K. P.; Brown, E. C. *Organometallics* **2000,** *19*, 944-946.
16. Gable, K. P.; Brown, E. C. *Synlett* **2003**, 2243-2245.
17. Arterburn, J. B.; Liu, M.; Perry, M. C. *Helv. Chim. Acta* **2002**, *85*, 3225-3236.
18. Roth, W.; Pigman, W., in *Methods in Carbohydrate Chemistry,* Whistler, R. L.; Wolfrom, M. L.; BeMiller, J. N., Eds. Vol. 2, pp. 405 408. Academic Press: New York, 1963.
19. Fürstner, A.; *Top. Catal.* **1998**, *4*, 285-299.
20. Voight, E. A.; Rein, C.; Burke, S. D. *J. Org. Chem.* **2002**, *67*, 8489-9499.

Chapter 12

Iron-TAML® Catalysts in the Pulp and Paper Industry

Colin P. Horwitz[1,*], Terrence J. Collins[1,*], Jonathan Spatz[1],
Hayden J. Smith[2], L. James Wright[2,*], Trevor R. Stuthridge[3,*],
Kathryn G. Wingate[2,3], and Kim McGrouther[3]

[1]Department of Chemistry, Carnegie Mellon University,
Pittsburgh, PA 15213
[2]Department of Chemistry, The University of Auckland,
Auckland, New Zealand
[3]Forest Research Limited, Rotorua, New Zealand

Here we report on a novel catalytic approach to removing
color from the effluent derived from the bleaching process of
bleached kraft pulp mills. First we show how an Fe-TAML
catalyst with hydrogen peroxide, after laboratory simulation,
can be used at a pulp mill as an effective end-of-pipe treatment
to decolorize the effluent water from the caustic extraction
stage of a chlorine-based bleaching process. Then we show
laboratory experimentation for applying the Fe-TAML catalyst
in the El extraction stage of the bleaching process itself. The
goal of this second treatment type is to obviate the need for the
previously described end-of-pipe treatment. The details
regarding the benefits that a Fe-TAML/H_2O_2 system could
have on mill operations will be discussed.

The pulp and paper industry is an essential component for growth in the world's economy. U.S. pulp and paper production accounted for $151.5 billion in 2002 and employed 531,000 workers (*1*). The Confederation of European Paper Industries accounts for about €73 billion in combined turnover and employs 250,000 people (*2*). Overall, this industry used approximately 11% of the total volume of water used in industrial activities in the thirty member OECD countries (http://www.oecd.org/home/). The worldwide water use by the industry has been projected to grow from 11 billion m^3 in 1995 to 18 billion m^3 by 2020. Wastewater produced during the manufacture of paper products contains organic materials with a wide variety of structures resulting from the degradation, modification and dissolution of cellulosic and lignin based materials. This results in wastes that are high in biochemical oxygen demand (BOD), 10–40 kg/t of air-dried pulp (ADP), total suspended solids (TSS), 10–50 kg/t ADP, and chemical oxygen demand (COD), 20–200 kg/t ADP (*3*). Worldwide the industry spends $5 billion/yr (est.) to treat these wastewaters.

The principles of a sustainable society make us consider water-borne discharges from industries like pulp and paper as sources of useful chemicals rather than as wastes to be treated and disposed (*4*). By converting what is now considered as waste material into useful products, the waste producers, chemical manufacturers, and regulators are offered new opportunities.

The most noteworthy limitation to this approach is the complexity and multi-functionality that exist in these waste streams. For example, a bleached kraft pulp mill is estimated to produce greater than 400 compounds including tannins, resins, and lignin degradation products (*5*). However, as only a relatively small number of chemical transformations offer benefits to an end-user, i.e. polymerisation, selective oxidation or reduction, C-H bond activation, and carbon-halogen bond cleavage, the problem is simplified to some extent.

The most appropriate point to install a treatment system is a key issue when developing waste utilization technologies. Targeting individual sources of organic materials prior to biological treatment appears to be the most effective approach. A major source of this material in bleached kraft pulp mills is the bleach plant where lignin residues are oxidized and dissolved into the water.

The colored effluent is one of the most obvious indicators of waste discharge from a pulp and paper mill as it can darken receiving waters with adverse effects to benthic and planktonic life forms. Thus, targeting color reduction is a starting point for developing a technology that utilizes waste to derive useful organic materials. Once it is understood how to control the color centers in the waste, the technology can be shifted toward obtaining useful chemicals from the waste. The studies described here focused on removing color from the first caustic extraction stage effluent either as an end-of-pipe treatment where color is treated after it is formed or as an in-process treatment where color is removed as it is generated during extraction.

Colored Material Composition in Pulp and Paper Wastewater

Cellulose is a long linear polysaccharide comprised of approximately 10,000 anhydroglucose units linked by β-1,4-glycosidic bonds, and has the general formula $(C_6H_{10}O_5)_n$. Hemicelluloses are short chain, branched polysaccharides. They are composed of 100 to 200, 5- and 6-carbon sugar monomers linked in various combinations. Lignin is an amorphous three-dimensional polymeric network containing substituted phenylpropane units. Its stucture has not been elucidated due to the random polymerisation that occurs during its biosynthesis (6,7). The complex structure of lignin allows for considerable double bond resonance, which leads to its dark color. The aim of pulping and bleaching is to remove the colored lignin from the cellulosic fiber material required for papermaking. The oxidized lignin fragments produced by the bleaching process are discarded but these could prove to be valuable chemical precursors.

The dominant chemical pulping technology is called the kraft process. During the process most of the lignin becomes solubilized and is carried off in the aqueous solution called 'black liquor'. The black liquor is burned in the mill's recovery boiler to generate energy and to return the majority of the pulping salts for reuse. The energy generation and salt recovery are important components of the financial feasibility of kraft mills.

Oxidative bleaching follows the kraft process to remove the remaining lignin. Chlorine dioxide now is the most common bleaching chemical as it is highly selective in oxidising and solubilizing the remaining lignin without degrading cellulose (8). Unfortunately, since chlorine salts are corrosive, the wastewater can not be burned for energy. Thus, the bleach plant is the dominant source of undesirable wastewater in terms of both color and toxicity (9).

Reactions of lignin model compounds with either Cl_2 or ClO_2 followed by a sodium hydroxide extraction stage have been used to characterize the structure of wastewater chromophores (10). Model phenolic compounds reacted with ClO_2 gave a mixture of muconic acid esters, lactones and maleic acids as the major products. Reactions of the lactones with sodium hydroxide resulted in saponification of the methyl esters and ring opening to form muconic acids.

Studies of bleach mill wastewater have found that the majority of the phenolic content in lignin-derived material is destroyed during bleaching. The phenolic content is less that 5 % of the total organic material (5). These residual structures are generally highly conjugated, and can be very colored. It has been estimated that 40 to 50 % of the color of bleach plant wastewater may be due to quinones and about 20 % due to other carbonyl structures (11). The remaining 30 to 40 % of the colored material may result from other structural by-products of lignin as well as sugar degradation products (12). The precise nature of the dissolved organic material in bleaching wastewaters and the chromophoric structures therein, remain largely unidentified.

The majority of past research on pulp and paper mill wastewater impacts and treatment has focused on the low molecular weight (LMW) compounds. High molecular weight (HMW) compounds have been considered to be biologically benign (*13*) and are resistant to traditional biological treatment (*9*). Recently the HMW component of wastewater has been shown to be the principal source of the color and COD (*14*). This HMW material has been characterised as being virtually non-aromatic with a low methoxyl and phenolic hydroxyl content and a high carbonyl and carboxyl content (*7*).

Color is often seen as an aesthetic concern, however, it is also reputed to be responsible for several negative environmental impacts. Colored material adsorbs sunlight, thereby reducing the photic depth of the receiving water which may then result in decreased photosynthetic rates and affect aquatic productivity. Also, the accumulation of precipitated colored particles in lower velocity regions can cause benthic inhibition. For example, the river bed community of the Tarawera River, in which pulp and paper wastewater is discharged, has decreased diversity (from 14 species to 5) and size (from 35 % of the river bed to just 3 %) due to color load reducing light penetration (*15*). Furthermore, possible nutrients bound to the color-causing compounds may increase algal or bacterial activity and phenolics can be a taste problem.

Processes for Color Removal

Aerobic biological treatment systems such as aerated stabilisation basins and activated sludge are most commonly used to treat pulp mill wastewaters. A considerable reduction in BOD and TSS occurs in these systems, but color remains relatively unaltered. For example, the secondary treatment system of an integrated bleached kraft mill showed a 92 % reduction in BOD, a 60 % reduction in TSS but only a 1 % reduction in color (*16*).

To find an economically feasible color removal technology, cost (chemical and operational), treatment time, by-product generation, compatibility with pulping, bleaching and treatment processes, and sludge handling and disposal options are all factors to be considered. Biological (*17*), chemical (*18,19*), physico-chemical means (*20*), and process conditions (*21*), have been applied to this problem, yet there is no single approach for color removal from pulp mill wastewaters. Recently, wastewater treatment by Advanced Oxidation Processes (AOPs) has gained popularity, leading to biodegradable materials.

The most common oxidants for AOPs are O_3 and H_2O_2. Laboratory-scale treatments of pulp and paper mill wastewater have been undertaken using O_3, O_3 with UV, and O_3 with UV and ZnO. Little difference between rate and extent of treatments when using O_3 alone, or simultaneously with UV radiation were observed (*22*), However, when ZnO was added, the system was more efficient at treating wastewater. The amount of O_3 required varied between 20 and 300 mg L^{-1} (23) and resulted in 83 % color reduction (24). In addition to removing color from wastewater, O$_3$ can also convert non-biodegradable substances into biodegradable ones (25). Unfortunately, O$_3$ is costly and often needs to be manufactured on-site. H$_2$O$_2$ has also been applied in AOP systems. It is considerably cheaper than O$_3$, however the treatment times are long and result in incomplete decolorization (26). UV radiation combined with H$_2$O$_2$ can result in a greater reduction of the wastewater color (19,27).

Fe-TAML Oxidation Catalysts

The family of catalysts called TAML® catalysts (TetraAmido Macrocyclic Ligand) are powerful activating agents of H$_2$O$_2$. Figure 1 shows the prototype iron TAML catalyst (Fe-TAML) and the one used in this study. Fe-TAML catalysts are the first catalysts that are highly suitable for commercial applications in wastewater treatment. The catalysts have been developed over a twenty year period (28) and possess some unique properties. They are formed from biochemically common elements and have biodegradable functionalities so they have little or no toxicity (29). They are water-soluble and stable to high and low pH conditions (30). They often are effective at minute concentrations, 0.1 to 4 ppm (31-33) and their H$_2$O$_2$ disporportionation activity is much slower than the catalyzed multi-step oxidation of most substrates. The major oxidation mechanism does not appear to be non-selective as found in the Fenton-type chemistry. Finally, they can be modified to capture novel selectivity – more than 20 variants have been synthesized thus far.

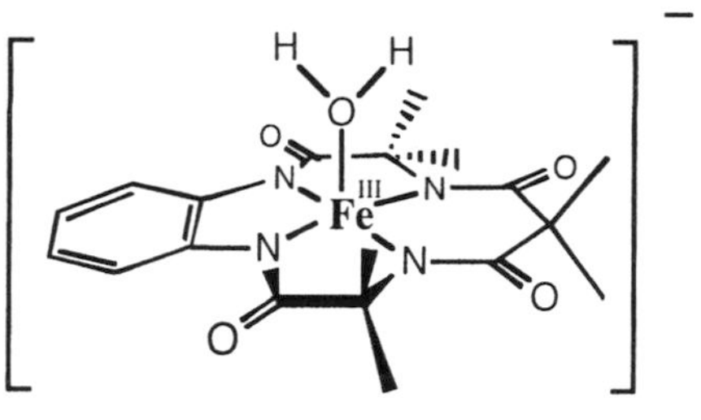

Figure 1. Fe-TAML catalyst used in these studies

Experimental Procedures

Source of samples. Samples were obtained from an integrated bleached kraft pulp and paper mill using a DE$_{op}$DED bleaching process (D = ClO$_2$, E =

NaOH extraction, $P = H_2O_2$, $O = O_2$). Total mill discharge color is 37.5 tonne CPU/day.

End-of pipe effluent source. Effluent samples were taken from the E1 (first extraction stage) sewer of the bleach plant while the mill was processing *Pinus radiata*. Samples were stored at 4 °C. Treatment trials were carried out on filtered wastewater samples (25 mL) which had been adjusted to pH 11. The Fe-TAML catalyst (0.5 – 10 μM) was added simultaneously with H_2O_2 (2 – 30 mM) and stirred at 50 °C for a prescribed amount of time 30 – 300 minutes (*34*).

E1 stage pulp source. Pulp was taken from the screen of the first D-stage washer. The pulp was squeezed to remove the liquor, the liquor was then filtered through GF/C filter paper and stored at 4°C for the duration of the experimental work. The remaining pulp was washed at 1% consistency (cty) in hot tap water for 1.5 h prior to dewatering to 26% cty, then washed at 1.5% cty in distilled water for 45 min prior to dewatering to 26% cty.

Color. The standard method for color determination, chloroplatinate standard (*35*), in pulp and paper effluents (pH 7.6, 465 nm) was employed.

Absorption spectra. Absorption spectra were recorded from 200 to 700 nm on a Philips PU 8740 spectrophotometer using a path length of 10 mm, a bandwidth of 1.0 nm, and a scan speed of 250 nm min^{-1}. Effluent samples were diluted tenfold using either NaOH (0.6 M) or pH 6.5 buffer ($NaH_2PO_4.2H_2O$).

Catalytic hydrogenation. A 5 % Pd on carbon sample (3 mg) was added to deoxygenated solutions of untreated and Fe-TAML/H_2O_2 treated effluents (5 mL, diluted fivefold with water). The mixture was stirred for 17 h under H_2 in a balloon. 1 mL was then filtered through a 0.45 μm PTFE filter and diluted fivefold with pH 7 buffer before determining the absorption spectra.

Reaction with sodium borohydride. Untreated and Fe-TAML/H_2O_2 treated effluents (100 mL) were diluted tenfold with water. NaBH$_4$ (200 mg) was added and stirred at room temperature. Further portions of sodium borohydride (200 mg) were added after two and four days. On day seven, samples (7 mL) were neutralised with 5 % sulfuric acid (2 drops) and centrifuged (3000 rpm, 15 min) to remove boric acid salts before spectral analysis.

Sodium periodate treatment. Effluent (5 mL, diluted fivefold with water) was added to 5 mg sodium periodate in a closed vessel flushed with nitrogen at room temperature. Samples (1 mL) were then taken at timed intervals (1, 4 and 24 h) and diluted with pH 7 buffer (4 mL) before UV/Vis analysis.

Treatment of effluent using Fe-TAML. Treatment of the effluent samples was performed in an agitating incubator as adequate sample mixing was found to be vital for reproducible color removal. Treatments involved the addition of different concentrations of Fe-TAML and H_2O_2 to E1 effluent samples that had been adjusted to pH 11 and maintained at 40-60 °C.

Treatment of pulp in an E1 stage simulation. The requisite amount of D-stage pulp and liquor or pH 11 distilled water (NaOH) were combined in a 100 mL glass bottle. The suspension was diluted to 1% cty, 0.75 g of pulp in 75 mL total pulp-suspension volume, and adjusted to pH 11 with 0.2 M NaOH. Fe-TAML catalyst was mixed in if required, (0.5μM). The bottle was sealed and flushed with N_2 or O_2 before pressurizing to 0.17MPa. H_2O_2 was added, 1% on pulp. The bottle was placed into an incubation shaker pre-heated to 80°C. The contents were gently swirled for 90 min. The reaction was quenched by placing the bottle in a 25°C water bath. Once cooled, the pulp suspension was filtered through a GF/C filter. The filtrate was analysed for color and the pulp was made into a brightness pad for brightness measurements after washing.

Results and Discussion

Model Bleach Plant Wastewater Decolorization using Fe-TAML/H_2O_2

For the current study, all experiments were carried out using caustic bleach plant wastewater processing pine. This wastewater was discharged at pH ~ 11 and T ~50 °C. At the time of discharge, the wastewater contained residual H_2O_2 (approximately 0.37 mM), however additional H_2O_2 was added for the trials as its concentration in the wastewater was insignificant by the time the wastewater was treated in the laboratory.

Initial trials with the Fe-TAML catalyst (0.5 μM) and H_2O_2 (16 mM) showed that the Fe-TAML/H_2O_2 methodology removed 41 % of color from the bleach plant wastewater after 30 minutes at 50 °C. Initial treatments also showed that adequate mixing of the wastewater was required for effective and consistent color removal. For example, when no mixing was used, color reductions varied from 9 to 42 % over seven repetitions, final color 495 ± 68 CPU, whereas when mixing was used, the final color was 348 ± 8 CPU.

Aqueous solutions of Fe^{2+} salts and H_2O_2 produce hydroxyl radical. To test the possibility that color removal using Fe-TAML/H_2O_2 could have resulted from reaction with free Fe^{2+}/Fe^{3+} derived from catalyst breakdown, reactions were investigated in which $FeCl_3$ replaced the Fe-TAML catalyst. The concentration of Fe^{3+} was the same as the Fe-TAML catalyst concentration, 1 μM in this case. No effect from the added Fe^{3+} was observed - control (0 μM Fe-TAML) 592 CPU, 1 μM Fe-TAML 348 CPU, and 1 μM $FeCl_3$ 592 CPU all 26 mM H_2O_2. Thus the Fe-TAML-H_2O_2 combination must be decreasing the color.

Optimization of catalyst and H_2O_2 concentrations and reaction time were examined. For the initial experiments, H_2O_2 was fixed at 16 mM and the catalyst was added to give concentrations ranging from 0 to 50 μM. The maximum color removal was 44 % and was achieved at 1 μM catalyst (Figure 2a). A substantial increase in color removal occurred within a very small range of catalyst concentrations (0.05 to 0.1 µM) demonstrating the efficiency of the system. Catalyst concentrations greater than 5 µM resulted in lesser color removal probably because Fe-TAML now catalysed disproportionation of H_2O_2. By employing a similar set of experiments, it was found that the optimum H_2O_2 concentration was 13 mM at 1 µM Fe-TAML concentration.

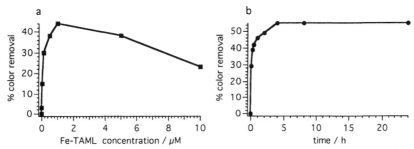

Figure 2. Color removal dependence on (a) Fe-TAML concentration (30 min) and on (b) time (1 µM Fe-TAML), pH 11, 50 °C, 16 mM H_2O_2.

For the process to be feasible industrially, color removal must be rapid. To monitor the progress of color removal over time, samples were collected at 5, 15 and 30 minutes, and at 1, 2, 4, 8, 16, 24 and 48 hours. The experiments were performed under the previously determined most favorable condtions catalyst (1 µM) and H_2O_2 (16 µM). Color removal rapidly reached a maximum (55%) within approximately four hours (Figure 2b). This result implied that the active oxidant formed quickly and then reacted rapidly with the chromophores present in the wastewater. A four hour reaction time is realistic industrially.

Treatments of wastewater using the conditions, pH 11, 4 h, 50 °C, 16 mM H_2O_2, and 1 µM Fe-TAML, were carried out using caustic bleach plant wastewater samples collected over several weeks. It was found that the mean color removal was 47% (593 CPU original to 279 CPU final with a 95% confidence interval of 11 CPU). These data show that color removal was very reproducible. This was an important finding for mill implementation, especially if the final wastewater has a color level close to the discharge limit.

The complicated nature of the effluent makes it difficult to determine with a high degree of certainty the exact nature of the chemistry occuring during treatment with the Fe-TAML/H_2O_2 system. However, some spectroscopic methods and chemical analyses do provide some insights. First, ionization difference spectra revealed phenolic groups and phenolic α-carbonyl structures in the effluent samples (*36*) were removed following a Fe-TAML/H_2O_2 treatment. Second, subjecting a Fe-TAML/H_2O_2 treated effluent to a sodium periodate oxidation produced in no additional spectral changes so most if not all of the oxidizable groups in the effluent were oxidized in the first treatment.

Third, hydrogenation of control and Fe-TAML/H_2O_2 treated effluents indicated that aliphatic double bonds like those found in stilbenes were removed in the treated sample (*37*). Finally, reduction using sodium borohydride, which reduces carbonyl groups in structures such as α-carbonyls and quinones (*37*), revealed that both types of structures were removed in the treated samples.

Pilot Plant Field Trial for the Treatment of E1 Effluent

Based on the laboratory results described above a pilot facility was built at the Tasman mill for the treatment of E1 effluent (*Pinus radiata* feedstock) to test a real time application of the chemistry. The pilot facility consisted of two stirred reactor vessels where the effluent was pumped into the first vessel (200 L capacity, hydralic retention time (HRT) 1 h) at a rate of 3.3 L/min (4800 L/day), before flowing into the second one (800 L capacity, HRT 4 h). Temperature and pH were continuously monitored but were not adjusted. Solution samples were regularly collected from each vessel along with untreated effluent. The Fe-TAML catalyst and hydrogen peroxide were dosed into treatment vessel 1.

Color reductions on the order of 78% could be achieved using 2 μM Fe-TAML and 22 mM H_2O_2 (Figure 3) (*38*). The color reduction was only slightly reduced, 67%, when 1 μM Fe-TAML and 11 mM H_2O_2 were used. The color reduction was also determined to be complete within 1 h. Over the twelve day duration of the trial, it was found that the color entering the reaction vessel varied from as low as 15 kg/ADT to as high as 45 kg/ADT. The amount of color removed by the Fe-TAML/H_2O_2 system to some degree tracked the variation of the input color (Figure 3). Following the Fe-TAML/H_2O_2 treatment, the color level was never above 10 kg/ADT. Thus, although the E1 effluent color could vary widely from day to day, the Fe-TAML/H_2O_2 treatment substantially evened out the color remaining after treatment. This is important for bleach plants that must maintain tight control over their color discharges.

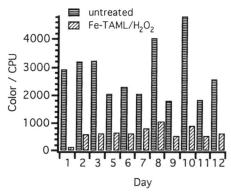

Figure 3. Color of E1 effluent before and after 1 h Fe-TAML treatment on a production basis, 2 μM catalyst, 22 mM H_2O_2, pH 11.8, 60 °C.

D-State Pulp Treatment with Fe-TAML/H$_2$O$_2$ - An E1 Stage Simulation

While the Fe-TAML/H$_2$O$_2$ system was effective at decolorizing E1 effluent, the addition of the Fe-TAML into the E stage tower where color generation occurs might obviate the need for effluent treatment. Furthermore, reaction conditions within the tower are relatively consistent between mills with respect to pH (10 – 11), H$_2$O$_2$ concentration (0.5 – 1% on pulp), T (~60 °C), residence time (60 – 90 min) and pulp consistency (10 – 14%). These are favorable reaction conditions for the Fe-TAML catalysts. Treating the effluent needs to be individualized for each mill as catalyst and H$_2$O$_2$ doses will depend on shower flows and counter-current wash operations at each mill. In order to investigate the utility of the Fe-TAML catalyst as an in-process treatment for E$_p$ and E$_{op}$ stages, a sample of pulp and associated liquor was obtained from the D1-stage washer screen. The pulp and liquor were separated in order to investigate the influence of Fe-TAML on pulp and liquor independently.

In order to have a pulp slurry that was managable in terms of mixing of catalyst and H$_2$O$_2$, the consistency was reduced to 1%. This allowed for placing the pulp into glass bottles which could then be placed in a shaking incubator at 80 °C. In a mill situation, the Fe-TAML catalyst could be introduced just prior to the re-pulper where efficient mixing with the pulp could occur.

In order to determine the impact of the Fe-TAML catalyst on the possible sources of color generated in the E1 process, the experiments outlined in Table 1 were devised.

Table 1: Experimental plan and results for Fe-TAML/H$_2$O$_2$ treatment on simulated E$_p$ amd E$_{op}$ bleaching in glass reactors.

T rial	P ulp	Liq uor	Fe-TAML	% color reduction[a], N$_2$	% color reduction[a], O$_2$
1	*	-	-	48	49
2	-	*	-	32	25
3	*	*	-	37	32
4	*	-	*	41	33
5	-	*	*	61	44
6	*	*	*	51	33

[a] color reduction refers to the percent color reduction in the presence of H$_2$O$_2$ (1% on pulp). Conditions for each component when present (*): pH 11, 80 °C, 90 min reaction, 0.75 g of pulp, 3.05 mL liquor, 75 mL total pulp-suspension volume (1% cty), Fe-TAML catalyst (0.5 μM), 0.17MPa N$_2$ or O$_2$. An (-) denotes an absent component.

Each experiment was performed in the absence then presence of H$_2$O$_2$ (1 % on pulp) under 0.17 MPa of N$_2$ or O$_2$. In the case where liquor was absent (Trials 1 and 4), pH 11 distilled water was used. When liquor was present, the liquor that had been squeezed from the pulp was added back to the pulp (3.05 mL to 0.75 g od pulp) and the remainder of the liquid was pH 11 water.

The results of the treatments under N_2 are shown graphically in Figure 4a and the data are presented in Table 2. The data in Figure 4a are displayed in terms of the Trials depicted in Table 1 in the absence and presence (*) of H_2O_2. It is noteworthy that the color generated by the caustic extraction process is lower than the color of the liquor carried over from the D-stage washer (Trial 1 vs 2; 0% H_2O_2) and this continues to be the case in the presence of 1% H_2O_2. The data also reveal that in the absence of Fe-TAML, H_2O_2 is capable of lowering the color from both the liquor and the alkaline extracted color. Upon addition of Fe-TAML, further reductions are observed for the samples containing liquor (Trials 2 and 5) and liquor/pulp (Trials 3 and 6). There appears to be little or no difference for pulp alone (Trials 1 and 4). The results suggest that the chromophore containing species that reside on the pulp after treatment with chlorine dioxide and are then solubilized in the alkaline extraction tower are not susceptible to the Fe-TAML/H_2O_2 treatment as applied here. In contrast, the colored species that are present in the liquor of the D-stage pulp, are oxidizable by the Fe-TAML catalyst beyond what H_2O_2 alone is capable of doing. The reasons for this are not yet obvious.

The results of the treatments under O_2 are shown in Figure 4b. Like Figure 4a, the data are displayed in pairs in terms of the Trials depicted in Table 1. There are some significant differences between some of the color values under N_2 and O_2. First, the color values from the simulated E_p and E_{op} treatments in the absence of the Fe-TAML catalyst (Trials 3 in Figures 4a and 4b), are 177 and 221 CPU, respectively (Table 2). E_{op} effluents therefore are darker than E_p effluents by 20%. The majority of this additional color comes from the liquor which is 27 CPU higher when O_2 is present (114 CPU N_2; 141 CPU O_2).

However, when Fe-TAML is added, the treated effluents differ by only 7% with E_{op} being slightly darker (137 vs 148 CPU; Trials 6 in Figures 4a and 4b). This leveling of the colors appears to be the result of the Fe-TAML catalyst playing a role in lowering all of the starting color values even in the absence of H_2O_2. The lower color values under oxygen suggest that the Fe-TAML catalyst reacts directly with O_2 to produce a species capable of oxidizing some of the chromophores generated during the extraction process. Rapid and direct O_2 activation by this Fe-TAML catalyst has been described in non-aqueous chemistry. It is not yet known in this case whether the decolorization is the result of direct reaction of the Fe-TAML catalyst with O_2 followed by oxidation of the chromophores or if some species in solution reduces the Fe-TAML catalyst which then reacts with oxygen to oxidize other chromophores. Investigations are ongoing to understand the oxygen chemistry. The key point is that a bleach plant operating under E_{op} conditions, could have an easier time reaching regulatory compliance if absolute color differences are employed.

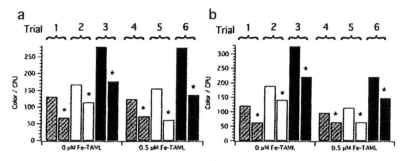

Figure 4. Comparison of color values in the presence or absence of Fe-TAML and presence and absence of H_2O_2, (a) 0.17 Mpa N_2 and (b) 0.17 Mpa O_2. The Trial numbers refer Table 1 and the (*) is reaction in presence of H_2O_2.

□ = □ = ■ = pulp +

Table 2. Color values in CPU for simulated E1-stage pulp and liquor treated with and without H_2O_2 and with and without Fe-TAML.

Trial[b]	N_2, Color (0% H_2O_2)	N_2, Color (1% H_2O_2)	O_2, Color (0% H_2O_2)	O_2, Color (1% H_2O_2)
1	131	68	121	62
2	168	114	190	141
3	281	177	327	221
4	124	73	96	64
5	156	61	114	64
6	278	137	221	148

Conclusions

An Fe-TAML/H_2O_2 based technology provides a unique opportunity to achieve regulatory compliance for color reduction in the pulp bleaching industry in a cost effective manner. The catalyst cost at full scale production is projected to be well within the budget for a mill, the quantities of catalyst required are very small per unit of water decolorized (0.5 – 4 ppm catalyst), and the catalyst as well as its decomposition products are safe and non-toxic. The process can potentially use the residual peroxide present in the effluent streams from the alkaline peroxide bleaching stage or the catalyst can be added directly into the caustic extraction stage of the bleaching process, and no additional heating (energy) needs to be applied for deep decolorization to occur. The ability of Fe-TAML catalysts to function as simple additives early on in the pulping process provides a tremendous savings for the mills with this treatment approach as compared to those that treat whole mill effluent.

References

1. Administration, I. T. "Nafta - 10 Years Later," U.S. Department of Commerce http://www.ita.doc.gov/td/industry/OTEA/nafta/nafta-index.html.
2. CEPI "The European Paper Industry on the Road to Sustainable Development," Confederation of European Paper Industries, 2003 http://www.cepi.org.
3. Group, W. B. In *Pollution Prevention and Abatement Handbook*; The World Bank: Washington, DC, 1999, pp 395-400.
4. Stevens, C. V. In *Renewable Bioresources Scope and Modification for Non-Food Applications*; Stevens, C. V. and Verhe, R. G., Eds.; Wiley: West Sussex, 2004, pp 160-188.
5. McKague, A. B.; Carlberg, G. In *Pulp Bleaching: Principles and Practice*; Dence, C. W. and Reeve, D. W., Eds.; TAPPI Press: Atlanta, 1996, pp 749-765.
6. Pettersen, R. C. *Adv. Chem. Ser.* **1984**, *207*, 57-126.
7. Capanema, E. A.; Balakshin, M. Y.; Kadla, J. F. *J. Agric. Food Chem.* **2004**, *52*, 1850-1860.
8. Gellerstedt, G. In *Pulp Bleaching: Principles and Practice*; Dence, C. W. and Reeve, D. W., Eds.; Tappi Press: Atlanta, 1996, pp 91-112.
9. Owens, J. W.; Lehtinen, K.-J. In *Pulp Bleaching: Principles and Practice*; Dence, C. W. and Reeve, D. W., Eds.; TAPPI Press: Atlanta, 1996, pp 767-798.
10. McKague, B.; Reeve, D. W. *Environ. Sci. Technol.* **1994**, *28*, 573-577.
11. Momohara, I.; Matsumoto, Y.; Ishizu, A.; Chang, H. M. *Mokuzai Gakkaishi* **1989**, *35*, 1110-1115.
12. Ziobro, G. C. *J. Wood Chem. Tech.* **1990**, *10*, 151-168.
13. Kringstad, K. P.; Lindstroem, K. *Environ. Sci. Technol.* **1984**, *18*, 236A-247A.
14. Dahlman, O. B.; Reimann, A. K.; Stromberg, L. M.; Moerck, R. E. *Tappi J.* **1995**, *78*, 99-109.
15. Dell, P.; Power, F.; Donald, R.; McIntosh, J.; Park, S.; Pang, L. *Environmental Fate and Effects of Pulp and Paper Mill Effluents, [International Conference on Environmental Fate and Effects of Bleached Pulp Mill Effluents], 2nd, Vancouver, B. C., Nov. 6-10, 1994* **1996**, 627-636.
16. Nicol, C. J. MSc, Waikato, Waikato, 1992.
17. Dilek, F. B.; Taplamacioglu, H. M.; Tarlan, E. *Appl. Microbiol. Biotechnol.* **1999**, *52*, 585-591.
18. Wingate, K. G.; Robinson, M. J.; Stuthridge, T. R.; Collins, T. J.; Wright, L. J. *International Environmental, Health & Safety Conference and Exhibit, Charlotte, NC, United States, Apr. 22-25, 2001* **2001**, 274-283.

19. Martin, C. A.; Alfano, O. M.; Cassano, A. E. *Water Sci. Technol.* **2001**, *44*, 53-60.
20. Dilek, F. B.; Bese, S. *Water SA* **2001**, *27*, 361-366.
21. Algehed, J.; Stromberg, J.; Berntsson, T. *Tappi J.* **2000**, *83*, 55.
22. Mansilla, H. D.; Yeber, M. C.; Freer, J.; Rodriguez, J.; Baeza, J. *Water Sci. Technol.* **1997**, *35*, 273-278.
23. Oeller, H. J.; Demel, I.; Weinberger, G. *Water Sci. Technol.* **1997**, *35*, 269-276.
24. Amoth, A. R.; Miller, J. P.; Hickman, G. T. In *TAPPI Environmental Conference Proceedings*: VA, 1992, pp 339-346.
25. Cecen, F.; Urban, W.; Haberl, R. *Water Sci. Technol.* **1992**, *26*, 435-444.
26. Prat, C.; Vicente, M.; Esplugas, S. *Water Res.* **1988**, *22*, 663-668.
27. Mantzavinos, D.; Psillakis, E. *J. Chem. Tech. Biotech.* **2004**, *79*, 431-454.
28. Collins, T. J. *Acc. Chem. Res.* **1994**, *27*, 279-285.
29. Sen Gupta, S.; Stadler, M.; Noser, C. A.; Ghosh, A.; Steinhoff, B.; Lenoir, D.; Horwitz, C. P.; Schramm, K.-W.; Collins, T. J. *Science (Washington, DC, United States)* **2002**, *296*, 326-328.
30. Ghosh, A.; Ryabov, A. D.; Mayer, S. M.; Horner, D. C.; Prasuhn, D. E., Jr.; Sen Gupta, S.; Vuocolo, L.; Culver, C.; Hendrich, M. P.; Rickard, C. E. F.; Norman, R. E.; Horwitz, C. P.; Collins, T. J. *J. Am. Chem. Soc.* **2003**, *125*, 12378-12379.
31. Horwitz, C. P.; Fooksman, D. R.; Vuocolo, L. D.; Gordon-Wylie, S. W.; Cox, N. J.; Collins, T. J. *J. Am. Chem. Soc.* **1998**, *120*, 4867-4868.
32. Collins, T. J.; Horwitz, C. P. *TAPPI Pulping Conf.* **1999**, *2*, 703-710.
33. Collins, T. J.; Gordon-Wylie, S. W.; Bartos, M. J.; Horwitz, C. P.; Woomer, C. G.; Williams, S. A.; Patterson, R. E.; Vuocolo, L. D.; Paterno, S. A.; Strazisar, S. A.; Peraino, D. K.; Dudash, C. A. *Green Chem.* **1998**, 46-71.
34. Hall, J. A.; Vuocolo, L. D.; Suckling, I. D.; Horwitz, C. P.; Allison, R. W.; Wright, L. J.; Collins, T. J. *Appita Annual General Conference Proceedings* **1999**, *53rd*, 455-461.
35. Cook, D.; Frum, N. In *Methods Manual*; National Council for Air and Stream Improvement, Inc.: Research Triangle Park, NC, 2000.
36. Sarkanen, K. V.; Ludwig, C. H.; Editors *Lignins: Occurrence and Formation, Structure, Chemical and Macromolecular Properties, and Utilization*, 1971.
37. Pasco, M. F.; Suckling, I. D. *Appita Journal* **1998**, *51*, 138-146.
38. Wingate, K. G.; Stuthridge, T. R.; Wright, L. J.; Horwitz, C. P.; Collins, T. J. *Water Sci. Technol.* **2004**, *49*, 255-260.

Chapter 13

New Heterogeneous Catalysts Derived from Chitosan for Clean Technology Applications

Duncan J. Macquarrie[1,*], Jeff J. E. Hardy[1], Sandrine Hubert[1,4],
Alexa J. Deveaux[1], Marco Bandini[2], Rafael Luque Alvarez[1,3],
and Marie Chabrel[1]

[1]Department of Chemistry, University of York, Heslington,
York YO10 5DD, United Kingdom
[2]Dipartimento di Chimica "G. Ciamician", Università di Bologna,
Via Selmi, 2 40126 Bologna, Italy
[3]Departamento de Química Orgánica, Facultad de Ciencias, Universidad de
Córdoba, Edificio Marie Curie (C–3) Ctra Nnal IV,
Km 496, CP 14004 Córdoba, Spain
[4]Current address: U.F.R. Sciences Exactes et Naturelles, Université
de Reims Champagne-Ardennes, Moulin de la Housse, B.P. 1039, 51687
Reims Cedex 2, France

Supported complexes of palladium and nickel on chitosan are
reported in this overview of work by our group. Palladium
catalysts are immobilized using a pyridylimine functionalized
chitosan, and are active in C-C bond-forming reactions. Nickel
(II) is immobilized either directly or via a salicylaldimine
ligand. These materials are active in the Baeyer-Villiger
lactonisation.

The drive towards cleaner manufacturing of chemicals – green chemistry (*1-3*) - is a major theme in synthetic chemistry, with all sectors of industry actively participating in the development of cleaner technologies. Several approaches and themes are being actively researched – these cover most of the key parameters which have to be considered in cleaning up a process. New reactions are being developed which operate with lower energy requirements, have greater atom economy (i.e. more of the atoms put into the reactor end up in the product, rather than as waste) and which require simpler isolation procedures. Novel solvents with reduced toxicity, volatility and with improved safety are being developed, as are better catalysts to improve reaction rates and, more crucially, selectivities. These catalysts can also remove the requirements for toxic stoichiometric reagents, replacing them with cleaner alternatives. As an example of this, new oxidation methods have been developed which utilise hydrogen peroxide or air as oxidant (with an appropriate catalyst) rather than the traditional stoichiometric high oxidation state transition metals such as Cr or Mn (*4-6*). Much of the waste produced in processes comes from the separation stages, which are often less well optimised than the reaction itself, and thus solvent systems such as supercritical fluids, which can readily be vented and recovered separately are exciting alternatives to traditional solvents which are harder to remove and recover (*7-9*). Similarly, liquid phase reactions benefit from the use of heterogeneous catalysts, as these are simple to remove from the reaction mixture, and are thus easier to recover and reuse, further reducing waste. A further benefit of these catalysts, often based on existing homogeneous systems and tethered to an insoluble support, is that they can be fixed inside continuous reactors and used in intensive processing systems such as microreactors (*10,11*) and spinning disc reactors (*12*).

The major thrust of research into green chemistry in the Centre for Clean Technology at York has always been the development of improved heterogeneous catalysts for the production of fine chemicals. More recently, we have begun to look at the use of renewable resources, both as alternative feedstocks for chemical production, and as novel catalyst supports. This paper will focus on one aspect of this work, the use of chitosan as a novel catalyst support for synthetic chemistry.

Chitosan **1** is derived from chitin **2**, which in turn is a component of the shells of crustaceans, and of the cell walls of certain fungi. Its isolation involves the removal of calcium carbonate (from the shells) and proteins, followed by the hydrolysis of the acetamide groups. This process involves the use of strong acids and bases, and leads to chitosan with varying degrees of deacetylation (DA) – typically DAs of 70-90% are achieved (*13*). Thus, it is more correct to speak of chitosan as a copolymer consisting of 2-amino-glucose and 2-acetamido-glucose units.

A second parameter, molecular weight, is likely to be affected by the nature and extent of the acid treatment. The acetal linkage is acid sensitive, and it is well known that the polymer molecular weight decreases in acidic conditions

(*14*). Indeed, the position of cleavage is influenced by the presence of amino, or acetamido linkages (*14*) meaning that the process is not entirely random, and may itself be influenced by the DA. Catalytic applications reported in the literature rarely report the DA and molecular weight, and it is therefore difficult to assess the variability that either parameter might induce in catalytic activity.

1   2

There are many features of chitosan which makes it attractive as a catalyst support. Firstly, it is cheap and readily abundant, and it is currently a waste stream from the fishing industry as well as from some parts of the biotechnology industry. Secondly, it has functionality through which one can attach catalytic groups; along with other cheap, abundant biopolymers such as starch and cellulose, it has many hydroxyl groups, but crucially, chitosan also has amino groups. These are the workhorses of modified silicas, and everything from protons to proteins have been attached to silica using amine functions. Clearly then, the rich field of silica modification (via aminopropylsilica) can serve as a model for further functionalisation of chitosan.

Chitosan has already been used as a support in several catalytic applications, including some which exploit its chirality (*15-19*). Finally, and a major benefit, is the ability of chitosan to form films and fibres readily, (*13*) allowing the coating of reactor walls and the forming of catalyst systems with controlled architecture, e.g. films, membranes, porous beads. Such behaviour relies on the ability of chitosan to dissolve in dilute aqueous acid (via protonation of the amino function) to give solutions from which films can be cast. Chitosan is completely insoluble in virtually all other solvent systems.

Chitosan-based Palladium Catalysts for C-C Coupling

Our interest in chitosan-metal systems derived originally from the pronounced ability of chitosan to adsorb heavy metals from seawater – this might indicate a strong affinity for such metals which could lead to a series of transition metal-based catalysts. While chitosan does adsorb such metals well, it does not generally hold them tightly enough, and leaching of metals such as Pd is relatively rapid during catalytic reactions. This is likely to be due to the formation of soluble Pd complexes as part of the reaction pathway, and without sufficient anchoring of the Pd to the support (e.g. via a bidentate ligand system) these soluble species will leach from the support. We have already found that Pd, correctly adsorbed on a bidentate (N,N) ligand, is stable to the conditions of the Heck and Suzuki reactions (*20-22*), but without correct complexation, Pd is rapidly lost during reaction. However, we have found that nickel (II) can be adsorbed and retained much better than Pd by chitosan in the absence of a coordinating ligand, in at least one reaction type. Here, chitosan's natural ability to coordinate via O and N groups is sufficient to retain virtually all the metal, with only a very small amount of Ni lost during reaction. A coordinating N,O ligand system improves the system even further.

An area which is currently generating a lot of interest relates to the development of mild C-C bond forming reactions using Pd catalysis. Several different variations exist for such reactions, but several commonalities exist. These reactions are normally catalysed by phosphine-Pd complexes, and require a base to remove the HX formed. The general outline of the two reactions are given in Figure 1.

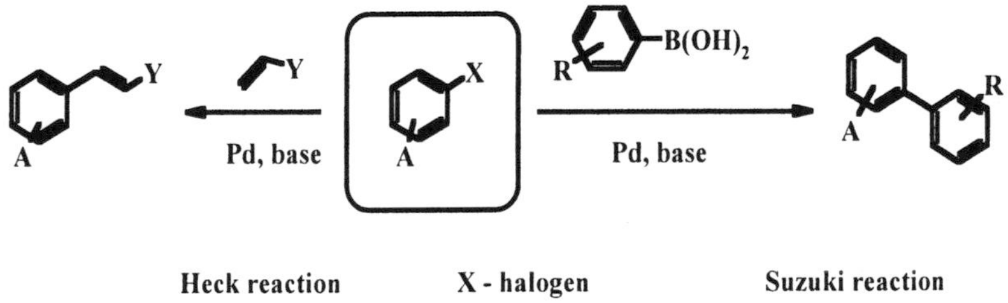

| Heck reaction | X - halogen | Suzuki reaction |

Figure 1. Heck and Suzuki coupling reactions

With the aim of avoiding the use of phosphorus-based ligands, we had previously developed catalysts based on silica supported bidentate iminopyridine ligand systems. These coordinate Pd well, and the catalysts are active and stable in both the Heck (*20,22*) and Suzuki (*21,22*) reactions. Since the catalysts are based on the functionalisation of an aminopropyl groups attached to the silica, these are ideal candidates to translate to a chitosan-based support, and indeed this was achieved in a straightforward manner (Figure 2).

Reaction of 2-pyridine aldehyde **2** with 80% DA chitosan led to a 38% conversion of the amine groups to imines, and this material adsorbed a total of 0.18mmol/g Pd(OAc)$_2$, after thorough conditioning to remove any loosely bound Pd species. (It was found for the SiO$_2$ materials that a conditioning with 3x refluxing acetonitrile, 3x refluxing ethanol, and 3x refluxing toluene led to a catalyst that was stable under the reaction conditions – i.e. no Pd was lost (ICP analysis of supernatant) and the supernatant liquid, after filtration did not continue to catalyse the reaction.) This conditioning was found to work equally well (and to be necessary) for the chitosan-based catalyst.

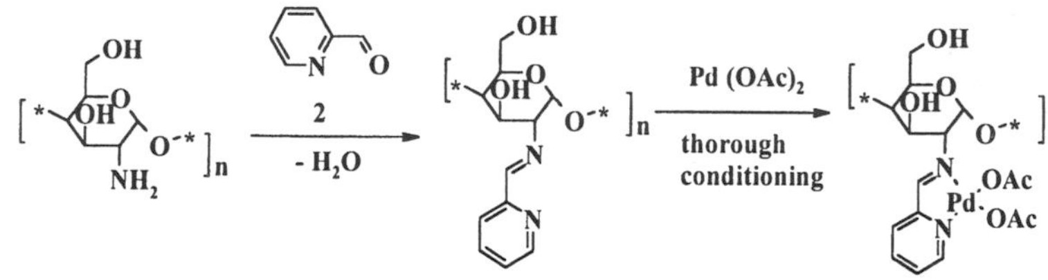

Figure 2. Preparation of chitosan-Pd catalyst 4

Chitosan –Pd complexes in the Heck reaction

Results of the Heck reaction are given in Table I.

Table I. Summary of the Heck results

A	X	Y	4(mmol/g)	Time (h)	Temp (^{o}C)	Yield (%)	TON
H	I	CO_2Bu	0.74	42	100	82	1090
H	I	CO_2Bu	1.47	20	100	87	545
H	Br	Ph	0.70	42	100	88	1250
H	Br	Ph	1.41	20	100	87	620

A, X, and Y as defined in Figure 1

TON defined as mol product / mol Pd.

As can be seen the Heck reactions proceed well, and in a similar manner to those catalysed by the silica-based systems previously studied. Reaction times and yields are very similar, and turnover numbers are comparable.

Chitosan-Pd complexes in the Suzuki Coupling Reaction

The Suzuki reaction also proceeds well at the elevated temperature of 143°C (Table II), with excellent yields being obtained in relatively short times for a range of substrates with a variety of functional groups being tolerated. Reuse of the catalyst is possible by simple filtration without washing or reactivation, up to a fifth reuse (Table II, 2nd entry). After this point, catalyst activity drops significantly, but can be restored by washing with methanol. The results obtained here have comparable turnover numbers (TON, mol product/mol Pd) to the silica-based materials (if anything, the chitosan system is slightly more efficient in this respect). However, significantly higher temperatures are required, with the silica-based catalysts performing well at 95°C, almost 50°C lower than the chitosan system. This is discussed in more detail below. A good range of substrates can be converted in high yields in relatively short times, including bromopyridines and bromopyrones. Nitro aromatics perform poorly, and it is clear from the rapid darkening of the catalyst that significant decomposition occurs which destroys the activity of the catalyst. As might be expected, chloroaromatics are essentially inert, reflecting their generally more sluggish reactivity in this reaction type.

Table II. Results of Suzuki Reaction

A	X	4(mmol/g)	Conditions	Yield (%)	TON
H	Br	0.83	143°C, 1h (6h)	78% (86%)	1032
H[a]	Br	0.83	143°C, 1h (6h)	74% (81%)	6028[b]
4-Me	Br	0.83	143°C, 1h	89%	1068
4-MeO	Br	0.83	143°C, 1h	92%	1104
4-CN	Br	0.83	143°C, 1h (3h)	68% (78%)	936
2-NO$_2$	Br	0.83	143°C, 1h (6h)	35% (36%)	432
4-NO$_2$	Br	0.83	143°C, 1h (6h)	3% (3%)	36
H	I	0.83	143°C, 1h	85%	1020
H	Cl	0.83	143°C, 1h (18h)	2% (3%)	36
2-bromopyridine		0.83	143°C, 6h	74%	888
		0.83	143°C, 4h	81%	972

A and X as in Figure 1, R = H.

(a)   5th reuse (b) cumulative TON after 6 uses.

Sodium carbonate base, p-xylene solvent.

TON defined as mol product / mol Pd.

Curiously, at lower temperatures, a second reaction is dominant – raising the temperature eliminates this side reaction almost entirely. The second reaction, protodeborylation, converts the boronic acid reagent to boric acid and the arene; such a reaction has been observed in certain Suzuki reactions (23-26), but rarely to the extent encountered here. The changeover in selectivity observed on raising the temperature is also highly unusual. Protodeborylation is usually explained by the attack of water on the boronic acid (27-30), but is minimal with the relatively hydrophilic silica-based systems, and is prevalent with the much more hydrophobic chitosan-based catalysts. The exact mechanism of this important side-reaction is currently under investigation.

Thus the Suzuki reaction works well for a wide range of substrates, although nitro-containing systems caused rapid decomposition of the catalyst. As expected, the relatively strong C-Cl bond was inert under the conditions used. Reuse of the catalyst was possible, and results after 5 reuses were almost identical to the initial run. Simple washing with methanol is sufficient to retain activity by removing the inorganics formed during the reaction. As found with the Heck reaction, no Pd leaching was observed (ICP) and filtration of the catalyst under reaction conditions stopped the reaction completely, indicating that the reaction proceeds in a genuine heterogeneous fashion.

Supported Nickel Catalysts in the Baeyer-Villiger Oxidation

Supported nickel (II) catalysts were prepared either by directly contacting chitosan with a solution of $NiCl_2$ in either water or ethanol. Stirring for 24h at room temperature led to catalysts **5** with loadings up to 0.4mmol/g $NiCl_2$. As might be expected, higher loadings were achieved at higher concentrations of $NiCl_2$, but the proportion of metal salt adsorbed decreased with concentration. Thus, 100ml of a 12.5mM aqueous solution led to 0.4mmol/g loading (**5a**), corresponding to adsorption of 32% of the available salt, whereas a 1.25mM led to a 0.084mmol/g material, adsorbing 67% of the available salt. In ethanol, 20% uptake of the 12.5mM solution was found (**5b**), but no adsorption was noted (within experimental error) for the 1.25mM solution.

A second type of supported catalyst was prepared based on the functionalisation of chitosan with salicylaldehyde **6** to give an analogous ligand system **7** to that for the palladium catalysts described earlier. (Figure 3).
Selected catalysts were then evaluated in the Baeyer-Villiger oxidation (*31*) of cyclohexanone to caprolactone with air and a sacrificial aldehyde (*32*). The results are summarised in Figure 4. Catalyst stability was measured by carrying out atomic absorption studies on the supernatant solution after reaction. Catalytic activity varied dramatically, but high yields of caprolactone were obtained with the best catalysts.

As can be seen from the Figure below, three catalysts are particularly effective at carrying out the reaction: **5a**, **8c**, and **8d**. After 6h at room temperature, under one atmosphere of oxygen, with dichloroethane as solvent, all three reached a 60% conversion to caprolactone. Other catalysts were significantly slower, although some appear to continue producing caprolactone slowly for longer periods of time. In all cases, caprolactone was the only product formed during the reaction.

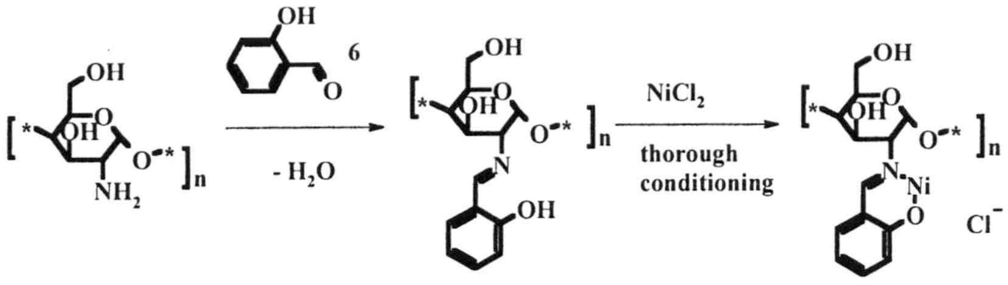

OH

OH
6
O

$[*$ ... OH ... O$^{-*}]_n$
NH$_2$

$- H_2O$ →

$[*$ ... OH ... O$^{-*}]_n$
N
OH

NiCl$_2$
thorough
conditioning
→

$[*$ ... OH ... O$^{-*}]_n$
N-Ni
O
Cl$^-$

1   7   8

Catalyst	Ligand attachment methodology	Ni attachment methodology	Loading of Ni (mmol/g)
5a	No ligand	Water 20°C	0.40
5b	No ligand	EtOH 20°C	0.25
8a	EtOH 80°C	EtOH 20°C	0.35
8b	EtOH 20°C	Water 20°C	0.18
8c	EtOH 20°C	EtOH 20°C	0.34
8d	Toluene 110°C	Water 20°C	0.24

Figure 3. General scheme for the preparation of supported NiCl$_x$ catalysts

Samples were taken after 3h reaction and upon completion of reaction to analyse for leached Ni. Filtration of the samples to remove solid catalyst and subsequent extraction of the liquid phase by heating in the presence of 2M HCl should extract any soluble Ni species into the aqueous phase. For catalysts 8c and 8d no Ni was detected during or after the reaction. For 5a, 1.2% leaching was detected midway through the reaction, and ca. 4% leaching at the end of the 6h was found. Thus, the ligand system aids in the stabilisation of the catalyst, even though it does not enhance its activity. Prolonged exposure of catalysts 8 to the oxidising conditions of the reaction does eventually lead to some loss of Ni (after 18h, ca 3% losses were noted). Whether this is related to oxidative degradation of the support is not clear as yet. However, recovered catalysts had low activity, despite essentially identical metal contents, and this may also be related to a slow degradation of the support, although other factors such as adsorption and site blocking cannot be ruled out at this stage. Analogous silica-supported Ni-salicylaldimine complexes displayed similar activities to the chitosan systems (33).

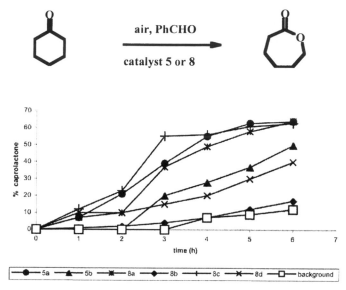

Figure 4. Conversion of cyclohexanone to caprolactone using chitosan-based Ni catalysts.

Catalytic Film Formation

Initial attempts at producing a catalytic chitosan film were carried out using a glass substrate and an Al substrate. A 1wt% solution of chitosan was made up in 1% aqueous acetic acid and was cast onto the substrates and allowed to dry, initially at room temperature, and then at 60°C. Clear films were successfully formed in both cases, and were moderately resistant to abrasion, and completely resistant to soaking in hot solvents (water, ethanol, toluene, THF at 50°C for 72h had no effect on the film). Diffuse reflection infrared studies indicated the presence of chitosan on the substrates before and after immersion in solvents.

Aluminum proved to be a better model in terms of infrared studies, since the cut-off for glass occurred at ca. 2000cm^{-1}. Reaction of the film by heating with a concentrated solution of pyridine-2-aldehyde gave the imine intermediate, as evidenced by the development of a band at 1649cm^{-1}, and subsequent reaction with Pd(OAc)$_2$ produced the final catalyst. Given the low occupancy of ligand sites by Pd (as detailed for the powdered catalyst above) and the relatively small changes effected by complexation of Pd to the ligand, no significant changes in the IR were noted. Extra activity around 1640-1550cm^{-1} was seen, consistent with the presence of acetate groups. However, the colour of the film changed to a distinct yellow-orange, consistent with the adsorption of some Pd.

Formation of chitosan films containing either silica particles or carbon was also successful, and incorporation of ca. 10wt% of silica or carbon was possible without significantly altering the robustness of the film. This observation indicates that it should be possible to prepare films from existing silica-based and carbon-based catalysts by forming a composite film on either glass or aluminum (or indeed other surfaces).

A catalytic film based on 5wt% Pd-C in chitosan, prepared from a 1% aq. AcOH solution was made on an Al plate (40mm x 23mm) with a shallow well in the centre (well dimensions 30mm x 17mm x 1mm). A second plate was placed on top of this, separated by a 1mm thick Viton seal. The second plate had an inlet and an outlet. To test the catalytic activity of the plate, a solution of 0.25g benzophenone in 8ml limonene was passed over the film at a rate of 1ml/h at 80°C. GC analysis of the product indicated a 17% conversion of benzophenone to diphenylmethane, along with the formation of some *p*-cymene from limonene (Figure 5). The activity displayed is significantly slower than a batch reduction carried out with Pd-C (85% reduction in 4h at 80°C), but is similar to that displayed by an analogous powdered chitosan-Pd/C composite in a batch reactor (26% reduction in 8h). Interestingly, a control experiment using Pd-C, which had been subjected to the same process as for making the chitosan composites (i.e., stirring in dilute acetic acid, and subsequent drying, gave 28% reduction in 8h reaction under the same conditions. Therefore, it appears that the main reason for the lower activity in this case is a partial deactivation of the Pd-C catalyst, rather than occlusion in the composite. This result, although on a rudimentary continuous reactor, does augur well for the potential of more sophisticated devices based on catalytic chitosan films.

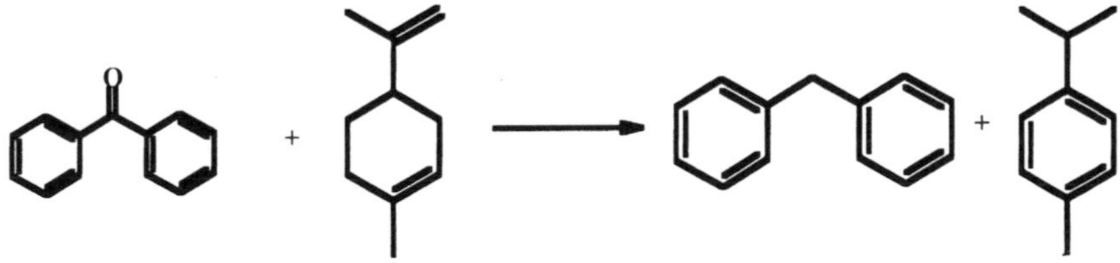

Figure 5. Transfer hydrogenation of benzophenone catalysed by chitosan-Pd/C

Experimental

Modification of Chitosan

2-Pyridinecarboxaldehyde (**2**, 0.643g, 6mmol) was dissolved in ethanol (50mL) and chitosan (1g) added. The mixture was then heated to reflux for 18h. After cooling, the resultant mixture was filtered. The solid was washed with ethanol and dried under vacuum for 6h to give **3**.

The production of the salicylaldehyde-functionalised material **7** was carried out using the same approach from chitosan and salicylaldehyde. In this case, three variations of the functionalisation were investigated, using ethanol at 20°C, reflux and toluene at reflux. In each case the material was bright yellow, consistent with salicylaldimine units attached to the backbone

3 (1.0g) was then stirred with a solution of palladium acetate (0.112g, 0.5mmol) in acetone (50mL). After 18h, the solid was filtered and washed with acetone. The catalyst, as isolated, has Pd bound to the ligand system, and loosely bound Pd. The catalyst was therefore washed thoroughly to remove any loose Pd species. The washing procedure developed involved 3x3h refluxes in each of ethanol, toluene, and acetonitrile. Finally the catalyst was dried at 90°C for 18h.

Typical Heck reaction

Catalyst **4** (25mg) was suspended in anhydrous dioxane (30mL). Iodobenzene (1.03g, 5.3mmol), *n*-butyl acrylate (0.68g, 5.3mmol) and triethylamine (0.98g, 10mmol) were added and the mixture was refluxed and followed by GC, again using *n*-dodecane as internal standard. In a second reaction, styrene and bromobenzene were coupled, using the same ratios.

Typical Suzuki Reaction

The catalyst **4** was suspended in *p*-xylene (20mL) and benzene boronic acid (0.914g, 7.5mmol) bromobenzene (0.805g, 5.1mmol), potassium carbonate (1.382g, 10mmol) and *n*-dodecane (0.749g, 4.4mol) as GC internal standard were added. The reaction was then heated to reflux and followed by GC.

Preparation of Nickel-Containing Catalysts

Chitosan (1g) or salicylaldehyde-modified chitosan **7** was added to either an aqueous or an ethanolic solution of $NiCl_2$ (100mL, 12.5mM) and stirred for 12 h. The solid was then filtered, and washed several times with the solvent used in the adsorption part of the experiments. The ethanol-derived catalysts were also finally washed with water. The resultant solids were then dried at 90°C before use.

Nickel loadings were determined by refluxing the catalyst in 2M HCl for 4h, filtering and repeating with fresh 2M HCl. The combined HCl fractions were combined and the nickel content determined by Atomic Absorption Spectrometry (AAS). Loading was also determined using UV-visible spectroscopy for catalysts **5** – results were in excellent agreement with those from AAS. For catalysts **8** interferences were evident which were ascribed as being due to breakdown of the salicylaldehyde during sample preparation.

Baeyer-Villiger Oxidation Reactions

The chitosan catalyst (**5** or **8**, 50mg) was suspended in 1,2-dichloroethane (50mL) and cyclohexanone (0.98g, 10mmol) and benzaldehyde (3.18g, 30mmol) were added. *n*-Dodecane (0.100g) was added as internal standard. The reaction mixture was magnetically stirred under an atmosphere of oxygen, maintained by a balloon. Samples were taken for regular GC analysis. Samples of the liquid phase were also taken for leaching studies. These were filtered and then treated with 2M aq. HCl to extract any Ni species by vigorous stirring at 50°C for a minimum of 4h. The aqueous phase was then analysed by AAS as described previously.

Conclusion

Chitosan has been explored as a catalyst support for two different metals, palladium and nickel. In both cases, stable catalysts have been produced, which display good activity in chosen reactions. Generally, activity appears to be similar to existing silica-based analogues, which should allow more extensive replacement of silica systems with renewable supports. Initial attempts to form films of catalyst appear to be very promising, and further work is planned in this direction.

References

1. *Green Chemistry, Theory and Practice*; Anastas, P. T.; Warner, J. C., Oxford University Press, Oxford, 1998.
2. *Handbook of Green Chemical Technology*; Clark, J. H.; Macquarrie, D. J., Eds.; Blackwell Science, Oxford, 2002.
3. Anastas, P. T.; Lankey, R. L. *Green Chem.* **2000**, *2*, 289.
4. Sheldon, R. A. *Chemtech* **1994**, 38.
5. Raja, R.; Sankar, G.; Thomas, J. M. *Chem. Commun.* **1999**, 829.
6. Dias, C. R.; Portela, M. F.; Ganán-Fereres, M.; Bañares, M. A., Granados, M. L.; Peña, M. A.; Fierro, J. L. G. *Catal. Lett.* **1997**, *43*, 117.
7. Furno, F.; Licence, P.; Howdle, SM.; Poliakoff, M. *Actualite Chimie* **2003** *4-5*, 62.
8. Kawanami, H.; Ikushima, Y. *Tetrahedron Lett.* **2004** *45* 5147.
9. Oku, T.; Arita, Y.; Tsuneki, H.; Ikariya, T. *J. Amer. Chem. Soc.* **2004**, *126*, 7368.
10. Zampieri, A.; Colombo, P.; Mabande, G. T. P.; Selvam, T.; Schwieger, W.; Scheffler, F. *Adv. Mat.* **2004**, *16*, 819.
11. Jackson, T.; Clark, J. H.; Macquarrie, D. J.; Brophy, J. *Green Chem.* **2004**, *6*, 193.
12. Brechtelsbauer, C.; Lewis, N.; Oxley, P.; Ricard, F.; Ramshaw, C. *Org. Proc. Res. Dev.* **2001**, *5*, 65.
13. Ravi Kumar, M. N. V. *React. Funct. Polym.* **2000**, *46*, 1.
14. Vårum, K. M.; Ottøy, M. H.; Smidsrød, O. *Carb. Polym.* **2001**, *46* 89.
15. Yin, M.-Y.; Yuan, G.-L.; Wu, Y.-Q.; Huang, M.-Y.; Liang, Y.-Y. *J. Mol. Cat., A* **1999**, *147*, 93.
16. Quignard, F.; Choplin, A.; Domard, A. *Langmuir* **2000**, *16*, 9106.
17. Buisson, P.; Quignard, F. *Aust. J. Chem.* **2002**, *55*, 73.
18. Sun, W.; Xia, C.-G.; Wong, H.-W. *New J. Chem.* **2002**, *26*, 755.
19. Chang, Y.; Wang, Y. P.; Su, Z. X. *J. Appl. Polym. Sci.* **2002**, *83* 2188.
20. Mubofu, E. B.; Clark, J. H.; Macquarrie, D. J. *Green Chem.* **2000**, *2*, 53.
21. Mubofu, E. B.; Clark, J. H.; Macquarrie, D. J. *Green Chem.* **2001**, *3*, 23.
22. Hardy, J. J. E.; Hubert, S,; Macquarrie, D. J.; Wilson, A. J. *Green Chem.* **2004**, *6*, 53.
23. Muller, D.; Fleury, J.-P. *Tetrahedron Lett.* **1991**, *32*, 2229.
24. Charette, A. B.; Giroux, A. *J. Org. Chem.* **1996**, *61*, 8718.
25. Anctil, E. J.-G.; Snieckus, V. C. *J. Organomet. Chem.* **2002**, *653*, 150.
26. Christoforou, I. C.; Koutentis, P. A.; Rees, C. W. *Org. Biomol. Chem.* **2003**, *1*, 2900.
27. Matteson, D. S., in *The Chemistry of the Metal-Carbon Bond*; Hartley, F. R., Ed.; Wiley, Chichester, 1987; p307.
28. Kuivala, H.; Nahabandian, K. V. *J. Amer. Chem. Soc.* **1961**, *83*, 2159.
29. Kuivala, H.; Nahabandian, K. V. *J. Amer. Chem. Soc.* **1961**, *83*, 2164.

30. Kuivala, H.; Nahabandian, K. V. *J. Amer. Chem. Soc.* **1961,** *83*, 2167.
31. Renz, M.; Meunier, B. *Eur. J. Org. Chem.* **1999,** 737.
32. Yamada, T.; Rhode, O.; Mukaiyama, T. *Chem. Lett.* **1991** 5.
33. Rafelt, J. R., *Selective Oxidation using Immobilised Metal Complexes*, D.Phil Thesis, University of York, 2000.

Chapter 14

Genetic Engineering of *S. cerevisiae* for Pentose Utilization

Peter Richard, Ritva Verho, John Londesborough, and Merja Penttilä

VTT Biotechnology, Tietotie 2, Espoo, P.O. Box 1500, 02044 VTT, Espoo, Finland

Recombinant *Saccharomyces cerevisiae*, able to ferment the pentoses D-xylose and L-arabinose, was modified for improved fermentation rates and yields. Pentose fermentation is relevant when low cost raw materials such as plant hydrolysates are fermented to ethanol. The two most widespread pentose sugars in our biosphere are D-xylose and L-arabinose. S. *cerevisiae* is unable to ferment pentoses but has been engineered to do so; however rates and yields are low. The imbalance of redox cofactors (excess NADP and NADH are produced) is considered a major limiting factor. For the L-arabinose fermentation we identified an NADH-dependent L-xylulose reductase replacing the previously known NADPH-dependent enzyme. For D-xylose fermentation we introduced an NADP-dependent glyceraldehyde 3-phospate dehydrogenase to regenerate NADPH.

Using biomass as a feedstock for renewable fuel production is desirable for several reasons: (i) it contributes to reducing the greenhouse gas CO_2, (ii) it increases the use of agricultural commodities and is therefore of economic advantage for rural areas, and (iii) it makes economies less dependent on the oil market.

The fermentation of hydrolyzed biomass to ethanol is an existing technology and therefore easy to implement. The biggest fraction in the hydrolyzed biomass consists of hexose sugars which are currently fermented to ethanol using existing technologies. The pentose sugars D-xylose and L-arabinose are a major fraction of the hydrolysate but are currently not fermented by yeast. Bacteria can be used for the fermentation of pentose sugars, (for review see (1)), but have drawbacks. One disadvantage associated with bacteria which were considered for ethanol fermentation are the low tolerances to inhibitors like acetic acid. Another disadvantage is that the biomass which is generated during the fermentation process might not be used as animal feed as is done in the existing process with yeast (1,2). Yeasts are often considered the more suitable microorganisms for ethanol fermentation. There are some yeast species which naturally ferment the pentoses D-xylose (3,4) and L-arabinose (5), however these species often require aeration or have poor inhibitor and ethanol tolerances which limits their utility. The yeast *Saccharomyces cerevisiae* is very efficient in ethanol fermentation. It has advantages such as high ethanol and inhibitor tolerance and high fermentation rates under anaerobic conditions and it has GRAS (Generally Regarded As Safe) status, i.e. it can be used as a food and feed additive. *S. cerevisiae* is also the yeast which is mainly used for ethanol fermentation in biotechnology; however it does not ferment pentoses. This has been the motivation to genetically engineer *S. cerevisiae* for pentose fermentation in order to generate a strain which can ferment, in addition to hexose sugars, the pentose sugars D-xylose and L-arabinose (for review see (6)). In order to engineer *S. cerevisiae* for pentose utilization, the genes coding for the enzymes of the corresponding pathways were introduced. There are several possible pathways for the conversion of L-arabinose and D-xylose as indicated in Figure 1. To engineer *S. cerevisiae* for the utilization of D-xylose two pathways are available, a fungal and a bacterial. In both pathways D-xylose is converted to D-xylulose 5-phosphate, which is a metabolite of the pentose phosphate pathway. In bacteria the pathway consists of with two reactions catalyzed by xylose isomerase and xylulokinase. The intermediate metabolite in this pathway is D-xylulose. In fungi the pathway has three reactions catalyzed by xylose reductase, xylitol dehydrogenase and xylulokinase. The intermediate metabolites are xylitol and D-xylulose. Since *S. cerevisiae* has xylulokinase activity the easiest approach would be to express the xylose isomerase of the bacterial pathway in *S. cerevisiae*. This approach has been followed (7) but only with limited success (for review, see (6)).

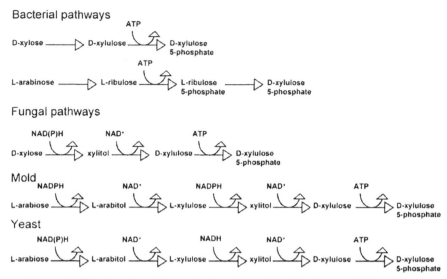

Figure 1. The fungal and bacterial pathways for D-xylose and L-arabinose catabolism. All pathways have in common that D-xylulose 5-phosphate is produced. The enzymes in the bacterial pathways are xylose isomerase and xylulokinase for the D-xylose pathway and L-arabinose isomerase, ribulokinase and L-ribulosephosphate 4-epimerase for the L-arabinose pathway. The fungal D-xylose pathway has the enzymes aldose reductase, xylitol dehydrogenase and xylulokinase. The enzymes in the L-arabinose pathways of mold and yeast are aldose reductase, L-arabinitol 4-dehydrogenase, L-xylulose reductase, xylitol dehydrogenase and xylulokinase. The differences between the mold and yeast pathway are in the cofactor requirements.

So far only a xylose isomerase from an anaerobic fungus showed activity when expressed in *S. cerevisiae* (*8*). The expression of the fungal pathway including xylose reductase and xylitol dehydrogenase in *S. cerevisiae* led to a strain able to grow on D-xylose and to produce ethanol under oxygen limited conditions (*9*). It is only required to express these two genes since *S. cerevisiae* can grow on D-xylulose (*10,11*) and has xylulokinase (*12,13*). However overexpression of xylulokinase can have a beneficial effect (*12,14*). The D-xylose fermentation in recombinant *S. cerevisiae* is slow and a major fraction of the D-xylose is converted to xylitol. A reason for that is that the xylose reductase uses NADPH and the xylitol dehydrogenase NAD, i.e. a cofactor imbalance is induced (*15*). NADPH is mainly regenerated by the oxidative part of the pentose phosphate pathway where the reduction of NADP is linked to CO_2 production so that the redox neutral conversion of D-xylose to equimolar amounts of ethanol and CO_2 is disabled. Yeast species which are efficient in D-xylose fermentation have xylose reductases that are unspecific for the cofactor, i.e. can use NADH and NADPH, as in the case of *Pachysolen tannophilus* (*16*) and *Pichia stipitis* (*17*). This might indicate that avoiding the redox cofactor imbalance is beneficial for D-xylose fermentation. A xylose reductase with a preference for NADH from the yeast *Candida parapsilosis* was described (*18*), however this enzyme was never tested in D-xylose fermentation. Most other xylose reductases from yeast and all known xylose reductases of filamentous fungi are NADPH specific.

L-arabinose pathways

For the catabolism of L-arabinose, similar to the catabolism of D-xylose, two distinctly different pathways are known, a bacterial pathway and a fungal pathway. Both pathways convert L-arabinose to D-xylulose 5-phosphate. In the bacterial pathway the three enzymes L-arabinose isomerase, L-ribulokinase and L-ribulose-5-phosphate 4-epimerase act as shown in Figure 1. A eukaryotic pathway was first described by Chiang and Knight (*19*) for the mold *Penicillium chrysogenum*. Here the enzymes aldose reductase, L-arabinitol 4-dehydrogenase, L-xylulose reductase, xylitol dehydrogenase and xylulokinase are used. In this pathway the reducing enzymes aldose reductase and L-xylulose reductase use NADPH as a cofactor, while the oxidizing enzymes L-arabinitol 4-dehydrogenase and xylitol dehydrogenase use NAD as a cofactor (Figure 1). This pathway was also described for the mold *Aspergillus niger* (*20*). After all genes coding for the enzymes of this pathway were identified the pathway was expressed in the yeast *S. cerevisiae*. The genes for L-arabinitol 4-dehydrogenase and L-xylulose reductase were from the mold *Hypocrea jecorina* (*Trichoderma reesei*), the other genes originated from yeast. The pathway was shown to be functional, i.e. the resulting strain could grow on and ferment L-arabinose, however at very low rates (*21-23*).

Information about the L-arabinose pathway in yeast is rare. It is probably similar to the mold pathway. It requires a xylitol dehydrogenase as shown by Shi et al. (*24*). In a mutant of *Pichia stipitis*, which was unable to grow on L-arabinose, overexpression of a xylitol dehydrogenase could restore growth on L-arabinose. In a study of Dien et al. (*5*) more than 100 yeast species were tested for L-arabinose fermentation. Most of them produced arabinitol and xylitol indicating that the yeast pathway is indeed similar to the pathway of molds and not to the pathway of bacteria. There is only little knowledge of the enzymes in the yeast pathway. Aldose reductases which are active with L-arabinose and D-xylose were described e.g. for the yeasts *S. cerevisiae* (*25*) and *P. stipitis* (*17*). The enzymes have similar affinity toward D-xylose and L-arabinose and convert both sugars with a similar rate. The *S. cerevisiae* enzyme however is strictly NADPH-dependent, while the *P. stipitis* enzyme can use both NADH and

NADPH, but has a preference for NADPH. Xylitol dehydrogenase and xylulokinase are also used in the xylose pathway and the enzymes of *S. cerevisiae* (*13,26*) and *P. stipitis* (*27,28*) have been characterized. There are no reports about L-arabinitol-4-dehydrogenases in yeast. We identified an L-xylulose reductase and the corresponding gene from the L-arabinose fermenting yeast *Ambrosiozyma monospora* (*29*). This L-xylulose reductase is different from previously described L-xylulose reductases since it is specific for NADH. All the previously described enzymes were specific for NADPH. The *ALX1* gene encoding the NADH-dependent L-xylulose reductase is strongly expressed in *A. monospora* during growth on L-arabinose as shown by Northern analysis indicating that it is indeed active in L-arabinose catabolism (*29*). The expression of the NADH-dependent L-xylulose reductase instead of the NADPH-dependent enzyme in *S. cerevisiae* facilitated the L-arabinose catabolism. Due to the difference in the cofactor specificity of the L-xylulose reductase one can distinguish between L-arabinose pathways for yeast and mold (Figure 1). In mold all reductions are strictly NADPH-dependent. In yeast the reduction at the L-xylulose reductase is strictly NADH-linked and the reduction at the aldose reductase is, dependent on the yeast species, specific for NADPH or unspecific so that it can use both cofactors, NADH and NADPH. The pathway is redox neutral, however in mold an imbalance of cofactors is generated, i.e. NADPH and NAD are consumed and NADP and NADH produced. In yeast this imbalance of cofactors is less pronounced.

For the genetic engineering of *S. cerevisiae* the fungal or the bacterial pathways can be used.

The bacterial L-arabinose pathway has been successfully expressed in *S. cerevisiae* by using genes from *E.coli* and *B. subtilis* (*30*). The resulting strain was able to grow on and ferment L-arabinose at high rates.

The imbalance of redox cofactors

Pentose fermentation to ethanol with recombinant *S. cerevisiae* is slow and has a low yield. One reason for this is that the catabolism of the pentoses D-xylose and L-arabinose through the corresponding fungal pathways creates an imbalance of redox cofactors. The process is redox neutral but requires NADPH and NAD. The cofactors have to be regenerated in separate processes. NADPH is normally generated through the oxidative part of the pentose phosphate pathway by the action of glucose 6-phosphate dehydrogenase (*ZWF1*) and 6-phosphogluconate dehydrogenase. This part of the pentose phosphate pathway is the main natural path for NADPH regeneration. Yeast and mold growing on D-xylose have an elevated glucose 6-phosphate dehydrogenase activity (*20,31*) indicating that this route is indeed the preferred route for NADPH regeneration. However use of this pathway causes wasteful CO_2 production and creates a

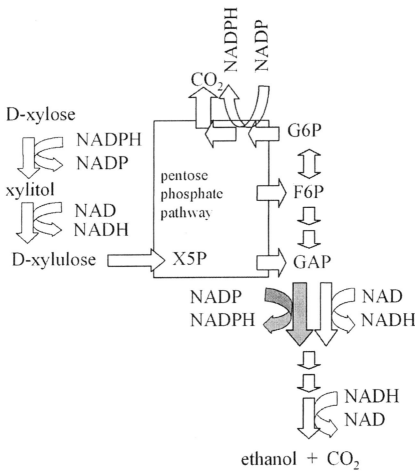

Figure 2: Anaerobic D-xylose fermentation with the fungal pathway. D-xylose fermentation with the fungal pathways requires regeneration of NADPH and NAD. The main path for regeneration of NADPH is the oxidative part of the pentose phosphate pathway. In this pathway glucose 6-phosphate is converted to ribulose 5-phosphate and CO_2. In this reaction two NADPH are regenerated for each glucose 6-phosphate. The solid arrows show the action of the introduced NADP utilizing glyceraldehyde 3-phosphate dehydrogenase. Abbreviations are G6P, glucose 6-phosphate; F6P, fructose 6-phosphate; X5P, xylulose 5-phosphate; GAP, glyceraldehyde 3-phosphate.

redox imbalance on the path of anaerobic pentose fermentation to ethanol because it does not regenerate NAD. An alternative to regenerate NADPH and NAD without the loss of CO_2 would be a transhydrogenase. Such an enzyme catalyses the reaction from NADH and NADP to NADPH and NAD. Jeppsson et al. (32) expressed the soluble transhydrogenase from the bacterium *Azotobacter vinelandii* in an *S. cerevisiae* strain with the xylose pathway. The expression resulted in a lower xylitol and higher glycerol yield; however the ethanol production was not affected. The soluble transhydrogenase might not be the best choice for resolving the imbalance of redox cofactors since it will not work effectively in the required direction if it is allosterically activated by NADPH and inactivated by NADP as described for the enzyme from *Pseudomonas aeruginosa* (33) and suggested for *E. coli* (34).

To facilitate NADPH regeneration we expressed the recently discovered gene *GDP1* coding for a fungal NADP dependent D-glyceraldehyde 3-phosphate dehydrogenase, NADP-GAPDH (EC 1.2.1.13) (35), in a *S. cerevisiae* strain with the D-xylose pathway. The resulting strain fermented D-xylose to ethanol with a higher rate and yield, i.e. the unwanted side products xylitol and CO_2 were lowered. The deletion of the gene *ZWF1* coding for the glucose 6-phosphate dehydrogenase, in combination with overexpression of *GDP1* further stimulated D-xylose fermentation with respect to rate and yield presumably by forcing the recombinant strain to use its NADP-linked GAPDH rather than its endogenous NAD-linked GAPDH, thus correcting both the NADP/NADPH and the NAD/NADH balances. Through genetic engineering of the redox reactions, the yeast strain was converted from a strain that mainly produced xylitol and CO_2 from D-xylose to a strain that produced mainly ethanol in anaerobic conditions (36).

References

1. Dien, B. S.; Cotta, M. A.; Jeffries, T. W. *Appl Microbiol. Biotechnol.* **2003,** *63,* 258-266.
2. du Preez, J. C. *Enzyme Microb. Technol.* **1994,** *16,* 944-956.
3. Jeffries, T. W.; Kurtzman, C. P. *Enzyme Microb. Technol.* **1994,** *16,* 922-932.
4. Hahn-Hägerdal, B.; Jeppsson, H.; Skoog, K.; Prior, B. A. *Enzyme. Microb. Technol.* **1994,** *16,* 933-943.
5. Dien, B. S.; Kurtzman, C. P.; Saha, B. C.; Bothast, R. J. *Appl. Biochem. Biotechnol.* **1996,** *57-58,* 233-242.
6. Jeffries, T. W.; Jin, Y. S. *Appl. Microbiol. Biotechnol.* **2004,** *63,* 495-509.
7. Sarthy, A. V.; McConaughy, B. L.; Lobo, Z.; Sundstrom, J. A.; Furlong, C. E.; Hall, B. D. *Appl. Environ. Microbiol.* **1987,** *53,* 1996-2000.
8. Kuyper, M.; Harhangi, H. R.; Stave, A. K.; Winkler, A. A.; Jetten, M. S.; de Laat, W. T.; den Ridder, J. J.; Op den Camp, H. J.; van Dijken, J. P.; Pronk, J. T. *FEMS Yeast Res.* **2003,** *4,* 69-78.

9. Kötter, P.; Ciriacy, M. *Appl. Microbiol. Biotechnol.* **1993,** *38,* 776-783.
10. Chiang, L. C.; Gong, C. S.; Chen, L. F.; Tsao, G. T. *Appl. Environ. Microbiol.* **1981,** *42,* 284-289.
11. Suihko, M.-L.; Drazic, M. *Biotechnol. Lett.* **1983,** *5,* 107-112.
12. Chang, S. F.; Ho, N. W. *Appl. Biochem. Biotechnol.* **1988,** *17,* 313-318.
13. Richard, P.; Toivari, M. H.; Penttilä, M. *FEMS Microbiol. Lett.* **2000,** *190,* 39-43.
14. Toivari, M. H.; Aristidou, A.; Ruohonen, L.; Penttilä, M. *Metab. Eng.* **2001,** *3,* 236-249.
15. Bruinenberg, P. M.; de Bot, P. H. M.; van Dijken, J. P.; Scheffers, W. A. *Eur. J. Appl. Microbiol. Biotechnol.* **1983,** *18,* 287-292.
16. Verduyn, C.; Jzn, J. F.; van Dijken, J. P.; Scheffers, W. A. *FEMS Microbiol. Lett.* **1985,** *30,* 313-317.
17. Verduyn, C.; Van Kleef, R.; Frank, J.; Schreuder, H.; Van Dijken, J. P.; Scheffers, W. A. *Biochem. J.* **1985,** *226,* 669-677.
18. Lee, J. K.; Koo, B. S.; Kim, S. Y. *Appl. Environ. Microbiol.* **2003,** *69,* 6179-6188.
19. Chiang, C.; Knight, S. G. *Biochem. Biophys. Res. Commun.* **1960,** *3,* 554-559.
20. Witteveen, C. F. B.; Busink, R.; van de Vondervoort, P.; Dijkema, C.; Swart, K.; Visser, J. *J. Gen. Microbiol.* **1989,** *135,* 2163-2171.
21. Richard, P.; Londesborough, J.; Putkonen, M.; Kalkkinen, N.; Penttilä, M. *J. Biol. Chem.* **2001,** *276,* 40631-40637.
22. Richard, P.; Putkonen, M.; Väänänen, R.; Londesborough, J.; Penttilä, M. *Biochemistry* **2002,** *41,* 6432-6437.
23. Richard, P.; Verho, R.; Putkonen, M.; Londesborough, J.; Penttilä, M. *FEMS Yeast Res.* **2003,** *3,* 185-189.
24. Shi, N. Q.; Prahl, K.; Hendrick, J.; Cruz, J.; Lu, P.; Cho, J. Y.; Jones, S.; Jeffries, T *Appl. Biochem. Biotechnol.* **2000,** *84-86,* 201-216.
25. Kuhn, A.; van Zyl, C.; van Tonder, A.; Prior, B. A. *Appl. Environ. Microbiol.* **1995,** *61,* 1580-1585.
26. Richard, P.; Toivari, M. H.; Penttilä, M. *FEBS Lett.* **1999,** *457,* 135-138.
27. Rizzi, M.; Harwart, K.; Erlemann, P.; Bui-Thanh, N.-A.; Dellweg, H. *J. Ferment. Bioeng.* **1989,** *67,* 20-24.
28. Jin, Y. S.; Jones, S.; Shi, N. Q.; Jeffries, T. W. *Appl. Environ. Microbiol.* **2002,** *68,* 1232-1239.
29. Verho, R.; Putkonen, M.; Londesborough, J.; Penttilä, M.; Richard, P. *J. Biol. Chem.* **2004,** *279,* 14746-14751.
30. Becker, J.; Boles, E. *Appl. Environ. Microbiol.* **2003,** *69,* 4144-4150.
31. Alexander, M. A.; Yang, V. W.; Jeffries, T. W. *Appl. Microbiol. Biotechnol.* **1988,** *29,* 282-288.
32. Jeppsson, M.; Johansson, B.; Jensen, P. R.; Hahn-Hägerdal, B.; Gorwa-Grauslund, M. F. *Yeast* **2003,** *20,* 1263-1272.

33. Widmer, F.; Kaplan, N. O. *Biochem. Biophys. Res. Commun.* **1977,** *76,* 1287-1292.

34. Sauer, U.; Canonaco, F.; Heri, S.; Perrenoud, A.; Fischer, E. *J. Biol. Chem.* **2004,** *279,* 6613-6619.

35. Verho, R.; Richard, P.; Jonson, P. H.; Sundqvist, L.; Londesborough, J.; Penttilä, M. *Biochemistry* **2002,** *41,* 13833-13838.

36. Verho, R.; Londesborough, J.; Penttilä, M.; Richard, P. *Appl. Environ. Microbiol.* **2003,** *69,* 5892-5897.

Chapter 15

Microbial Formation of Polyhydroxyalkanoates from Forestry-Based Substrates

Thomas M. Keenan[1], Stuart W. Tanenbaum[2], and James P. Nakas[1,*]

Departments of [1]Environmental and Forest Biology and [2]Environmental Chemistry, State University of New York, College of Environmental Science and Forestry, Syracuse, NY 13210

This study reports on the bench-scale production of poly-β-hydroxybutyrate-*co*-β-hydroxyvalerate (P(3HB-*co*-3HV)) by *Burkholderia cepacia* using detoxified aspen-derived hemicellulosic hydrolysates and levulinic acid as renewable, forestry-based platform substrates. Shake-flask cultures were found to produce P(3HB-*co*-3HV) at yields of 0.5-2.0 g/L and compositions of 16-52 mol % 3HV were achieved by adding 0.25-0.5 % (w/v) levulinic acid as a cosubstrate. Physical characterizations revealed a relatively wide thermal processing range for the copolymers, with a 105°C average differential separating the respective melting and decomposition temperatures. Molecular mass determinations also supported the potential of these polyesters for commercial processing, with viscosity-derived M_v values averaging 557 kDa.

Introduction

Analogous to petroleum refineries, biorefineries are envisioned to produce power, fiber, fuel, or commodity and specialty chemicals from a variety of renewable feedstocks. These substrates may include agricultural wastes, selected whole crops, municipal solid wastes, or lignocellulosic sources. Alternate feedstock-mixes, process-mixes, and product-mixes attendant to the underlying principles behind the biorefinery concept have been delineated in depth by Kamm and Kamm (*1*). Scenarios for implementation of such large-scale biotechnology transfers have been put forth at the recently concluded, inaugural "World Congress on Industrial Biotechnology and Bioprocessing" held in Orlando, FL, and have been critically summarized by Ritter (*2*). Work for over a decade has brought to fruition a dedicated, short-rotation willow growth program to supply forest-related biomass towards projected biobased industries, primarily for the northeastern United States (*3*). In parallel, improvements in fractionation of lignocellulosic biomass into relatively clean process streams containing cellulose, hemicellulose, and lignin have evolved (*e.g.* NREL-Clean Fractionation, Purevision, etc.). This paper reports on bench-scale investigations regarding the conversion of residual hemicellulosic and cellulosic process streams into value-added poly-β-hydroxyalkanoates (PHAs) by way of a pentose-rich hydrolysate and the platform intermediate, levulinic acid.

PHAs represent a unique class of biopolymers, synthesized as an intracellular carbon and energy reserve by a variety of microorganisms when carbon sources are provided in excess and growth is limited by the lack of at least one other nutrient (*4,5*). Due to the wide-range of thermoplastic and elastomeric properties, which can be modulated as a function of polymeric composition (*6*), these environmentally degradable (*7,8,9*) and biocompatible (*10,11*) microbial polyesters are receiving increased attention as alternatives to conventional plastic resins. With the appropriate choice of microorganism, carbon source(s), cosubstrate(s) and culture conditions, a variety of homopolymers and copolyesters can be produced with properties similar to those of commodity plastics such as polypropylene and polyethylene (*12,13*), while avoiding many of the environmentally negative characteristics of petroleum-based plastics.

A major limiting factor in the development of biodegradable polyesters is the expense of the carbon substrate used in the fermentation, which can account for up to 50 % of the overall production cost of PHAs (*14,15,16,17*). Production based on relatively inexpensive renewable substrates, including the variety of carbon feedstocks available in underutilized agricultural, forestry, and food wastes, could make PHA-derived thermoplastics more competitive with plastic products derived from future dwindling petroleum reserves. Reductions in the overall cost of production could be attained by utilization of the hemicellulosic component of wood, which represents a vastly underutilized carbon source that is often burned within the pulping/paper plant or simply released as an environmental effluent. The use of hemicellulosic hydrolysates as the primary carbon substrate for the production of poly-β-hydroxybutyrate (PHB) has been reported to reduce expenditures by 2-4 fold as compared to glucose (*14,18,19*). Recent work by Keenan et al. (*20*) has demonstrated the ability of *Burkholderia* (formerly *Pseudomonas*) *cepacia* to convert xylose and levulinic acid to poly-β-hydroxybutyrate-*co*-β-hydroxyvalerate (P(3HB-*co*-3HV)). With lignocellulosic biomass comprising approximately 50 % of the global biomass (*21*), the eminently fermentable heteroglycans of the hemicellulosic fraction (20-50 % w/w) may therefore hold significant potential for relatively low-cost microbial conversions to value-added products, such as PHA polymers. In addition, the use of levulinic acid as a cosubstrate for P(3HB-*co*-3HV) production has been shown to result in stimulatory effects on growth and PHA formation (*20,22*), and has proven superior to propionate and valerate as a 3HV precursor in *Wautersia* (formerly *Ralstonia*) *eutropha*-based fermentations (*23*).

In this communication we extend our prior observations and demonstrate the use of xylan-rich, hemicellulosic residual fractions of wood for the production of P(3HB-*co*-3HV) by *B. cepacia*. Levulinic acid, the secondary carbon source utilized in this bioconversion process, can be produced cost-effectively from a vast array of renewable carbohydrate-rich resources including cellulose-containing forest and agricultural waste residues (*24,25*). This five-carbon cosubstrate (4-ketovaleric acid) serves as a precursor to the 3-hydroxyvalerate (3HV) component of the *B. cepacia*-derived P(3HB-*co*-3HV) copolymer (Figure 1). Further, the mol % 3HV composition and associated physical/mechanical properties of the copolymer can be manipulated as a function of the substrate concentrations provided in the fermentation. Physical-chemical characterizations of such PHA copolymers are reported herein, as evidence supporting the potential of these biodegradable thermoplastics to serve as viable replacements for conventional, environmentally recalcitrant commodity plastics.

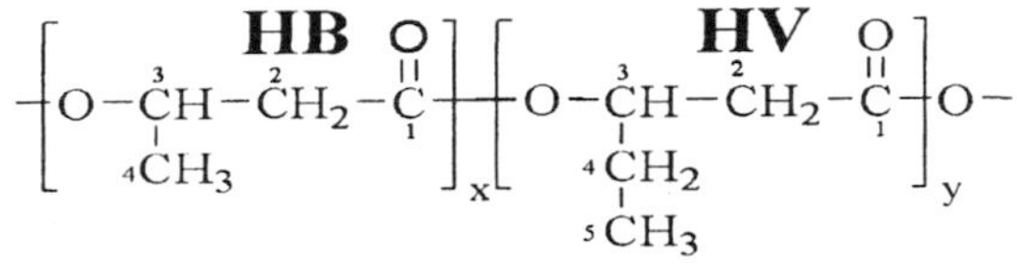

Figure 1. Chemical structure of poly(3-hydroxybutyrate-co-3-hydroxyvalerate) (P(3HB-co-3HV)).

Experimental

Culture and fermentation conditions. PHA production experiments were conducted in shake-flasks using cultures of *Burkholderia cepacia*, obtained from the American Type Culture Collection (ATCC No. 17759, Bethesda, MD). Cultures of *B. cepacia* were maintained at 25°C and transferred weekly on agar slants containing xylose (2.2 % w/v) and levulinic acid (0.3 % w/v) as carbon sources. Levulinic acid was added to cultures as a concentrated solution (pH adjusted to 7.2 via NaOH prior to sterilization). Inocula were prepared as 500-1000 ml cultures in 2,800 ml Fernbach flasks containing 2.2 % (w/v) xylose and 0.07 % (w/v) levulinic acid, incubated at 28°C and 150 rpm for 72 hours. For PHA production, shake-flasks were inoculated with a 5 % (v/v) aliquot of the above-described seed culture, incubated at 28°C and 150 rpm for 20 hours, and were then supplemented with a second dose (0-0.5 % w/v) of levulinic acid. Flasks were subsequently incubated for an additional 62-115 hours of growth in order to assess time-related differences in polymer production, and then harvested for biomass and PHA extraction, as previously described (*20*).

PHA extraction and sample preparation. Cell pellets from the shake-flask cultures were collected by centrifugation at 16,000 x g for 10 minutes, resuspended and washed in 100 ml of distilled water, and then lyophilized overnight at −50°C under vacuum. Dried biomass pellets were weighed, ground to powder, mixed with chloroform, and incubated at 55°C in sealed glass jars for 24 hours. Cellular debris was removed from the resultant viscous chloroform solutions by mixture with distilled water, followed by phase separation in 150 ml separatory funnels. White PHA cakes were obtained by precipitating the polymer from chloroform solutions with cold 95% ethanol (1:10 mixture; $CHCl_3$:EtOH) and subsequent filtration (Whatman #1 paper). Residual debris was removed from these PHA samples by secondary and tertiary extractions in chloroform, before solvent-casting in 10 cm Pyrex glass petri dishes. Translucent films were dried for 24 hours in a vacuum oven at 65°C, in order to remove residual chloroform from the PHA samples prior to thermal characterization.

Detoxification of hemicellulosic hydrolysates. The hemicellulosic hydrolysates were prepared at the National Renewable Energy Laboratory (NREL, Golden, CO) according to the NREL Clean Fractionation™ (NREL CF) procedure (*26*), using aspen as the source of lignocellulosic biomass. Using methyl isobutyl ketone (MIBK), ethanol, and H_2SO_4 in the hydrolysate mixture, the NREL CF process produces a solid cellulosic fraction, as well as an organic and aqueous phase containing lignin and hemicellulosic sugars, respectively (*26*). Concentrations of reducing sugar present in the hydrolysate through the detoxification procedure were estimated by the dinitrosalicylic acid (DNS) assay (*27*). In order to reduce levels of volatile organics in the aqueous phase to microbially sub-toxic concentrations, detoxification of the C5 stream was carried out by rotary-evaporation. Typically, 500-1200 ml quantities of crude hydrolysate were evaporated under vacuum at a bath temperature of 75°C to 250-600 ml, restored to original volume with distilled water, and evaporated again to half-volume. The subsequent steps were based on the detoxification procedures of Stickland and Beck (28), Perego et al. (29), Martinez et al. (30), and Mussatto and Roberto (31). The pH of the hydrolysate was adjusted to 10.0 with Ca(OH)$_2$ and stirred for 1.5 hours at 35°C, to remove phenolics, furfuraldehydes, and sulfates. This "overliming" step was then followed by filtration through Whatman #1 paper for removal of the Ca-based precipitate. The pH of the hydrolysate was then lowered to 5.5 with concentrated H$_2$SO$_4$, sodium sulfite was added (1g Na$_2$SO$_3$/1000 ml hydrolysate), and after 1 hour of stirring at 30°C the resulting precipitate was removed by filtration. Activated charcoal was added to the above-described filtrate (1g charcoal/40 ml hydrolysate) and mixed at 30°C for 1.5 hours, in order to complete removal of phenolics and to decolorize the growth medium. The translucent-yellow filtrate obtained following filtration and removal of the activated carbon was then mixed with essential growth/PHA production salts and trace elements (12). The final pH of the growth medium was adjusted to 7.0-7.2 with concentrated NaOH and then sterilized by passage through a 0.22 μm membrane. Levulinic acid and seed cultures of B. cepacia were then added to shake-flasks containing the sterile hydrolysate. Growth and polymer accumulation were monitored spectrophotometrically (via OD at 540 nm) and by the chloroform extraction of lyophilized cell pellets, respectively.

Compositional and structural analyses of PHAs. Compositions of PHA copolymers were established by ^{1}H and ^{13}C nuclear magnetic resonance (NMR) spectroscopy using Bruker BioSpin Avance 300 or 600 NMR spectrometers (Bruker BioSpin NMR Corp.) operating at 300 MHz (^{1}H) and 150 MHz (^{13}C), respectively. Sample preparation and calculation of monomeric composition (32) were accomplished as previously described (20). For quantitative ^{13}C NMR analyses, fully relaxed conditions were created by gated decoupling and inclusion of chromium acetylacetonate to the solution-state samples. Field locations and splitting patterns of the relevant chemical shifts obtained in these analyses were correlated with those obtained using authentic P(3HB-co-3HV) standards (Sigma-Aldrich Chemical Co.) and compared to ^{1}H and ^{13}C spectra available in the literature (32,33,34,35).

Physical-chemical characterization. Thermal characterizations of the P(3HB-co-3HV) samples, including glass transition temperature (T$_g$) and melting temperature (T$_m$), were made using a TA Instruments model 2920 Modulated DS calorimeter over a -50°C to 200°C temperature range. Solvent-cast film samples were prepared for thermal analyses as previously described (20). Under a dry nitrogen purge, PHA samples were first heated at a rate of 20.0°C/min. from 25°C to 200°C (i.e. above T$_m$) and allowed to equilibrate for 5 min. at 200°C. The samples were then recrystallized by decreasing the temperature to -50°C at a rate of 5°C/min. Following a second isothermal period at -50°C, the samples were analyzed for thermal transitions occurring during the second heating cycle, with the temperature increased at 5°C/min up to 200°C. Calibrations were made using indium (T_m=156°C) and the thermal data were recorded and analyzed using Universal Analysis 2000 software. The T_g and T_m values were determined according to earlier protocols (20). Crystallization temperatures (T_c) were also recorded along the second heating cycle in a similar fashion, as the peak temperature of the corresponding exotherm. In order to determine the temperature corresponding to the onset of thermogravimetric decomposition (T_{decomp}, arbitrarily defined as the loss of 0.032% of the original weight), samples were subjected to thermogravimetric analyses using a TA Instruments model 2950 Hi-Res TG analyzer. Thermogravimetric data were recorded and analyzed using Universal Analysis 2000 software with a 10°C/min. scanning rate, over temperatures from 25°C to 400°C. PHA film sample preparation and instrument calibrations for thermogravimetric analyses were completed as previously described (20).

Molecular mass approximation. Dilute solution viscometry was used to determine the intrinsic viscosity ([η]) of each PHA sample according to earlier protocols (20). Flow times of dilute P(3HB-*co*-3HV) solutions (2-0.25 mg/ml) were determined using a Ubbelohde capillary viscometer (Cannon Instrument Co., model 50 N35) immersed in a constant temperature bath (Cannon Instrument Co., model CT-2000) and controlled to within 0.02°C of the 30°C set temperature. From the corresponding average [η] values and the Mark-Houwink constants reported for the PHB homopolymer, K= 1.18×10^{-4} dL/g and α = 0.78 (36), viscosity average molecular weights (M_v) were approximated for each PHA sample according to a rearranged format of the Mark-Houwink-Sakurada equation: $M_v = ([η]/K)^{1/\alpha}$.

Results

Samples of unprocessed, aspen-derived xylan hydrolysate obtained via the NREL CF process contained 1.8 % (w/v) reducing sugar, which was lowered to approximately 1.4 % (w/v) in the final detoxified hydrolysate. As shown in Figure 2, microbial dry biomass and associated PHA yields from shake-flask cultures of *B. cepacia* containing the detoxified hydrolysate-based medium and 0.45 % (w/v) levulinic acid reached maximum levels of 5.1 g/L and 2.0 g/L, respectively, when harvested at 112 hours post-inoculation. Corresponding levels of PHA accumulation over this harvest period (82-135 hours post-inoculation) ranged from 40 % to 18 % (w/w) of the dry cell pellet weights. Yields of polymer prior to 82 hours of fermentation were generally below 1 g/L and these data were therefore not included in this experiment (Figure 2).

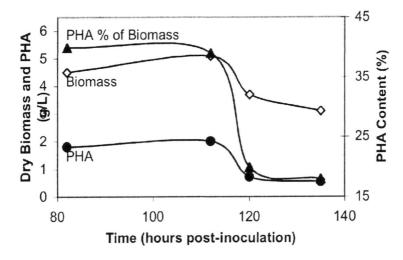

Figure 2. Time profiles of dry biomass, P(3HB-co-3HV) yields, and PHA contents for shake-flask cultures of Burkholderia cepacia *grown on detoxified aspen-derived hemicellulosic hydrolysate and 0.45 % (w/v) levulinic acid.*

Compositional analyses of P(3HB-*co*-3HV) samples produced from aspen-derived hemicellulosic hydrolysate supplemented with increasing concentrations of levulinic acid (0.25-0.5 % w/v), showed products to contain a range of 16 to 52 mol % 3HV, corresponding to the ratio of cosubstrate to reducing sugar in the fermentation medium. Definitive structural determination of the hemicellulose-based PHA by 300 MHz ^{1}H and 150 MHz ^{13}C NMR showed the copolymers to be composed of 3HB and 3HV monomers, with no detectable 4-hydroxyvalerate (4HV) fraction. Chemical shift locations and structural assignments of individual carbon atoms in the 3HB and 3HV monomers (Figure 3) demonstrated the absence of characteristic 4HV ^{13}C peaks at 31, 70, and 172 ppm (*37*). Under fully-relaxed and high-resolution operating conditions, ^{13}C peak area integrations of the relevant carbon atoms in the 3HB and 3HV monomers also correspond to the expected molar ratios present in the 24 mol% 3HV hemicellulose-based sample (Figure 3). Integrations of the three distinct carbonyl peaks located at 169.4, 169.2, and 169.1 ppm, correspond to expected diad sequence areas (3HV-3HV, 3HB-3HV, 3HB-3HB, from left to right in the expanded inset of Figure 3) for a statistically random P(3HB-*co*-3HV) sample composed of 24 mol % 3HV (*33,35,38*). The random nature of the sequences of comonomer subunits in the P(3HB-*co*-3HV) polymer chains is also supported by the ^{13}C carbon resonance signals splitting into multiple peaks (*38*). Proton resonance peak locations and splitting patterns (Figure 4) are characteristic of those reported for other bacterial P(3HB-*co*-3HV) samples *(20,33,35)*. In particular, the resonance locations, splitting patterns, and peak area integrations for the methyl group protons of the 3HB and 3HV monomers (1.27 ppm doublet and 0.88 ppm triplet, respectively, as shown by the expanded inset in Figure 4) in the 24 mol% 3HV sample, correspond to previously reported results for P(3HB-*co*-3HV) *(32)*.

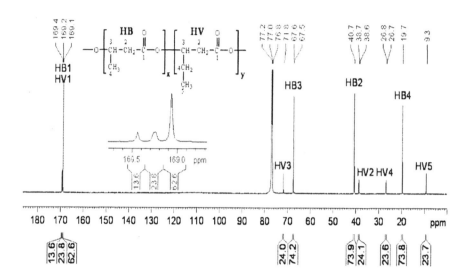

Figure 3. Fully relaxed, 150 MHz ^{13}C NMR spectrum for a P(3HB-co-24 mol % 3HV) sample produced by Burkholderia cepacia *from detoxified (aspen-based) NREL CF hemicellulosic hydrolysate with 0.3 % (w/v) levulinic acid. The expanded inset displays the field locations and relative intensities of the three carbonyl diad sequences (3HV-3HV, 3HB-3HV, 3HB-3HB, left to right).*

Thermal and viscosity-based characteristics of the P(3HB-*co*-3HV) samples produced from detoxified aspen hemicellulosic hydrolysate and levulinic acid are presented in Table I. The hemicellulosic hydrolysate-based samples displayed melting temperatures that decreased from 157°C to 98-103°C, with an increase in the mol % 3HV fraction of the copolymers. This T_m profile is relatively consistent with the pseudoeutectic behavior previously reported for other isodimorphic, P(3HB-*co*-3HV) copolymers *(32,33)*. The isodimorphic properties displayed by these short-chain length copolymers relate to the similar size and crystalline lattice conformations of the monomers, which permit crystallization of polymer chains within either the 3HB or 3HV lattices *(33)*. The T_m pattern determined for *W. eutropha*-derived P(3HB-*co*-3HV) samples showed T_m values to decrease from 175-180°C at 0 mol % 3HV to an 80°C minimum at 30 mol % 3HV (approximate "pseudoeutectic point"), and to then increase to 100°C at 47 mol % 3HV (33, 34). The 98°C and 103°C melting temperatures reported for our 50 and 52 mol % 3HV samples, respectively, are consistent with the T_m reported for the *W. eutropha*-derived 47 mol % 3HV sample. Additional hemicellulose-based samples containing 25-40 mol % 3HV are required for definitive identification of the crystalline lattice transition. As described in Table I, the molecular masses of these *B. cepacia*-derived PHA samples (487-588 kDa) are higher than the *W. eutropha*-derived samples described above (113-526 kDa), which may contribute to the higher T_m values determined for P(3HB-*co*-3HV) films characterized in this study.

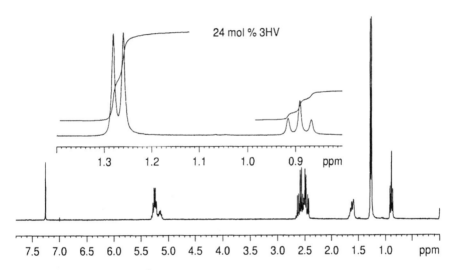

Figure 4. The 300 MHz ^1H NMR spectrum and methyl resonance chemical shift expansions of a P(3HB-co-24 mol % 3HV) sample.

Glass transition temperatures (T_g) for the hemicellulose-based polymers listed in Table I are consistent with those reported for P(3HB-*co*-3HV) (*39*), which exhibit a linear decline with increasing mol % 3HV. In general, the thermal characteristics of these hemicellulose-based films are comparable to those of commercial-grade xylose- (Sigma Aldrich Chemical Co.) and levulinic acid-based copolymers produced in separate experiments by *B. cepacia* and analyzed according to the same protocols (Table II). The onset temperature for thermal decomposition ($T_{decomp.}$) of the polymer chains and thus, expected mechanical property failure, was 53-152°C above the corresponding T_m for the hemicellulose-based polymers (Table I). Since the thermal decomposition temperatures do not decrease to the same extent as the corresponding T_m values over the range of increasing mol % 3HV (40), a relatively wide temperature differential exists (averaging 105°C) between the T_m and $T_{decomp.}$ values.

Table I. Physical characteristics of P(3HB-co-3HV) copolymers produced by Burkholderia cepacia from detoxified NREL CF/aspen-based hemicellulosic hydrolysate and levulinic acid.

Mol % 3HV	T_m^a (°C)	T_g (°C)	$T_{decomp.}^b$ (°C)	$[\eta]^c$	M_v^c (KDa)
16	157.1	-0.8	223.8	3.73	588
25	156.5	-0.9	209.6	3.67	575
50	98.1	-9.1	245.7	3.22	487
52	102.9	-10.3	254.5	3.68	578

[a]T_m (Melting temperature) determined to be the peak of the second, higher temperature endotherm in DSC cycles with multiple melting peaks. [b]$T_{decomp.}$ (Thermal decomposition temperature) represents the onset temperature for thermogravimetric decomposition, arbitrarily defined as the loss of 0.032 % of the original sample weight via analysis of the first derivative weight (%/°C) curve. [c]M_v (Viscosity average molecular mass) calculations based on the corresponding $[\eta]$ (intrinsic viscosity) values and Mark-Houwink constants (K, α) published for the PHB homopolymer and thus represent approximations.

Intrinsic viscosities and M_v calculations (average M_v = 557 kDa) for these hemicellulose-based copolymers (Table I) are similar to those of xylose- (Sigma Aldrich Chemical Co.) and levulinic acid-derived samples, as previously reported (20), and are well-within the range of molecular masses determined for other bacterial PHAs (13). Using the K= 1.36×10^{-4} dL/g estimate proposed by Bloembergen et al. (33) for P(3HB-co-3HV) with relatively high mol % HV fractions (41), viscosity average molecular masses (M_v) determined for the hemicellulose-based P(3HB-co-3HV) copolymers are reduced by 17 % to 406-490 kDa.

Table II. Thermal characteristics of P(3HB-*co*-3HV) copolymers produced by shake-flask cultures of *Burkholderia cepacia* from 2.2 % (w/v) xylose and 0.07-0.57 % (w/v) levulinic acid.

Mol % 3HV	T_m (°C)	T_g (°C)	$T_{decomp.}$(°C)
0.4	175.3	1.6	270.4
2.6	173.5	1.9	254.7
17	161.3	0.2	253.7
28	154.3	-1.1	259.3
56	94.9	-12.7	258.9

NOTE: Thermal data obtained and interpreted as noted in Table I.

Discussion

Using xylose as the principal carbon source and levulinic acid as the cosubstrate in a defined mineral salts medium, shake-flask cultures of *B. cepacia* ATCC 17759 produced 2-4 g/L PHA yields (42-51 % of cellular dry weight), with maximal PHA contents attained from 66 to 70 hours post-inoculation (*20*). Using detoxified, aspen-derived hemicellulosic hydrolysate and levulinic acid as carbon sources, shake-flask cultures of *B. cepacia* produced P(3HB-*co*-3HV) at yields ranging from 0.5-2 g/L (18-40 % of cellular dry weight), following 82-135 hours of cultivation. In some instances, detoxified hemicellulosic hydrolysates have been found to confer a stimulating effect on P(3HB) production, when used as a supplement to or compared with xylose-based cultures (*42,43*). Silva et al. (*43*) used *B. cepacia* strain IPT 048 (a soil-isolate) to obtain a 3.4 g/L yield of dry biomass containing 15.4 % P(3HB) from shake-flask cultures containing detoxified sugarcane bagasse hydrolysate. In bioreactor studies, the level of P(3HB) accumulation by *B. cepacia* IPT 048 increased from 37 % of the cellular dry weight on xylose and glucose to 53 % on sugarcane bagasse hydrolysate. Lee (*42*) found that the P(3HB) content level achieved by recombinant *Escherichia coli* increased from 35 % on xylose to 62 % and 74 % when the same synthetic medium was supplemented with cotton-seed and soybean hydrolysates, respectively.

The use of levulinic acid as a cosubstrate in PHA fermentations has also been shown to exhibit a growth and PHA enhancing effect in shake-flask cultures of *Alcaligenes* sp. SH-69 and *B. cepacia* ATCC 17759 with glucose or xylose as primary carbon sources, respectively (*22,20*). For cultures of *Alcaligenes* sp. SH-69, concentrations of 0.05 % (w/v) levulinic acid resulted in significant enhancements of cell growth and PHA accumulation, with polymer yields 27 % greater than those observed with 1.5 % (w/v) propionic acid. Using xylose as a principal carbon source, flask cultures of *B. cepacia* ATCC 17759 were found to increase cell and PHA yields over a range of 0.07 to 0.52 % (w/v) levulinic acid, but were inhibited by concentrations exceeding this level (*e.g.* 0.62-0.67 %) (*20*). Levulinic acid is known to inhibit biosynthesis of tetrapyrrols, including cytochromes (*44*), and may therefore act to suppress growth and polymer production at these concentrations by disruption of general metabolic processes. In order to circumvent such cosubstrate-associated inhibitory effects, Shang et al. (*35*) used a sequential feeding of glucose followed by valerate in fed-batch cultures of *W. eutropha* to produce P(3HB-*co*-3HV). This feeding strategy resulted in PHA yields of 37.8 g/L containing 62.7 mol % 3HV. Further, with 0.05-0.4 % (w/v) concentrations of levulinic acid as a 3HV precursor cosubstrate, it was found to be more effective than propionate or valerate for P(3HB-*co*-3HV) production by *W. eutropha* KHB-8862 on fructose as the principal carbon source (*23*). Thus, the relatively favorable biomass and PHA yields reported for *B. cepacia* utilizing hemicellulosic hydrolysates and levulinic acid may in part be due to a stimulatory effect of this cosubstrate. Amendment of the detoxified NREL CF hydrolysate with 0.25-0.5 % (w/v) concentrations of levulinic acid also allows for production of P(3HB-*co*-3HV) by *B. cepacia* with a wide-range of 3HV compositions (16 to 52 mol %).

Definitive structural determinations of these copolymers confirmed the absence of 4HV as a polymer constituent. These compositional findings are in contrast to those reported by Gorenflo et al. (*45*), in which recombinant strains of *Pseudomonas putida* and *W. eutropha* were found to produce a polyester containing 15.4 mol % 4HV from levulinic acid and octanoic acid (with significant fractions of 3HB and 3HV, as well as traces of 3-hydroxyhexanoic and 3-hydroxyoctanoic acids). The use of levulinic acid as a 4HV precursor cosubstrate (*45,46*) for these genetically engineered strains, relates to fundamental differences in metabolic carbon fluxes and expression of the biosynthetic genes. Such substrate-based methods for controlling the monomeric composition of PHAs allow for the production of copolymers with predictable physical and chemical properties that resemble those of conventional, petroleum-based polymers as displayed in Table III. Substitution of these petroplastics with PHAs such as the hemicellulose-based P(3HB-*co*-3HV) listed in Table I has potential since the thermal processing window, created by the widely separated T_m and $T_{decomp.}$ values (105°C average differential), would allow for a variety of conventional, polymer melt-related processes (*e.g.* extrusion, molding, fiber spinning, etc.) without jeopardizing the molecular integrity of the polymer. In addition, the viscosity average molecular masses (M_v) determined for P(3HB-*co*-3HV) produced from hemicellulosic hydrolysate and levulinic acid by *B. cepacia* are comparatively high (487-588 kDa) for bacterially derived PHA (13), which provide appropriate material characteristics for a variety of end-use applications.

Table III. Comparison of physical properties of selected polyhydroxyalkanoate polymers and petroleum-based plastics.

Polymer	T_m (°C)	T_g (°C)	% crystallinity
P(3HB)[a]	177	4	60
P(3HB-co-20 % 3HV)[a]	145	-1	56
Polypropylene[b]	176	-10	50-70
Polyethylene (LDPE)[b]	130	-36	20-50

[a] Melting temperature (T_m), glass transition temperature (T_g), and degree of crystallinity (% crystallinity) data for P(3HB) and P(3HB-co-20 % 3HV) obtained from Madison and Huisman (13). [b] Data for polypropylene and low density polyethylene obtained from Tsuge (47).

Commercial development of such value-added PHAs and substitution for conventional petrochemical-based plastics in commodity and specialty applications (48) is hindered by the overall production cost for these biopolymers. Substrate costs are the most significant expenditure in PHA production processes (15,17), comprising up to 50 % of the total manufacturing expense (16). Therefore, the utilization of inexpensive, renewable carbon sources, optimization of PHA content as a function of dry cellular mass, and improvements in the downstream polymer recovery processes, can result in significant economies of scale. Using xylanase-treated, steam-exploded, aspen hemicellulosic hydrolysates as a substrate for P(3HB) production by *B. cepacia* ATCC 17759, Ramsay et al. (18) reported a substrate-specific yield of 0.11g P(3HB)/g and calculated a substrate cost of $US 0.69/kg of P(3HB) that was 3-4 fold less than costs associated with the use of glucose or sucrose (14). Even accounting for the relatively low substrate-specific yields using hemicellulosic hydrolysates (0.2 g/g) compared to more conventional PHA substrates such as glucose (0.38 g/g), an economic comparison study by Lee et al. (19) reported that the use of hemicellulose as a carbon source (raw material cost of $US 0.07/kg) would reduce the final substrate cost to $US 0.34/kg P(3HB), which was roughly 2-4 fold lower than that for glucose. Results reported in this study demonstrate that the wild-type, natural PHA producer, *B. cepacia*, can convert a detoxified hemicellulose-based medium and levulinic acid to P(3HB-co-3HV), with potential for improved yields in more controlled fermentor-based studies.

Considering that the P(3HB-co-3HV) copolymer is derived from two distinct substrates, the production economics calculations must include not only the cost of the principal carbon source, but also expenditures and substrate specific yields associated with the 3HV-contributing cosubstrate. Commercial production of P(3HB-co-3HV) as *Biopol* (*49*) utilized glucose and propionic acid as principal carbon source and cosubstrate, respectively, with *W. eutropha*, to produce a product at an estimated market price of $US 16.00/kg (*17*). Since the price of the 3HV-precursor, propionic acid was approximately twice that of glucose, production cost of the copolymer was found to increase linearly with an increase in the 3HV mol fraction (*50*). Levulinic acid is an alternative 3HV-providing cosubstrate that can be produced cost-effectively from diverse renewable carbohydrate-based feedstocks including cellulose-containing forest and agricultural waste residues, paper mill sludge, and cellulose fines from paper production processes (*24,25*). The levulinic acid used in this study as the 3HV-precursor was produced by the Biofine Corporation (South Glens Falls, NY) through a patented two-stage/reactor process that converts six-carbon sugars to levulinic acid with >60 % yields and minimal byproduct formation (*24*). Selling prices for refined chemical-grade levulinic acid produced by the Biofine process, under a 5 million kg/year production scale, range from $US 0.99 to 1.21/kg (*51*). In addition to cost-advantages and production from renewable feedstocks, the growth and PHA enhancing properties of levulinic acid have been determined to be more marked (*22,23*) and to occur at 3-fold lower concentrations when compared to propionate (*22*). These economic and physiological advantages could translate to substantial reductions in substrate requirements for larger scale P(3HB-co-3HV) fermentations. Considering the relative availability of forestry-related renewable resources, the additional costs of the preparative and detoxification steps required to generate the platform substrates are far outweighed by the economic advantages conferred by the utilization of hemicellulosic wastestreams and levulinic acid.

Conclusions

The dwindling supply of fossil reserves and problems associated with global pollution, coupled to a supply of vastly underutilized renewable feedstocks, gives impetus to the search for alternative, biodegradable polymers synthesized from relatively inexpensive forest and agricultural wastestreams. This study extends our prior investigations into co-utilization of xylose and levulinic acid and documents the ability of *B. cepacia* to convert detoxified, aspen-derived hemicellulosic hydrolysates and levulinic acid to P(3HB-co-3HV) in shake-flask cultures to levels comprising 40 % of the cell dry mass. Using 0.25-0.5 % (w/v) concentrations of the renewable-based levulinic acid cosubstrate, PHA copolymers were produced with a wide-range of controllable 3HV compositions (16-52 mol % 3HV). The T_m and $T_{decomp.}$ profiles of these hemicellulose-based polymers appear industrially favorable from a melt-processing perspective, because of a 105°C average temperature differential separating these values. Additionally, the relatively high molecular mass determinations (487-588 kDa) also support the biotechnological potential for these copolymers to substitute for conventional petroleum-derived resins in a variety of commodity (*52*) and expanding specialty applications (*53*) markets. Salient features of these PHAs (*54,55*), including biocompatibility and biodegradability (*7,8,9*) continue to unveil a variety of diverse biomaterial applications for these polymers.

Acknowledgements

This project was funded by the New York State Energy Research and Development Authority, under the research grant 22931-1019995-1. Mr. David Kiemle and Dr. Art Stipanovic are gratefully acknowledged for their assistance with the physical-chemical characterizations conducted in this study.

References

1. Kamm, B.; Kamm, M. *Appl. Microbiol. Biotechnol.* **2004**, *64*, 137-145.
2. Ritter, S. K. *Chem. Eng. News* **2004**, May 31, pp 31-34.
3. Heller, M. C.; Keoleian, G. A.; Mann, M. K.; Volk, T. A. *Renewable Energy* **2004**, *29*, 1023-1042.
4. Anderson, A. J.; Dawes, E. A. *Microbiol. Rev.* **1990**, *54*, 450-472.
5. Steinbüchel, A.; Füchtenbusch, B. *Trends Biotechnol.* **1998**, *16*, 419-427.
6. Holmes, P. A. in *Developments in Crystalline Polymers*; Bassett, D.C., Ed.; Elsevier: London, UK, 1988; vol. 2, pp 1-65.
7. Williams, D. F.; Miller, N. D. *Adv. Biomater.* **1987**, *7*, 471-476.
8. Yue, C. L.; Gross, R. A.; McCarthy, S. P. *Polym. Degrad. Stabil.*, **1996**, *51*, 205-210.
9. Tokiwa, Y.; Calabia, B. P. *Biotechnol. Lett.* **2004**, *26*, 1181-1189.
10. Brandl, H.; Gross, R. A.; Lenz, R. W.; Fuller, R. C. *Adv. Biochem. Eng. Biotechnol.* **1990**, *41*, 77- 93.
11. Reddy, C. S. K.; Ghai, R.; Rashmi; Kalia, V. C. *Bioresour. Technol.* **2003**, *87*, 137-146.
12. Bertrand, J. L.; Ramsay, B. A.; Ramsay, J. A.; Chavarie, C. *Appl. Environ. Microbiol.* **1990**, *56*, 3133-3138.
13. Madison, L.; Huisman, G. W. *Microbiol. Molec. Biol. Rev.* **1999**, *63*, 21-53.
14. Byrom, D. *Trends Biotechnol.* **1987**, *5*, 246-250.
15. Yamane, T. *FEMS Microbiol. Rev.* **1992**, *103*, 257-264.
16. Choi, J.; Lee, S. Y. *Bioprocess Eng.* **1997**, *17*, 335-342.

17. Lee, S. Y.; Park, S. J.; Park, J. P.; Lee, Y.; Lee, S. H. in *Biopolymers*; Steinbüchel, A. Ed.; Wiley-VCH: New York, NY, 2003; vol. 10, pp 307-338.
18. Ramsay, J.; Hassan, M.; Ramsay, B. *Can. J. Microbiol.* **1995**, *41*, 262-266.
19. Lee, S. Y. *Trends Biotechnol.* **1996**, *14*, 431-438.
20. Keenan, T. M.; Stipanovic, A. J.; Tanenbaum, S. W.; Nakas, J. P. *Biotechnol. Prog.* **2004**, *20*, 1697-1704.
21. Galbe, M.; Zacchi, G. *Appl. Microbiol. Biotechnol.* **2002**, *59*, 618-628.
22. Jang, J. H.; Rogers, P. L. *Biotechnol. Lett.* **1996**, *18*, 219-224.
23. Chung, S. H.; Choi, G. G.; Kim, H. W.; Rhee, Y. H. *J. Microbiol.* **2001**, *39*, 79-82.
24. Bozell, J. J.; Moens, L.; Elliott, D. C.; Wang, Y.; Neuenscwander, G. G.; Fitzpatrick, S. W.; Bilski, R. J.; Jarnefeld, J. L. *Resour. Conserv. Recycl.* **2000**, *28*, 227-239.
25. Cha, J. Y.; Hanna, M. A. *Ind. Crops Prod.* **2002**, *16*, 109-118.
26. Kulesa, G. http://www.oit.doe.gov/chemicals/factsheets/ch_cellulose. pdf , **1999**.
27. Miller, G. L. *Anal. Chem.* **1959**, i, 426-428.
28. Strickland, R. J.; Beck., M. J. 6th International Symposium on Alcohol Fuels Technology; Ottawa, Canada, 1984.
29. Perego, P., Converti, A.; Palazzi, E.; Del Borghi, M.; Ferraiolo, G. *J. Indust. Microbiol.* **1990**, *6*, 157-164.
30. Martinez, A.; Rodriguez, M. E.; York, S. W.; Preston, J. E.; Ingram, L. O. *Biotechnol. Bioeng.* **2000**, *69*, 526-536.
31. Mussatto, S. I.; Roberto, I. C. *Bioresour. Technol.* **2004**, *93*, 1-10.
32. Bloembergen, S.; Holden, D. A.; Bluhm, T. L.; Hamer, G. K.; Marchessault, R. H. *Macromolecules*, **1986**, *19*, 2865-2871.
33. Bloembergen, S.; Holden, D. A.; Bluhm, T. L.; Hamer, G. K.; Marchessault, R.H. *Macromolecules* **1989**, *22*, 1663-1669.
34. Bluhm, T. L.; Hamer, G. K.; Marchessault, R. H.; Fyfe, C. A.; Veregin, R. P. *Macromolecules* **1986**, *19*, 2871-2876.
35. Shang, L.; Yim, S. C.; Park, H. G.; Chang, H. N. *Biotechnol. Prog.* **2004**, *20*, 140-144.
36. Akita, S.; Einaga, Y.; Miyaki, Y.; Fujita, H. *Macromolecules* **1976**, *9*, 774-780.
37. Valentin, H. E.; Schönebaum, A.; Steinbüchel, A. *Appl. Microbiol. Biotechnol.* **1992**, *36*, 507-514.
38. Doi, Y.; Kunioka, M. K.; Nakamura, Y.; Soga, K. *Macromolecules* **1986**, *19*, 2860-2864.
39. Sudesh, K.; Abe, H.; Doi, Y. *Prog. Polym. Sci.* **2000**, *25*, 1503-1504.
40. Savenkova, L.; Gercberga, Z.; Bibers, I.; Kalnin, M. *Process Biochem.* **2000**, *36*, 445-450.

41. Van Krevelen, D. W.; Hoftijzer, P. J. *Properties of polymers. Their estimation and correlation with chemical structure.* Elsevier: Amsterdam, Germany, 1976, 2nd ed.
42. Lee, S. Y. *Bioprocess Eng.* **1998**, *18*, 397-399.
43. Silva, L. F.; Taciro, M. K.; Michelin Ramos, M. E.; Carter, J. M.; Pradella, J. G. C.; Gomez, J. G. C. *J. Ind. Microbiol. Biotechnol.* **2004**, *31*, 245-254.
44. Sasaki, K.; Ikeda, S.; Nishizawa, Y.; Hayashi, M. *J. Ferment. Technol.* **1987**, *65*, 511-515.
45. Gorenflo, V.; Schmack, G.; Vogel, R.; Steinbüchel, A. *Biomacromolecules* **2001**, *2*, 45-57.
46. Schmack, G.; Gorenflo, V.; Steinbüchel, A. *Macromolecules* **1998**, *31*, 644-649.
47. Tsuge, T. *J. Biosci. Bioeng.* **2002**, *94*, 579-584.
48. Khanna, S.; Srivastava, A. K. *Process Biochem.* **2004**, *40*, 607-619.
49. Mohanty, A. K.; Misra, M.; Hinrichsen, G. *Macromol. Mater. Eng.* **2000**, *276*, 1-24.
50. Choi, J.; Lee, S. Y. *Appl. Microbiol. Biotechnol.* **2000**, *53*, 646-649.
51. Fitzpatrick, S. *Personal communication* regarding potential levulinic acid production costs based on Biofine technology. Biofine Corporation, **2004**, South Glens Falls, NY.
52. Aldor, I.; Keasling, J. D. *Curr. Opin. Biotechnol.* **2003**, *714*, 475-483.
53. Luengo, J. M.; Garcia, B.; Sandoval, A.; Naharro, G.; Olivera, E.R. *Curr. Opin. Microbiol.* **2003**, *6*, 251-260.
54. Martin, D. P.; Williams, S. F. *Biochem. Eng. J.* **2003**, *16*, 97-105.
55. Shishatskaya, E. I.; Volova, T. G. *J. Mat. Sci.: Mat. In Med.* **2004**, *15*, 915-923.

Chapter 16

Hemicellulose from Biodelignified Wood: A Feedstock for Renewable Materials and Chemicals

Arthur J. Stipanovic[1], Thomas E. Amidon[2], Gary M. Scott[2], Vincent Barber[2], and Misty K. Blowers[3]

Faculties of [1]Chemistry and [2]Paper Science and Engineering, State University of New York–College of Environmental Science and Forestry (SUNY–ESF), Syracuse, NY 13210
[3]Air Force Research Laboratory, Information Directorate, Rome, NY 13440

The hemicellulose fraction of woody biomass, typically 20-35% of the dry weight of wood, is currently an underutilized renewable resource potentially useful for biobased fuels, chemicals and polymeric materials. In this study, a fungal biodelignification pretreatment stage was successfully employed to enhance the accessibility of fast growing willow wood chips to hemicellulose extraction using water at 140-160°C. Films of hardwood xylan hemicellulose blended with commercially available cellulose esters were prepared from solution in an initial effort to generate new, biodegradable hemicellulose-based materials.

Introduction

The 21st century is envisioned to become the "age of biology" as sustainable biomass resources replace non-renewable petroleum in energy and industrial product applications (*1,2*). Motivated by concerns over national energy security, global CO_2 reduction, an evolving need for biodegradable products, and enhanced rural economic development, the engineering and construction of "biorefineries" is now a critical national priority. The vision of a "wood-based" biorefinery becoming a commercial reality will require the development of new core competencies:

- Development of fast-growing woody species that can be used year round as the biorefinery feedstock. These feedstocks should enjoy "life cycle" energy benefits compared to alternative biomass resources such as agricultural crops or residues.
- Efficient pretreatment and separation scenarios for woody biomass that provide at least three major process streams for the biorefinery: cellulose, hemicellulose and lignin.
- Creation of a value-added portfolio of fuels, chemicals and materials from each of the three biorefinery process streams that mimic the diverse product slate characteristic of today's petroleum refineries. Although cellulose fiber is profitably exploited by the paper industry, the hemicellulose and lignin components of wood are generally underutilized.

In the present study, attention was focused on the ability of a fungal biodelignification pretreatment stage to enhance the accessibility of wood to hemicellulose extraction and the use of the resulting hemicellulose as a component in biodegradable polymer blends. More specifically, a fast growing species of shrub willow (*Salix sp.*) was selected as a feedstock based on its proven potential as a renewable energy crop grown in an agro-forestry environment (*3*) and the fact that willow is a hardwood relatively rich in hemicellulose, primarily glucuronoxylan (xylan; ≈ 22 wt% on a dry wood basis).

Biodelignification

Lignin-degrading fungi have been used for several applications within the paper industry. Biopulping is probably the most studied application in which the fungi are used as a pretreatment to the pulping of wood or as a pulping method itself. These fungi have also been used for the treatment of pulp for both pulping and bleaching applications with some success. The use of white-rot fungi for the biological delignification of wood was first seriously considered by Lawson and Still (*4*) at the West Virginia Pulp and Paper Company research laboratory (now

Westvaco Corporation). Subsequent work showed that paper strength properties increased with the extent of natural degradation of pine by white-rot fungi (5,6). Related work was done at a Swedish research laboratory (STFI) in Stockholm, and the first published report on biopulping demonstrated that fungal treatment could result in significant energy savings for mechanical pulping (7). Considerable efforts by the Swedish group were directed toward developing cellulase-less mutants of selected white-rot fungi to improve the selectivity of lignin degradation and, thus, the specificity of biopulping (8-10). Samuelsson and co-workers (11) applied the white-rot fungus *Phlebia radiata* and its cellulase-less mutant to chips and pulp. Eriksson and Vallander (12) treated wood chips (with or without added glucose) with white-rot fungi, in most cases a cellulase-less mutant of *Phanerochaete chrysosporium.*

Pearce and co-workers (13) screened 204 isolates of wild-type wood decay fungi for biomechanical pulping of eucalyptus chips. Some of these fungi saved 40 to 50% electrical energy in refining and resulted in greater brightness of unbleached pulp as compared with that of pulps from untreated control chips. Some of the strains were found to be effective on unsterilized wood chips. Other details on biopulping research have been described in review articles and the literature cited therein (14,15). Recent work has shown that the white-rot fungus *Ceriporiopsis subvermispora* is very effective for both mechanical (16) and chemical pulping (17). Scott (18) demonstrated the feasibility of fungally treating wood chips on a semi-commercial scale.

Hemicellulose Composition, Extraction and Utilization

Significant differences exist in the hemicellulose composition of hardwoods and softwoods. Typically, softwoods contain galactoglucomannan (15-20% of dry wood) and arabinoglucuronoxylan (5-10%) while hardwoods contain very little glucomannan (< 4%) and a preponderance of glucuronoxylan (20-35 wt%; 19). Hemicelluloses found in other plants such as grasses and cereal grain stalks may approach 50 wt % in some tissues (20) but the hemicellulose fraction typically contains two or more polymers. An idealized backbone molecular structure for hardwood xylan is shown in Figure 1. In nature, this polymer contains one 4-O-methyl-glucuronic acid side chain per 4-16 backbone xylose residues (21,22) while acetyl groups are found on 3.5-7.0 repeat units per 10 xylose residues (22,23). For hardwoods, the terms xylan and glucuronoxylan are used interchangeably, while "xylans" from other sources may contain arabinofuranose-, glucurono(arabino)-, and rhamno- sugars as side chains.

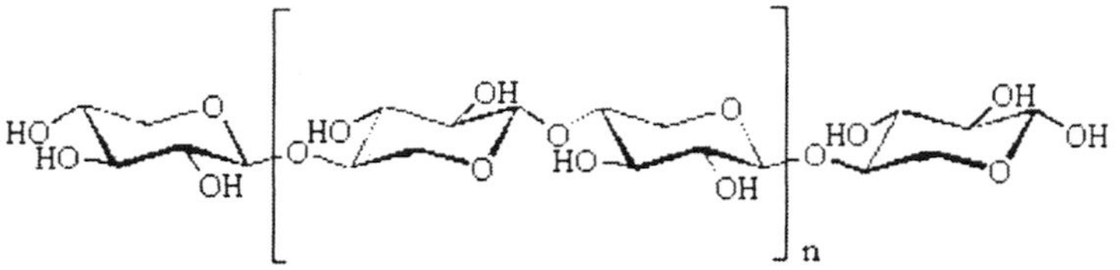

Figure 1: Idealized backbone structure of xylan (pyranose ring hydrogen atoms omitted for clarity).

The acetyl groups on xylan (not shown in Figure 1) are critically important in maintaining the water solubility of the native polymer since they block potential sites for hydrogen bonding. As a result, crystallization is inhibited. If they are removed by base-catalyzed hydrolysis (or acid hydrolysis) xylan becomes insoluble in water and can only be re-dissolved in alkali or other more polar organic solvents.

Despite the huge potential volume of xylan and other hemicelluloses available for commercial development, very few products currently exist which exploit this renewable resource. At least two reasons for this under-utilization are commonly cited: (1) most biological sources of hemicellulose and xylan produce a number of different heteropolysaccharides and the concentration of any single polymer is usually quite low [<10% dry weight of biomass; (23)] except in the case of certain hardwoods where; (2) isolation of the xylan component (up to 35% dry weight) is very difficult since wood is a dense "composite material" that also contains cellulose and lignin, both intractable polymers, which may actually be covalently linked to the hemicelluloses (20). In a recent review article, Ebringerova and Heinz summarize the various methodologies available to extract xylan from wood and other lignocellulosics (20). In general, the biomass is mechanically ground in some fashion followed by a delignification step typically involving aqueous NaOH and Cl_2 or H_2O_2 at 50-70°C. After filtration or centrifugation, the delignified material is then treated with additional aqueous 5% NaOH at lower temperature (20°C) to yield water-soluble (acetylated) and water-insoluble (deacetylated) xylan fractions. Solid xylan is obtained by precipitation with alcohol.

Other methods to "loosen up" the wood structure prior to alkaline extraction include the use of steam and / or treatment with concentrated ammonia although these processes were observed to produce xylan with significantly reduced degree of polymerization and a high residual lignin content (20). Nelson has shown that a "moderate" degree of matrix swelling is required to extract a maximum amount of xylan from Birch wood (24). He concluded that excessive swelling of the cellulosic component at high alkali concentrations restricts accessibility of the solvent (aqueous alkali) to the hemicellulose. Nelson also suggests that care must be taken during the swelling process to exclude oxygen which is capable of reducing the molecular weight of the extracted polymer fraction.

Assuming that biodelignification can enhance the ability of wood hemicellulose to be extracted under mild, aqueous conditions, applications can be sought for this potentially biodegradable, film-forming "plastic". Unfortunately, the molecular weight of hardwood xylan is relatively low (<10,000) and it typically forms very brittle, low strength films when cast from aqueous solvents. However, both naturally acetylated and synthetically modified xylans could be blended with other, higher performance polymers, such as commercially available cellulose esters and microbial polyesters. It is envisioned that these blends would exhibit both biodegradability and performance suitable for disposable plastic applications (packaging, paper coatings, etc).

Experimental Methods

Woody biomass feedstock

Willow wood chips (*Salix. sp*) were obtained from the SUNY-ESF Genetics Field Station in Tully, New York. The chips were from a single harvest at four years of age of a multi-clone trial and they were not de-barked. The chips were laid out for two weeks to air dry (AD) with a resulting oven-dry (OD) solids content of 92.3%. After air-drying, the chips were well-mixed and then divided and placed into large plastic bags for storage. It was important to bring the chips to a constant and low moisture content to ensure natural degradation did not take place during storage. When chips were needed for treatment, a 1625 gram AD chip sample (1500 g OD was needed for each bioreactor vessel) was brought up to a 50% moisture content by soaking overnight in distilled water.

The chips were then incubated in an aerated static bed-bioreactor consisting of 21-L polypropylene containers. The lid on the containers vented to the atmosphere through an exit tube. At the bottom of the polypropylene container, a 1-cm side opening provided for controlled inlet airflow. Prior to inoculation, the clean, empty bioreactors were autoclaved for twenty minutes. After the chips were added to the vessel, steam was injected for thirty minutes through latex tubing connection at the bottom of the reactor. The bioreactors' lids were left slightly ajar to prevent over pressurization. After steaming, the bioreactor was drained to remove the excess water that had condensed inside the vessel. The vessel and its contents were then cooled for two hours before inoculation, with the inlet and outlet of the vessel covered with aluminum foil to avoid contamination.

Preparation of the Inoculum

C. subvermispora strain L14807 SS-3 (Cs SS-3) was obtained from the USDA Forest Service, Forest Products Laboratory (FPL) in Madison WI. All stock culture slants were incubated at 26°C, stored at 4°C, and maintained at 2% (w/v) potato dextrose sugar plates. The samples were prepared and maintained as reported in Bartholomew (*25*). When needed for treatment, 2.31 ml of mycelium was added to 100 ml of sterile water and blended for 75 seconds. The blending was done in 15-second intervals followed by a 15-second pause to avoid heat build up, up to a total of 75 seconds of blending. The blended mycelium was transferred to a sterile beaker, additional makeup water was added directly to the chips to yield a 55% moisture content, and 0.5 % unsterilized corn steep liquor at 50% solids added to the beaker. The mixture was then poured over the chips in the bioreactor.

The bioreactors were then incubated at 27°C with an airflow of 7.87 cm^3/s (1.0 ft^3/h) per bioreactor. The air was humidified by flowing through two water-filled 2-L Erlenmeyer flasks and a fritted ground glass sparger. Humidified air was passed through a water trap, filtered through a 0.2 µm Millipore filter, and entered the base of the bioreactor. After the two weeks, the chips were frozen to prevent any further fungal growth prior to the analysis or subsequent extraction. The chips were kept frozen until 12 hours before they were used for xylan extraction.

Hot Water Hemicellulose Extraction

Hot water extraction was carried out in a 4-L capacity M&K digester equipped with indirect heating through heat exchangers with forced liquor re-circulation (*26*). The basket was filled with chips (1500 g OD) from air-dried willow samples for the control. For pretreated samples, the chips were removed from the freezer allowed to thaw for 12 hours. The basket was placed in the digester and distilled water was added to achieve a 4:1 liquor to wood ratio. The digester cover was then closed and the circulation pump turned on. The temperature was set (experiments were at 140°C, 145°C, 150°C, 155°C and 160°C) and the heaters were turned on. The chips were brought up to temperature in approximately 15 minutes and the two-hour extraction began.

After the two-hour extraction, the pump and heater were turned off and a bottom valve opened slowly to relieve the pressure and to withdraw the extract for analysis. The extract was collected through a valve and heat exchanger to cool the sample below the boiling point. The chips were washed thoroughly until a clear liquid was observed. The wash water was not collected. The chips were then placed in a drying oven at 105°C overnight to determine the mass loss of the chips. The % hemicellulose extracted was determined using a NMR technique to quantify the sugar distribution resulting from the acid hydrolysis of original wood and extracted samples (27).

Polymer Blend Preparation

In this preliminary study, extracted willow xylan was not immediately available for use in blending studies. As a result, native, partially acetylated birch xylan and a synthetically acetylated elm xylan were obtained from the sample collection of Professor T. E. Timell, an emeritus faculty member at our institution. Blends of these polymers with cellulose esters and bacterial polyesters were typically prepared by mixing solutions of the respective polymers dissolved in either dimethylformamide (DMF) or water followed by evaporation in a vacuum oven at 105°C yielding a thin film. A thermoplastic bacterial co-polyester (poly(hydroxybutyrate)) containing 30% hydroxyvalerate content) was purchased from the Aldrich Chemical Company (cat. no. 28,248-0). The glass transition temperature (T_g) of each polymer and blend was determined using a TA Instruments 2920 differential scanning calorimetry (DSC) instrument.

Results and Discussion

After two weeks of fungal treatment, bark-containing willow wood chips showed a very modest degree of % lignin reduction, typically less than 1% based on total original wood mass or 3-4% based on the initial lignin content (26). Previous work in our laboratory suggested that up to 5% of total lignin contained in Norway spruce chips could be removed by fungal delignification after two weeks and nearly 10% was removed after 4 weeks (25). A more significant reduction in lignin content was apparent for willow, however, after extraction with hot water as shown in Table I.

The results shown in Figure 2 demonstrate that fungal pretreatment increases hemicellulose yield in the hot water extraction. Comparing results at 150°C, approximately 20% of the available hemicellulose can be extracted from untreated "control" chips while the fungal treated chips provide extraction efficiency in the range of 40-50%. Variations between the treated "A" and "B" batches might have originated from microbial contamination which was observed to be more of a challenge with bark containing chips compared to debarked materials. Clearly, this preliminary work demonstrates that the level of extractable hemicellulose increases significantly with biodelignification, perhaps by cleaving linkages that may exist between hemicellulose and lignin (29). Alternatively, a scanning electron microscopy image of biodelignified wood chips, shown in Figure 3, illustrates that fungal hyphae penetrate through the cell walls of wood fibers potentially creating enhanced accessibility to the hemicellulose polysaccharides found there.

Table I: Lignin Content of Bark-Containing Willow Wood Chips (Lignin by TAPPI Method T-222; Original willow = 28.2 % lignin)

Biodelignification Pretreatment	After 140°C water extraction	After 155°C water extraction
A	23.2	22.2
B	23.1	26.4
G	-	25.5
H	-	25.0
None - Control	26.8	27.5

Note: A,B,G and H represent replicate delignification and extraction procedures using the same treatment protocol.

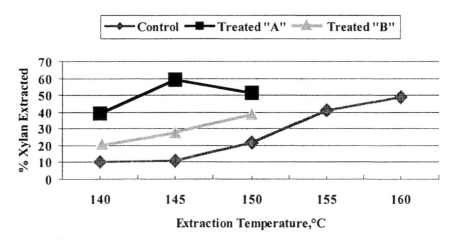

Figure 2. Hemicellulose removal efficiency of hot water extraction

As shown in Table II, blends of native and modified xylan hemicellulose were prepared with a compositionally diverse group of biobased and potentially biodegradable polymers and resulting T_g values, determined by DSC, were compared to the individual polymers. If two blended polymers form a true thermodynamic solution, then a single T_g is observed for the resulting blend

(*28*). If the polymers are incompatible, then the "apparent" blend is actually two distinct phases and two T_gs are seen. Only in the former case is the blend expected to show "hybrid" properties intermediate between the original polymers. Phase separated systems typically show poorer mechanical properties than true blends.

These DSC results provide evidence that mixtures of both native and acetylated xylan with cellulose acetate are indeed true blends based on the appearance of a single T_g. In the case of Blends 5+2 and 6+2, prepared in water, it is possible that a homogeneous blend was achieved despite the absence of an observed T_g. In these cases, differences in the ability of residual water and DMF to act as plasticizers for these polymers may effect molecular motion and apparent T_g. The higher boiling point of DMF, near 150°C, may enable this solvent to provide a plasticizer effect at higher temperatures than water, allowing a distinct T_g to be observed below the degradation temperature of the polymers (230-250°C). For Blend 7+2, the relatively hydrophobic nature of the copolyester may render this polymer incompatible with acetylated xylan which is more polar. Experiments to better characterize the mechanical properties of these blends are in progress.

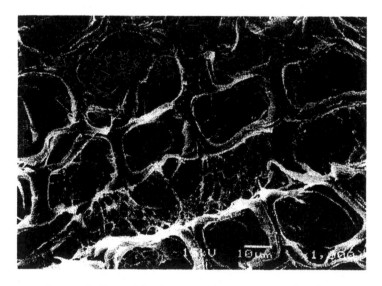

Figure 3. Fungal hyphae penetrating wood cell walls after biodelignification.

Table II: Blends of xylan with biobased polymers (50 wt% of each polymer. ND: Not detected – a distinct T_g was not observed.)

Film Composition	Casting Solvent	T_g (°C) by DSC
1. Cellulose Acetate	DMF	187
2. Native Birch Xylan	DMF	151
Blend: 1+2	DMF	164
3. Acetylated Elm Xylan	DMF	197
Blend 1+3	DMF	182
4. Cellulose Acetate Butyrate	DMF	102
Blend 4+2	DMF	98, 148 (2 T_g's)
5. Hydroxypropyl cellulose	Water	179
Blend 5+2	Water	ND (1)
6. Carboxymethyl cellulose	Water	> 250
Blend 6+2	Water	ND
7. Bacterial co-polyester	DMF	-7
Blend 7+2	DMF	-4, 139 (2 T_g's)

Conclusions

Hardwoods represent a unique opportunity as a feedstock for biobased products since they are especially rich in hemicellulose. In addition, hardwood hemicellulose is predominantly a single polymer, glucuronoxylan (xylan), which ultimately reduces the need for fractionation if a single product is desired. In this study, a fast growing species of willow was subjected to fungal biodelignification and subsequent extraction by hot water. Attempts to create new biodegradable polymer films from xylan by solution blending with other biobased polymers were partially successful. Specific findings are summarized below:

- Treatment of bark-containing willow wood chips with the fungus *C. subvermispora* for two weeks at 25-30°C resulted in a modest degree of delignification but a significantly higher level of water-soluble "extractables" compared to untreated controls.
- In general, the level of hemicellulose that can be extracted by water from biodelignified willow increases with temperature and reaches almost 60% of the total hemicellulose in the sample. At comparable

extraction temperatures, biodelignification increases the yield of soluble hemicellulose by 3 to 4 fold over untreated controls.

• DSC results suggest that native acetylated xylan and synthetically acetylated xylan form homogenous blends with commercially available cellulose acetate when cast from DMF solutions.

References

1. *Biobased Industrial Products – Priorities for Research and Commercialization,* National Research Council, National Academy Press, Washington, DC, 2000.
2. *The Biobased Economy of the Twenty-First Century: Agriculture Expanding into Health, Energy, Chemicals and Materials,* NABC Report 12, Eaglesham, A., Brown, W. F. and Hardy, R. W. E., Eds; National Agricultural Biotechnology Council, 2000.
3. Heller, M. C.; Keoleian, G. A.; Volk, T. A. *Biomass Bioenergy* **2003**, 25, 147-165.
4. Lawson, L. R.; Still Jr., C. N. *Tappi* 40:56A, **1957.**
5. Kawase, K. *J. Fac. Agri. Hokkaido Univ.* **1962**, 52, 186.
6. Reis, C. J.; Libby, C. E. *Tappi J.* **1960**, 43, 489.
7. Ander, P.; Eriksson, K.-E., *Svensk Papperstidning* **1975**, 18, 647.
8. Johnsrud, S.C.; Eriksson, K.-E. *Appl. Microbiol. Biotech.* **1985**, 21(5) 320-327.
9. Eriksson, K.-E.; Johnsrud, S. C.; Vallander, L. *Arch. Microbiol.* **1983**, 135, 161.
10. Eriksson, K.-E. *Wood Sci. Technol.* **1990**, 24, 79.
11. Samuelsson, L.; Mjober, P. J.; Harler, N.; Vallander, L.; Eriksson, K.-E. *Svensk Papperstidning* **1980**, 8, 221.
12. Eriksson, K.-E.; Vallander, L. *Svensk Papperstid.* **1982**, 85, R33.
13. Pearce, M. H.; Dunlop, R. W.; Falk, C. J.; Norman, K. *Proc. 49th Appita Annual General Conf.,* Australia, p. 347, 1995.
14. Kirk, T. K.; Koning Jr., J. W.; Burgess, R. R.; et al. *Res. Rep. FPL-RP-523,* Madison, WI., 1993.
15. Kirk, T. K.; M. Akhtar, M.; Blanchette, R. A. *Proc. 1994 TAPPI Biological Sciences Symposium,* Tappi, Atlanta, p. 57, 1994.
16. Akhtar, M., Blanchette, R. A.; Myers, G. C.; Kirk, T. K. R.A. Young and M. Akhtar, Eds, John Wiley & Sons, New York, p. 309, 1998.
17. Messner, K., Koller, K.; Wall, M. B.; Akhtar, M.; Scott, G. M. *Environmentally Friendly Technologies for the Pulp and Paper Industry,* R.A. Young and M. Akhtar, editors, John Wiley & Sons, New York, p. 385, 1998.

18. Scott, G. M., Akhtar, M.; Lentz, M. J.; Swaney, R. E. *Environmentally Friendly Technologies for the Pulp and Paper Industry*, R.A. Young and M. Akhtar, editors, John Wiley & Sons, New York, p. 341, 1998.
19. Sjostrom, E. *Wood Chemistry – Fundamentals and Applications*, 2nd Edition, Academic Press, 1993.
20. Ebringerova, A.; Heinze, T., *Macromol. Rapid Commun.* **2000**, *21*, 542-556.
21. Timell, T. E. *Adv. Carb. Chem.* **1964**, 19, 247.
22. Bolker, H. I. *Natural and Synthetic Polymers*, Marcel Dekker, Inc, New York, 1974.
23. Stenius, P. *Forest Products Chemistry*, TAPPI Press, 2000.
24. Nelson, R. *Factors Influencing the Removal of Pentosans from Wood By Alkali: Evidence of the Lignin-Carbohydrate Bond*, Ph.D. Thesis, SUNY-College of Environmental Science and Forestry at Syracuse, 1956.
25. Bartholomew, J. *Identification and isolation of lignolytic enzymes of Phlebia subserialis and an analysis of white-rot fungi on Picea abies for Mechanical Pulp*, M.S. Thesis, State University of New York- College of Environmental Science and Forestry, Syracuse, NY, 2003.
26. Blowers, M.K, *Xylan extraction from short rotation willow biomass*, M.S. Thesis, State University of New York- College of Environmental Science and Forestry, Syracuse, NY, 2003.
27. Kiemle, D. J.; Stipanovic, A. J; Mayo, K. E. *ACS Symp. Ser.* **2004**, *864*, 122-139.
28. Fried, J. R., *Polymer Science and Technology*, Prentice Hall PTR, p. 302, 1996.
29. Glasser, W. G.; Kaar, W. E.; Jain, R. K.; Sealey, J. E. *Cellulose* **2000**, 7, 299-317.

Chapter 17

Life-Cycle Assessment of Energy-Based Impacts of a Biobased Process for Producing 1,3-Propanediol

Robert P. Anex[1] and Alison L. Ogletree[2]

[1]Department of Agricultural and Biosystems Engineering, Iowa State University, Ames, IA 50011
[2]Archer Daniels Midland Company, 1001 North Brush College Road, Decatur, IL 62521

Life Cycle Assessment (LCA) is applied to biobased 1,3-propanediol (PDO) production. The cradle-to-factory gate system includes growing of corn, corn milling, converting cornstarch to glucose and fermentative production of PDO using a genetically modified biocatalyst. The LCA is of a preliminary biobased process design and is used to identify system improvement targets and benchmark life-cycle environmental performance. The analysis indicates that approximately 70% of nonrenewable energy is used in producing PDO from glucose. Of this, the largest part is used in purifying PDO. Biobased PDO is predicted to have significantly lower cradle-to-factory gate nonrenewable energy and climate change potential than PDO derived from ethylene oxide. PDO converted to polytrimethylene terephthalate (PTT) is predicted to outperform Nylon 6 in these same categories and also have lower NO_x emissions.

Introduction

From a historical perspective, it is only during the very recent past that humans have learned to decouple primary industrial productivity from biomass production. In this brief period, we have come to rely heavily on fossil biomass –petroleum, coal, and natural gas- to enable our modern way of life by magnifying our ability to do work and allowing us to synthesize new compounds with unique properties. Even agriculture is now predominantly practiced in an intensive manner that relies on petroleum in the form of fuels and chemicals to produce the large harvests that have allowed human population to grow beyond what was believed possible only 100 years ago (1).

Attempts to reverse this trend and return to satisfying significant amounts of the human appetite for power and materials using plant-derived raw materials will have complex social and environmental impacts. Past efforts to wean civilization from its ever-growing dependence on fossil fuels have mostly proved unsuccessful. In the United States, even vigorous efforts by such capable engineers and scientists as Henry Ford, George Washington Carver, and Thomas Edison during the early 20th century had little success (2). What may allow current efforts more success is the availability of advanced biotechnology and tightening environmental and resource constraints.

In terms of reducing humanity's heavy dependence on fossil resources, developing biobased polymers may not appear to be a key step. Worldwide, only eight percent of oil consumption is used in plastic production - four percent as feedstock and four percent as energy in production (3). Higher value biobased products such as polymers, however, improve the economics of multi-product biobased production (i.e., the "bio-refinery") and make feasible the production of significant quantities of fuel and power from abundant sources of lignocellulosic biomass (4). For example, producing high value chemical intermediates, such as polymers, polyols, enzymes, amino acids, vitamins, and carotenoids, from lignocellulose-dervied mixed sugar solutions will be made possible by understanding and engineering the metabolic pathways of biocatalysts (5).

Biopolymers may also have more resource and environmental significance when the biobased polymer has unique properties that make it possible to save or recover other resource streams. For example, due in part to the difficulty of separating the different polymers used in the caps and labels of polyethylene terephthalate (PET) bottles, more than eighty percent of recovered material is "down cycled" into terminal uses such as fibers. If the non-PET components were made of a polyhydroxyalkanoate (PHA) biopolymer that converted to carbon dioxide and water during an initial caustic step in the PET recycling process, much more bottle-to-bottle recycling would be possible. This type of recycling would, by using the same PET material many times over, result in significant reductions in emissions and savings of nonrenewable energy.

Biobased 1,3-Propanediol

One of the first biobased products being commercialized exemplifying the promise of combining metabolic engineering with advanced process design to achieve a sustainable product with superior properties at a low cost is biobased 1,3-propanediol (PDO). Because biobased PDO is not yet produced in commercial quantities, this is a good time to take early measure of how well the biobased pathway meets the objectives of sustainable and environmentally benign production.

Biomass-derived PDO is not new. The first record of PDO is from 1881 when it was discovered by August Freund in a glycerol-fermenting mixed culture. PDO is a monomer in a range of polycondensation reactions that produce polyesters, polyethers and polyurethanes. PDO can also be used to improve the properties of solvents, adhesives, laminates, resins, detergents and cosmetics. Until recently, PDO was limited by its high cost to a few solvent applications.

Commercial Production of PDO

There are three commercially interesting routes to PDO (6):
- hydration of acrolein followed by hydrogenation;
- hydroformylation and hydrogenation of ethylene oxide; and
- enzymatic processing of glucose.

The hydration of acrolein (obtained by catalytic oxidation of propylene) followed by hydrogenation route was initially investigated by Shell between 1948 and 1972 (7). Shell abandoned the route in the 1970s, but Degussa further developed the route until it could be commercialized and has supplied PDO to DuPont for product and market development since 1998 (8). Total cost of production in 27,000 t/yr plant using acrolein at a cost of $1700 US/t is estimated to be approximately $3.40/kg (9).

The hydroformylation and hydrogenation of ethylene oxide (prepared by oxidation of ethylene) did not become available until the early-1990s. Shell commercialized this technology in 1995. Total cost of production in 27,000 t/yr plant using ethylene oxide at a cost of $1010 US/t is estimated to be approximately $2.00/kg (9).

The third route to PDO is enzymatic conversion of glucose has been under development by DuPont since the early 1990s. Noyes and Watkin first noted the enzymatic conversion of glycerol to PDO in 1895 when these researchers found an impurity occurring in stored glycerol. The search for high-purity glycerol for the production of explosives led in 1914 to the identification (by Voisenet) of PDO as the impurity produced during the anaerobic fermentation of glycerol (7).

No naturally occurring organism can produce PDO directly from glucose. Genetically engineering a single organism to produce PDO from glucose, combining the conversion of glucose to glycerol with the conversion of glycerol to PDO is desirable because it reduces the complexity of the fermentation. In the early-1990s DuPont and its partners succeeded in genetically engineering a microorganism (*E. coli* bacterium) capable of producing PDO from glucose. Several studies expect the DuPont route to be competitive with the Shell route and potentially to be superior when made commercial (*8*). Total cost of production in a 27,000 t/yr first generation plant using glucose at a cost of $300 US/t is estimated to be approximately $2.00/kg (*9*).

In a joint venture with Tate and Lyle PLC, known as DuPont Tate & Lyle BioProducts LLC, DuPont plans to produce 1,3-propanediol (PDO) using their proprietary fermentation and purification process based on corn-derived glucose. Sample quantities of the biobased PDO are currently produced at a pilot plant in Decatur, Illinois and beginning in 2006, commercial-scale quantities will be produced at a manufacturing facility in Loudon, Tennessee (*10*).

Assessment Goals

The analysis reported in this paper is of a system similar to the one pioneered by DuPont, but based on a process design and simulations developed by the authors. The process for fermenting glucose to PDO was designed to match the performance of the DuPont process as defined by data publicly available at the time of the study. The performance of other unit processes, such as separation and purification of PDO, was predicted based on best practice engineering design. Life-cycle indicators for transport as well as production and processing of the corn feedstock were gathered from publicly available data sources.

This analysis represents one informed estimate of how a biobased PDO process could perform. It is the type of study that should be performed during preliminary design to benchmark performance, identify trade-offs and guide process improvement efforts. There are a variety of ways that the unit processes designed in this analysis could be improved as well as ways to improve system performance through integration. This analysis is thus not appropriate for making comparative assertions about the performance of mature technology of this type relative to other processes or products, but it does allow one to gauge the potential for sustainability of this type of biobased production.

Methodology of LCA

Life cycle assessments (LCAs) consist of four independent elements (*11*): 1) definition of goal and scope; 2) life cycle inventory analysis; 3) life cycle impact assessment; and 4) interpretation.

Element 1, the definition of goal and scope, includes choice of a functional unit which forms the basis of comparison, the product or service system boundaries, and allocation procedures. The goal of this assessment is benchmarking a preliminary design for producing PDO enzymatically from glucose. Because the function of 1 kg of PDO is the same regardless of whether the PDO was produced from fossil resources or corn, a cradle-to-factory gate system boundary is applicable. Although final product manufacture, use and end-of-life phases of the life cycle may involve significant impacts, they will be independent of the PDO production process. The functional unit is taken to be 1 kg of PDO. The geographic location is the corn-producing region of the Midwest United States, centered in Iowa.

The need for allocation arises when a unit process produces several products and the inputs and outputs must be allocated among them. Following ISO 14041 guidelines (*12*), analysis showed that allocation could not be avoided for some processes in the biobased PDO system. Four allocation and physical partitioning procedures were evaluated and allocation based on mass fraction was chosen and is used in all analyses presented here. Ogletree (*13*) presents detailed discussion of this comparison of allocation methods. Boustead (*14*) and Vink (*15*) make similar arguments recommending the mass fraction allocation method.

Element 2, the life cycle inventory, involves data collection and calculation to quantify the system's inputs and outputs that are relevant from an environmental point-of-view. The biobased PDO process under evaluation is a preliminary design with most performance data coming from simulations that are able to predict with accuracy only the most significant life-cycle flows. One of the reasons to investigate the potential use of biomass for producing chemicals is the reduction of greenhouse gas (GHG) emissions that goes along with the replacement of fossil fuels by biomass. In this context, the use of non-renewable fuels and the resulting GHG emissions are key indicators to evaluate the effectiveness of biomass use. The life cycle inventory categories are nonrenewable energy use, agricultural land use, and GHG and NO_x emissions.

Element 3, life cycle impact assessment, involves evaluating the significance of the life-cycle inventory. One goal is to aggregate flows with comparable effects using equivalence factors (e.g., all GHGs are converted to the common basis of kg of CO_2 equivalent).

Element 4, the life cycle interpretation, is where conclusions are drawn from the life cycle inventory and impact assessment, and recommendations are made for system improvement.

Biobased PDO System Description

The cradle-to-factory gate biobased PDO system is divided into six stages as shown in Figure 1. Life-cycle inventory data for the first four stages were taken from the Greenhouse gases, Regulated Emissions and Energy use in Transportation (GREET) model (*16*). GREET is a cradle-to-grave model that accounts for all emissions and energy consuming processes in various fuel-cycle paths, including corn-to-ethanol. The model tracks energy and material inputs and includes emissions of five air pollutants (volatile organic compounds [VOC], carbon monoxide [CO], nitrogen oxides [NO_x], particulate matter smaller than 10 microns [PM_{10}] and sulfur oxides [SO_x]) and three greenhouse gases (methane [CH_4], nitrous oxide [N_2O], and carbon dioxide [CO_2]).

The GREET study area for corn production is the four largest corn-producing states: Illinois, Iowa, Minnesota and Nebraska. All corn production-related inventory data are a weighted average of agronomic practices in these four states. The weighting factors are based on 1994-96 share of corn production for each state (*17*).

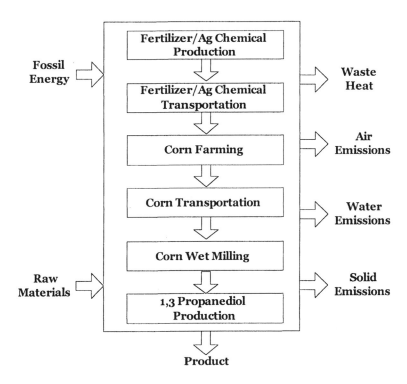

Figure 1. System boundaries for biobased PDO production.

Fertilizer/Ag Chemical Production

Field corn production usually involves application of nitrogen, phosphate and potash fertilizers and sometimes lime is incorporated into the soil to reduce acidity. Application rates vary with soil type and crop rotation practices. The GREET model estimates inventory data for three major types of nitrogen fertilizer - ammonia, nitrogen mixtures and urea – which represent the eight types of N-fertilizer used in significant amounts in growing corn in the Midwest region of the United States (*17*).

The pesticides tracked include the four herbicide agents most commonly applied in corn production: atrazine, metochlor, cyanazine, and acetochlor. Herbicide use is computed from a weighted average of application rates in the study region, based on 1996 data. The increasing use of genetically modified "plant pesticide" corn varieties may be reducing pesticide use, as has been suggested by recent data (*18*).

Corn Farming

U.S. corn yield per acre has increased over the last 30 years by over 50% due to improved corn varieties, better farming practices, and farming conservation measures (*19*). Corn farming inventory data represent the following subprocesses: land preparation, planting, plowing, weeding, chemical application, irrigation, harvesting, and drying. Corn production is estimated to be 146 bushels/acre (9173 kg/ha) based on 2002 production data for Iowa. In this analysis it is assumed that all CO_2 fixed through photosynthesis in the corn plant is re-emitted during processing or subsequent biodegradation of residual biomass. In this way, no climate change credit (charge) is taken for carbon sequestration in (release from) the soil. Emissions and resource use associated with the manufacture of farm equipment is assumed to be negligible.

The biological aspect of farming presents many difficult challenges for the LCA process. Environmental impacts of farming are generally specific to a given location because emissions are controlled by soil type, hydrology and the time-dependent processes involved in soil chemistry, and depend on the local conditions and ecosystem. For example, the amount of nitrogen released into surface waters and into the atmosphere as N_2O is heavily influenced by rainfall, local surface water conditions, and the partial pressure of oxygen, carbon, and pH of the soil. The GREET model utilizes static emission factors these complex natural processes. For example, the emission factor for N_2O is 1.5% of applied fertilizer-N (released as N_2O-nitrogen from combined soil direct emissions and leaching).

Corn Wet Milling

The purpose of the wet milling process is to break corn into its components, yielding a starch slurry of high purity with high value co-products. Corn grain is cleaned, steeped in a dilute sulfuric acid solution for 20 to 36 hours and then ground to separate the oil-bearing germ from the kernel. Germ is sold and will be processed to yield corn oil and animal feed. The germ-free kernel is milled and washed to separate starch and gluten meal from fiber. Gluten is filtered and sold as corn gluten meal. The fiber is mixed with steepwater and other residuals to produce corn gluten feed. The purified starch is enzymatically converted to glucose in a two-step liquefaction and hydrolysis process. In some instances starch may be converted to a variety of other products such as high fructose corn syrup. In this analysis the corn wet mill is assumed to have five product streams: germ, steepwater, gluten feed, gluten meal and glucose.

Corn wet mill energy use and emissions were estimated using industry data from Penford Corporation (20) and values reported in the literature (21). According to Vink, et al. (15), approximately 74% of corn wet mills use coal as their primary fuel and the others use natural gas. In this analysis natural gas is taken as the primary source of energy for the corn wet mill. Using mass fraction allocation, the gross fossil energy requirement of corn wet milling is 5.2 MJ/kg PDO. Other inventory data are summarized below.

Transport

Inventory of transport includes resource use and emissions produced moving feedstock for fertilizer and chemical production, fertilizers and chemicals to the farm, and corn from the farm to the mill. The PDO production facility is assumed to be collocated with the corn mill, so there is no transportation of glucose from the mill to the PDO facility.

Agricultural chemical transportation data are taken from the GREET model and include barge, rail, and two truck modes. Corn is moved from farms to mills in tractor-trailers ("semis"), wagons, and single and tandem axle trucks. Several studies of corn transport have produced similar results. For example, Graboski (18) finds that the average energy required to move corn from farms to the corn wet mill is 4600 Btu/ bushel and Wang (19) estimates 4100 Btu/bushel.

PDO Production

Life cycle inventory data for the production of PDO are generated from a detailed process model for bioproduction from glucose. The process design is developed from publicly available data on the biocatalyst developed by DuPont and its partners. They have constructed a strain of *E. coli* containing genes from *S. cerevisiae* for glycerol production and the genes from *K. pneumoniae* that can convert glucose to PDO (*22*). The biocatalyst process was designed to match the performance of the DuPont organism as defined by published data (*23*): concentration of PDO is 135 g/L; rate of production is 3.5 g/L·h; yield is 0.511 g PDO/g glucose.

The representative biomass for the process is taken to be *E. coli* ($CH_{1.78}O_{0.6}N_{0.19}$). The overall reaction in the fermentor is shown by the reaction (*24*):

$$37\ C_6H_{12}O_6 + 3\ NH_3 +\ 10\ O_2 \longrightarrow 60\ CO_2\ +\ 16\ H_2O +\ 50\ C_3H_8O_2\ +\ 3\ Biomass$$

Biocatalyst growth is represented by the standard Monod kinetic model using model coefficients for *E. coli* found in the literature. The specific growth rate, saturation constant, and cell maintenance coefficients are 0.50 h^{-1}, 4.0 mg/L, and 0.045 h^{-1} respectively (*25*).

Other unit processes, such as separation and purification of PDO, were designed using best engineering design practice with some guidance from DuPont (*24*) on applicable methods. The resulting process design for production of PDO from glucose is shown schematically in Figure 2.

Process Design

As shown in Figure 2, the feed products to the fermentor are glucose, ammonia, water, innoculum and air. A seed fermentor, not shown, is used to grow the organism innoculum before it is fed to the production fermentor. Energy and resource use in the seed fermentor are assumed to be negligible, a common assumption for systems of this type (*26*). The aerobic fermentation proceeds for 38 hours at 37°C, 1 atm with a working volume of 130,000 L. The biocatalyst excretes the PDO product extra-cellularly according to the standard Monod growth model.

The fermentation broth flows through a heat sterilization unit operated at 140°C to kill the genetically modified biocatalyst organisms complying with the Toxic Substances Control Act (TSCA) Part 725.422. After sterilization, the broth is moved to a storage tank used to accumulate product from the batch fermentor. The broth is centrifugally separated to remove the cell mass from the aqueous product. The aqueous product passes through a dewatering column while the cell mass is further washed to separate residual PDO. Biomass washing utilizes a diafiltration system using water as the rinse liquid. Water is added to the diafiltration feed tank at the same rate as the permeate flux to maintain constant feed volume. The aqueous product leaving the diafiltration advances to a centrifugal filtration system to remove residual biomass. The aqueous product from both centrifugal filtration systems are combined and passed through a dewatering column reducing water from 90% to 20%.

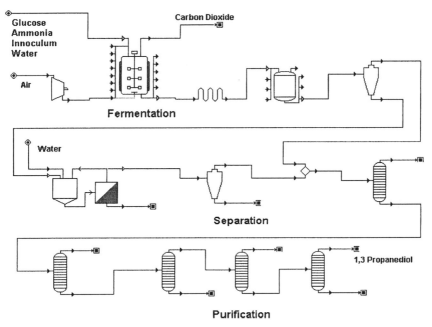

Figure 2. Bio-PDO process flowsheet.

Entering the distillation train the product stream has a pH greater than 8.5, is free of biomass, and contains less than 0.1% protein, total dissolved solids and glycerol. All of the columns in the distillation train are operated under a vacuum to reduce operating temperatures and to minimize unwanted reactions and product degradation. The first column of the distillation train reduces water to less than 3%. The bottom stream proceeds to a second distillation column where the heavy impurities, primarily glycerol and residual sugars, are removed. The third column removes the light components. The final distillation column purifies the PDO to a dry weight purity of 99.9%, with glycerol content less than 50 ppm, and pH between 5.5-7.0.

The process design was evaluated using the CHEMCAD process simulation software (27). The process was simulated at a scale of 3.1×10^6 kg/yr PDO. Total energy use in the PDO production process is 26 MJ/kg PDO. Electrical power is assumed to come from the Midwest U.S. electrical power grid.

Presentation of LCA Results

Cradle-to-factory gate life cycle indicators per functional unit of 1 kg of PDO are presented in Table I. Cradle-to-gate nonrenewable energy use is 37 MJ/kg of PDO. The largest fraction (70%) is used in the production of PDO from glucose, with 26% of this used in fermentation, 64% in purification and 10% in separation. Climate change potential and NO_x emissions are also largest in the PDO production stage, due primarily to the products of combustion related to power generation.

Table I. Key Life Cycle Indicators for Biobased PDO

Life Cycle Stage	Cradle-to-gate Nonrenewable Energy Use [MJ/kg]	Climate Change Potential [kg CO₂ eq./kg]	NOx Emissions [kg NOx/kg]
Ag Chemical Production & Transport	4.4×10^{-2}	1.6×10^{-1}	2.0×10^{-5}
Corn Farming	2.8	1.3×10^{-1}	9.4×10^{-4}
Corn Transport	4.6×10^{-1}	3.0×10^{-2}	1.1×10^{-4}
Corn Wet Milling	5.2	6.8×10^{-1}	6.2×10^{-4}
PDO Production	26	4.5	9.2×10^{-3}

Process Improvement

An obvious way to improve the process design is to reduce the amount of energy used, with the largest target being the energy required for purification and separation of PDO from the fermentation broth. Candidate techniques that could be investigated include solvent extraction and spray drying with evaporation. The next largest energy use is fermentation. This process would be more efficient if it could be made to operate in a continuous mode with PDO extracted from a slipstream and the biomass returned to the fermentor. This would increase average yield, reduce separation, and reduce the labor of operating the process.

Comparative Data

In addition to improvement of the preliminary process design, a goal of this LCA study is to benchmark the performance of this biobased PDO process against competing processes. The most direct comparison can be made with the process producing PDO from ethylene oxide (hereafter "EO").

In the EO-process, PDO is produced by the hydroformylation and hydrogenation of EO according to the two-stage reaction:

$$C_2H_4O \ + \ CO \ + \ H_2 \ \rightarrow \ C_3H_6O_2 \ + \ H_2 \ \rightarrow \ C_3H_8O_2$$

The first reaction is a hydroformylation process lengthening the carbon chain to three. This is accomplished by catalytic combination of ethylene oxide with carbon monoxide and hydrogen (synthesis gas) to yield hydroxypropanaldehyde (HPA). The hydroformylation process is performed under high pressure and temperature using a hydrophobic dicobaltoctacarbonyl catalyst (*6*). The second reaction is hydrogenation of HPA to PDO with a nickel-based catalyst occurring in three stages with a gradual temperature increase between stages (*6*).

Life cycle inventory data for the EO process were derived by combining SimaPro (*28*) data for hydrogen production with syngas production data (*29*). Key life cycle indicators for ethylene oxide-derived PDO and biobased PDO are presented in Table II. The biobased PDO life cycle fossil energy requirement is about 40% less than that of the petroleum-based (EO) process. Correspondingly, climate change potential of biobased PDO is less than that of the EO-derived PDO. There is no significant difference in NO_x emissions. In addition to NO_x emissions related to combustion, the biobased life cycle involves significant NO_x emissions associated with N-fertilizer production and application.

Table II. Key Life Cycle Indicators for Biobased PDO and Ethylene Oxide-derived PDO.

PDO Type	Cradle-to-gate Nonrenewable Energy Use [MJ/kg PDO]	Climate Change Potential [kg CO₂ eq./kg PDO]	NOx Emissions [kg NOx/kg PDO]
Bio PDO	37	5.5	0.01
EO PDO	61	7.0	0.01

There is significant uncertainty associated with this life cycle assessment since the biobased PDO process is only a preliminary design and inventory data are generated through simulation. The significantly lower fossil energy use and somewhat lower climate change potential of the biobased process are indications of the environmental potential of this process. However, biopolymer production technology is in its infancy and there are likely to be large improvements as it matures, whereas the petroleum-based processes are relatively more mature and thus may have less potential for dramatic improvement.

Evaluation of a Polymer Application

To better understand the potential impact of biobased PDO, it is useful to consider a large-scale application that could be made feasible if biobased production can significantly lower the cost of PDO. One such use is the manufacture of polytrimethylene terephthalate (PTT). In this process, PTT is formed by esterfication of PDO and terephthalic acid:

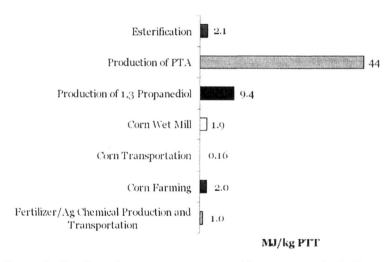

| PDO | Terephthalic acid | Polytrimethylene terephthalate |

Pure terephthalic acid (PTA) is formed by the oxidation of para-xylene. Life cycle inventory data for production of PTA are from Boustead (*30*). Data on the esterification process were estimated from a process model developed by SRI (*6*). The process includes two esterification reactors followed by a three column polycondensation reactor train.

The life cycle indicators for PTT are computed in terms of a new functional unit - 1 kg of PTT. Figure 3 shows the fossil energy required for the production of PTT from biobased PDO and PTA. The energy required for the production of PDO is dwarfed by that used in making PTA.

Esterification	2.1
Production of PTA	44
Production of 1,3 Propanediol	9.4
Corn Wet Mill	1.9
Corn Transportation	0.16
Corn Farming	2.0
Fertilizer/Ag Chemical Production and Transportation	1.0

MJ/kg PTT

Figure 3. Cradle-to-factory gate nonrenewable energy per kg PTT.

In several applications, biobased PTT has product performance similar to that of Nylon 6, so the life cycle performance of Nylon is a useful benchmark. Considering the use and end-of-life phases of PTT and Nylon 6 are beyond the scope of this study. However, some common functional unit must be found to serve as a basis of comparison. For this rough comparison, Nylon 6 life cycle data from Boustead (*31*) were converted to a functional unit of 1 kg of PTT by correcting for the density difference between the two polymers. Key life cycle indicators for biobased PTT, EO-derived PTT and Nylon 6 are presented in Table III.

Table III. Key Life Cycle Indicators for Biobased PTT, Ethylene Oxide-derived PTT and Nylon 6.

Polymer	Cradle-to-gate Nonrenewable Energy Use [MJ/kg PTT]	Climate Change Potential [kg CO₂ eq./kg PTT]	NOx Emissions [kg NOx/kg PTT]
Biobased PTT	61	4.1	0.013
EO-based PTT	68	4.4	0.011
Nylon 6	100	7.3	0.024

The differences previously seen between biobased PDO and EO-derived PDO are obscured in this PTT-Nylon comparison because of the large impacts associated with the PTA that is incorporated into both types of PTT. It is clear from Table III that, in the three impact categories examined, the PTT polymers are superior to Nylon 6 with the biobased PTT most preferred in two of three categories, and roughly equivalent in the third.

Discussion

The cradle-to-factory gate system boundary is justified on the basis of all forms of PDO have the same use and end-of-life impacts. However, life-cycle assessments of polymers are frequently sensitive to assumptions regarding end-of-life disposition (*32*). Additional analyses taking a cradle-to-grave perspective including use and waste management should be conducted.

There are many possible process improvements that could be conceived and included in future analysis. For example, the PTT life cycle assumes that PTA is newly manufactured. It is possible to significantly reduce energy use and CO_2 emissions by recovering PTA from PETE or PTT.

There are many PDO applications that should be examined. These analyses could be used to identify those applications where PDO would result in the largest environmental improvements. For example, some PDO compounds have very good thermal conductivity and stability properties that make it a good heat exchanger fluid (*33*). These properties should also lead to environmental and energy savings.

There remains a range of potentially important environmental impacts that have not been investigated. In particular the impacts on water quality and soil fertility of growing biomass for industrial uses are not known. Industrial demand for biomass will likely change the crops that farmers grow and how they grow them. How these changes will affect the environment should be further investigated.

Agriculture poses many difficult problems for LCA because of the complicated dynamic nature of the agroecosystem. For example, allocating life cycle flows among agricultural products is a problem with no clear solution. Crops are usually grown in a complementary rotation. For example, corn is usually grown in rotation with soybeans and N-fertilizer is applied only during the corn rotation. This rotation helps control pests and maintains soil fertility. Although soybeans are legumes that can fix nitrogen, in practice soybeans usually use more nitrogen than they fix, so they are utilizing fertilizer applied during the corn rotation. The allocation question then, is how much of the fertilizer applied during the corn rotation should be allocated to the soybean product? Or, since soybeans aid in corn production, should soybean inputs be partially allocated to corn? Just how to measure and allocate the impacts of agricultural practice in the LCA process remains an open question.

Conclusion and Outlook

This preliminary life cycle evaluation of energy impacts suggests that biobased production of PDO has the potential to deliver environmental performance superior to that of ethylene oxide-based processes. LCA was useful for benchmarking and evaluating preliminary process designs, providing insight as to where in the life cycle improvement efforts may be focused most usefully. Biobased production technologies of the type evaluated here are in their infancy and their performance is likely to improve rapidly as these technologies mature and are more widely used. Biobased production may help achieve sustainability by providing a source of chemicals from renewable resources with attractive life cycle environmental performance. LCA applied early in the design process can help fulfill that promise.

References

1. Cohen, J. E. *How Many People Can the Earth Support?* Norton: New York, NY, 1996.
2. Finlay, M. R. *J. Industrial Ecology* 2003, 7, 33-46.

3. *Plastics: Contributing to Environmental Protection*, URL http://www.apme.org/

4. Wyman, C. E. *Biotechnology Progress* **2003**, *19*, 254-262.

5. Realff, M. J.; Abbas, C. *J. Industrial Ecology* **2003**, *7*, 5-9.

6. *1,3-propanediol and Polytrimethylene Terephthalate*; SRI Consulting; Process Economics Program Report 227; SRI Consulting: Menlo Park, CA, 1999.

7. Andringa, L. R., M.Sc. thesis, Utrecht University, Utrecht, The Netherlands, 2004.

8. Zeng, A. P.; Biebl, H. *Adv. Biochem. Eng. Biotechnol.* **2002**, *74*, 239-259.

9. Held, M. Institute of Process Engineering, Swiss Federal Institute of Technology, Zurich, Switzerland. Unpublished results. 2004.

10. DuPont website. http://www.dupont.com/sorona/biobasedinitiative.html. Accessed October 25, 2004.

11. *Life-Cycle Assessment: Inventory Guidelines and Principles*; U.S. Environmental Protection Agency [USEPA]; EPA/600/R-92/245, USEPA, Office of Research and Development: Washington, D.C., 1993.

12. ISO 14041 Environmental management - Life cycle assessment - Goal and scope definition and inventory analysis; International Organization for Standardization: Geneva, Oct. 1998.

13. Ogletree, A. L., M.S. Thesis, Iowa State University, Ames, IA, 2004.

14. Boustead I. *An Introduction to Life Cycle Assessment*, Boustead Consulting Ltd.: West Grinstead, 2003.

15. Vink, E. T. H.; Hettenhaus, J.; O'Connor, R. P.; Dale, B. E.; Tsobanakis, P.; Stover, D. *The Life Cycle of NatureWorks™ Polylactide: The Production of Dextrose Via Corn Wet Milling*; Cargill Dow BV: Naarden, The Netherlands, 2004.

16. Wang, M. Q. *Development and Use of Greet 1.6 Fuel-Cycle Model for Transportation Fuels and Vehicle Technologies*; ANL/ESD/TM-163; Argonne National Laboratory: Argonne, IL, 2001.

17. Wang, M.; Saricks, C.; Wu, M. *Fuel-Cycle Fossil Energy Use and Greenhouse Gas Emissions of Fuel Ethanol Produced from U.S. Midwest Corn*; Argonne National Laboratory: Argonne, IL, 1997.

18. Graboski, M. *Fossil Energy Use in the Manufacture of Corn Ethanol*; National Corn Growers Association: Washington, DC, 2002.

19. Wang, M.; Saricks, C.; Santini, D. *Effects of Fuel Ethanol Use on Fuel-Cycle Energy and Greenhouse Gas Emissions*; ANL/ESD-38; Argonne National Laboratory: Argonne, IL, 1999.

20. Quarels, Jim; Personal communication. October 2003.

21. Galitsky, C.; Worrell, E.; Ruth, M. *Energy Efficiency Improvements and Cost Saving Opportunities for the Corn Wet Milling Industry*; Paper LBNL-52307; Lawrence Berkeley National Laboratory: Berkeley, CA, 2003.

22. Chotani, G.; Dodge, T.; Hsu, A.; Kumar, M.; LaDuca, R.; Trimbur, D.; Weyler, W.; Sanford, K.; *Biochim. Biophys. Acta* **2000**, *1543*, 434 – 455.
23. Nakamura, C. *Production of 1,3-propanediol by E. coli*; presented at Metabolic Engineering IV: Applied System Biology, Il Ciocco, Castelvecchio Pascoli, Italy, October 8, 2002.
24. Miller, R.; DuPont, Personal communication, October 28, 2003.
25. Bailey, J.; Ollis, D. *Biochemical Engineering Fundamentals*, 2nd ed.; McGraw Hill: New York, NY; 1986.
26. Vink, E.; Rabago, K.; Glassner, D.; Gruber, P.; *Polym. Degrad. Stab.* **2003**, *80*, 403-419.
27. *CHEMCAD Suite Simulator Software, Version 5.4*; Chemstations, Inc.: Houston, TX, 2003.
28. *SimaPro 5.4*; Pré Consultants: Amersfoot, the Netherlands, 2002.
29. Chauvel, A.; LeFebvre, G. *Petrochemical Processes; Vol. 1 Synthesis-Gas Derivatives and Major Hydrocarbons*; Gulf Pub Co: Houston, TX, 1989.
30. Boustead, I. *Eco-Profile: Polyethylene Terephthalate*, URL http://www.apme.org, 2002.
31. Boustead, I. *Eco-Profile: Nylon 6*, URL http://www.apme.org, 1995.
32. Patel, M.; Bastioli, C.; Marini, L.; Würdinger, E. In *General Aspects and Special Applications*; Steinbüchel, A., Ed.; Biopolymers; Wiley: Weinheim, Germany, 2003; Vol. 10, pp 409-452.
33. Zhang, Z.-Y.; Xu, Y.-P. *Solar Energy* **2001**, *70*, 299-303.

Chapter 18

Ecological Evaluation of Processes Based on By-Products or Waste from Agriculture: Life Cycle Assessment of Biodiesel from Tallow and Used Vegetable Oil

Anneliese Niederl and Michael Narodoslawsky

Institute for Resource Efficient and Sustainable Systems, Graz University of Technology, Inffeldgrasse 21B, A–8010 Graz, Austria

Life cycle assessment (LCA) is an increasingly important evaluation tool for decision making and stakeholder discussion. New methodological aspects arising with the application of LCA on products from renewable resources that are by-products or waste of other processes are reflected on the basis of a LCA of biodiesel from tallow (TME) and used vegetable oil (UVO). Two different impact assessment methods (Sustainable Process Index SPI and CML-method) show largely concordant results. The SPI gives "a bigger picture" of the environmental impacts and will be helpful in decision making. Depending on the setting of system boundaries the ecological footprint (SPI) is between -1,2 to 4,8 m^2a/MJ for UVO, between 0,85 to 8,3 m^2a/MJ for TME compared to 26,1 m^2a/MJ for fossil diesel.

Introduction

Processes based on renewable resources always have an "intrinsic" perception of being environmentally friendly and sustainable. However, as this property is a major advantage for products generated by these processes there is a necessity to prove their sustainability credentials in a rigorous manner that can withstand the scrutiny of a competitive market.

Life cycle assessment (LCA) has turned out to be a good basis for product and process evaluation in order to account for all environmental impacts incurred by the provision of the good in question. Although strict standards for LCA are laid down in ISO standards of the 14.00X family (*1*), it must be noted that these standards only provide a procedural guide to evaluation and do not prescribe a fixed evaluation method.

On top of this LCA of processes and goods on the base of renewable resources are bound to face special methodological challenges. Firstly, many industrial raw materials are by-products or surplus products from agricultural activities leading to other (more valuable) products. In contrast to conventional, for example fossil, resources we do not face linear value chains but more complex production networks. In these cases the general problem of allocating the pressures of the agricultural sector arise (as agricultural production is not driven by generating the by product that is utilised for these products). This may considerably influence the outcome of any valuation. In some cases the raw materials are even streams that are considered as waste, which makes a prudent valuation even more complicated. In the following case-study this challenge is reflected on the base of the production of biodiesel from tallow, a by-product from meat production, and biodiesel from used vegetable oil, that has already gone through a usage step and therefore represents a genuine waste-material.

The second challenge that has to be faced is the sustainability evaluation of processes leading to the same sort of goods on the base of different raw materials. This valuation must account for the different impacts from different raw material generations. Especially the difference between renewable and depletable raw material systems must be evaluated. Most of the LCA are based on the problem oriented approach to impact assessment (Centrum vor Milieukunde Leiden, CML-method) (*2,3*) resulting in various impact categories. They provide a reasonable communication and discussion tool for questions like which environmental problems are caused to which extent over the life cycle of a product. But in many "real world cases" improvements on one front, like reduction of greenhouse gas emissions, tend to lead to disadvantages in other areas. Decision makers or engineers want to know about these trade-offs and are therefore asking for a high degree of aggregation which allows to valuate one impact versus another. Instead of eco-indicators (e.g. EcoIndicator 99 (*4*)) that are based on weighing factors set by experts that are thus not undisputable, a tool for engineering purposes must have a more generally acceptable methodological base. The Sustainable Process Index (SPI) (*5*), a member of the ecological footprint family, offers this feature and has therefore been applied in the case study along with the CML-method to identify possible improvements along the life cycle.

Case Study - Scope Definition

In the last few years biodiesel has become a hot topic, which is impressively demonstrated by the large number of LCAs published to deal with this issue (*6-13*). The findings of these studies - although some of them compare the same life cycle, e.g. biodiesel from rape seed oil – are not always in accordance. One reason is that although these authors look at the same life cycle, they often cut out different scopes for their analysis.

When starting a Life Cycle Assessment the first task is the definition of goal and scope. As already mentioned the setting of system boundaries for products based on renewable resources is not straightforward. In order to exemplify the influence of the question "what shall be included?" on the result of the assessment three different scenarios will be compared for the life cycle of biodiesel from tallow and used vegetable oil (see Figure 1).

The function of the products biodiesel from tallow (TME – tallow methyl ester) and biodiesel from used vegetable oil (UVO) is to serve as fuel for combustion in (unspecified) motor vehicles. The functional unit used to quantify this function is the combustion energy (calorific value) of biodiesel. The reference flow is 1 MJ combustion energy.

The scope of this LCA study considers the provision of the product biodiesel from tallow (TME) and biodiesel from used vegetable oil (UVO) from raw material extraction to the usage of the finished product (resulting in vehicle emissions caused by fuel combustion). The scope here is therefore a "cradle-to-wheel" LCA. The production of energy, raw materials and auxiliary materials is included as is the waste disposal and the treatment of liquid and gaseous emissions during all steps of the life cycle. The production and operation of the infrastructure needed in the provision of the function is excluded from the LCA as it has turned out to be of minor influence (see Figure 1).

Starting point for the development of the different inventory scenarios is the transesterification process. The scenarios only differ in the generation of the raw material, since the provision with process chemicals and process energy as well as the end use (fuel delivery and fuel combustion) are assumed to be identical.

- Scenario I: This scenario addresses the production of biodiesel from used vegetable oil (UVO). The first step in the life cycle of the production of biodiesel from UVO is the collection of waste cooking oil, causing transport. Then this collected raw material is processed in a

242

transesterification step to yield biodiesel. A further transport unit (fuel delivery) is included before the biodiesel is burned in an engine. Scenario I also accounts for the life cycle of biodiesel from tallow, with the assumption that tallow, which comes out of the rendering plant, is a waste material. This means that, comparable to the life cycle of UVO a transport unit (here from the rendering plant to the biodiesel plant) stands at the beginning of the life cycle of biodiesel from tallow.

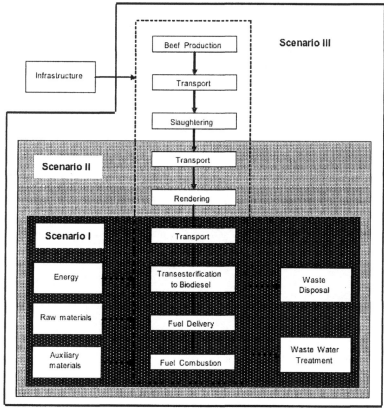

Figure 1. Different System Boundaries in the Life Cycle of Biodiesel from Tallow and Used Vegetable Oil

- Scenario II: In contrast to the previous setting, the system boundaries for biodiesel from tallow are set outside the gate of the slaughtering house. Rendering products leaving the slaughtering process are considered as waste-stream comparable to UVO in scenario I. These render products are further processed in the rendering plant to yield meat-and-bone-meal and tallow. Tallow is then transported to the biodiesel-plant and transesterified. Scenario III: The slaughtering process is above all carried out to produce meat. Without the returns from meat sale the slaughtering process and the upstream processes related to meat production would not be profitable. Without meat sales there would be no upstream processes and no slaughtering. Still, in order to investigate the influence of the agrarian chain for biodiesel from tallow, Scenario I includes the slaughtering process and the production of cattle, fodder and fertilizers together with all its connected processes in the agrarian sector. We will see that in this scenario allocation will define the outcome of the analysis.

Another tricky question arises when facing life cycles of multi-output processes. Inputs, outputs and the related environmental impacts must then be allocated to products. This can be done according to physical properties of the product flows (mass or energy flows). If this is not possible or not justifiable the usual way is to allocate according to the economic value of the products (prices). In our study, both mass and price allocation have been applied to highlight the influence of this allocation procedure on the outcome of the analysis.

Table I. Multi-Output Processes in the Production of Biodiesel from Tallow

Process	Outputs
Slaughtering Process	Meat after Slaughtering
	Render Products
	Hides
	Offal
	Rest
Rendering Process	Tallow
	Meat and Bone Meal
Transesterification	Biodiesel
	Glycerol
	K_2SO_4
KOH-Production	KOH
	Chlorine Gas
	Hydrogen

Table I shows the multi output process of the biodiesel from tallow model. Four processes have more than one economic output. For the life cycle of biodiesel from used vegetable oil only the two letter processes (transesterification and KOH-production) are relevant.

The slaughtering process produces meat (36.2 %), render products (22.5 %), hides (8.3 %), offal (3.2 %) and rest (29.8 %) (*14*). Mass allocation ("render products are held responsible for the environmental impact of the value chain to the same extent as meat") might seem unjust vis-à-vis the co-products of the slaughtering process. Therefore price-allocation has been applied. The following prices have been assumed for the products: meat: 3.6 €/kg, hides: 1.2 €/kg, offal: 1.3 €/kg, render products: 0.08 €/kg.

The share of the agrarian upstream processes allocated to biodiesel production is reduced considerably to 1.2 % when price allocation is applied compared to 32 % with mass allocation. The lion's share of the upstream environmental burden is then allocated to the main product meat (see Table II).

Table II. Comparison between different Allocation Models for the Slaughtering Process

	Mass Allocation [%]	Price [€/kg]	Price Allocation [%]
Meat	51.5	3.6	89.0
Hides	12.0	1.2	6.9
Render Products	**32.0**	0.08	**1.2**
Offals	4.5	1.3	2.8

The rendering process produces fat (24 %), meat and bone meal (21 %) and water (55 %) from 100 % render products at the input side. Upstream processes have to be allocated to the two products that may be further processed. At present in Europe, meat and bone meal (MBM) are not used to produce goods of economic value. With this in mind, one could argue that all upstream environmental impact should be allocated to the only output of market value, the fat. But the rendering process is not only carried out to gain fat. The production of MBM is desired – if not economically then at least from the hygienic point of view. This makes an allocation justifiable. As a market price is not available (or even negative for MBM), mass allocation has been carried out.

Potassium hydroxide is a by-product of chlorine production in the chlor-alkali process. Its low relative market value (chlorine 37.5 €/kg; H_2 123.8 €/kg; KOH 0.65 €/kg) makes economic allocation seem sensible. The process would run without the KOH by-product but not for KOH production exclusively.

Glycerol is a marketable co-product of the biodiesel transesterification process. Therefore, mass allocation can be justified. Due to similar prices for biodiesel and glycerol, economic allocation would only yield a slightly changed picture.

In literature it is sometimes suggested to avoid allocation by system boundary expansion. This method balances co-products through substitution processes (15). This means that the main-product biodiesel gets credit for the co-products glycerol and K_2SO_4, because these co-products can substitute equivalent products and therefore energy consumption and emissions, respectively. (see Figure 2)

The equivalent product to biodiesel-glycerol is synthetic glycerol that is petrochemically produced. Although the main market share of glycerol comes from natural sources, natural glycerol is an unsuitable equivalent to glycerol from the transesterification process, because most of it is - like biodiesel-glycerol – a by-product (for example from the production of organic tensides). Fossil based glycerol is produced to make up the difference between glycerol produced as by-product from various processes and would therefore be substituted by any new side product glycerol.

K_2SO_4 from the transesterification process is used as fertilizer. Therefore, the production of K-fertilizer was chosen as substitution process.

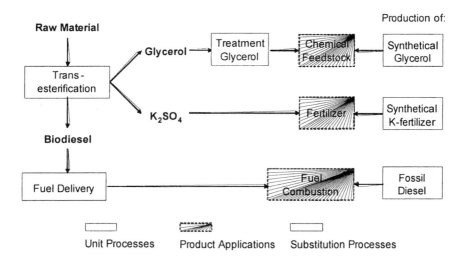

Figure 2. Substitution Processes

In order to be able to compare biodiesel with fossil diesel 1 MJ of combustion energy was chosen as reference flow. The data for fossil diesel were taken from (*16*). For details on the data used in this case study see (*17*).

Case Study - Impact Assessment

Two methods of impact assessment are applied in this study. First, the CML method of effect categories (problem oriented approach) is suggested in the ISO-standards. For the problem oriented approach only the characterisation step is carried out, normalisation and weighting are omitted.

As a second impact assessment method the Sustainable Process Index method (SPI) has been chosen. The SPI (*5,18*) a measure of ecological sustainability, expresses pressure on natural systems as area needed to embed the process sustainably into the ecosphere. It is a measure of the environmental footprint (EF) family, though methodologically different from the EF. According to the SPI methodology, pressure is put on ecosystems by flows of matter and energy transgressing the boundaries between natural and anthropogenic systems. Input flows of (renewable and non-renewable) resources as well as output flows of (solid, fluid and gaseous) emissions result in overall pressure. In addition to flows that are converted to area appropriation, "direct" area use in the form of sealed ground (due to e.g. installations, roads) is considered. The reference for ecological sustainability of the activities assessed is the (geographical) area available to embed the activities. In the particular case of national and regional accounting the reference value is given by the surface of a region or a nation. This surface can be used to generate the resources needed for the domestic and foreign (transboundary flows) productive and consumptive activities and to absorb the residuals from such domestic and foreign activities. The area available for the absorption of local and regional emissions is, in most cases, land surface. The absorption of global emissions, such as CO_2, is not limited to land area. Global emissions are absorbed by the oceans as well, and therefore, sea area has to be included in the calculation of the sustainability reference area for such emissions. The SPI assumes that when the area needed for the generation of resources and the absorption of residuals exceeds the area available, an activity or a number of activities is not sustainable. For details on the calculation procedure of the SPI see (*5*).

Case Study - Results

The comparison of the two different impact assessment methods is made on the basis of Scenario I, the production of biodiesel from waste cooking oil without the consideration of substitution processes. Since SPI and CML results cannot be directly compared the main focus of investigation is to discuss if both methods point to the same environmental problems arising over the life cycle and identifying the process steps contributing most prominently to the ecological impact.

Starting point for the analysis of the different stages of the biodiesel life cycle is again the transesterification process with its inputs, process energy and process chemicals. The other two stages of the life cycle are classified as transport, considering both collection of the raw material and fuel delivery, and the combustion of biodiesel (see Table III).

Table III. Comparison between different impact assessment methods of biodiesel from used vegetable oil

	fuel combustion	*process energy*	*process[1] chemicals*	*transport[2]*	*total*	*fossil diesel*
	[%]	[%]	[%]	[%]	\multicolumn{2}{c}{absolute values}
SPI	42.5	18.0	22.0	17.8	4.78	28.06
GWP	29.1	20.7	40.5	9.7	1.8E-2	9.0E-2
AP	85.8	4.2	7.8	2.21	2.1E-4	2.3E-4
EP	94.8	2.4	1.4	1.4	3.3E-5	2.2E-5
ADP	0.7	67.1	2.0	30.2	3.7E-5	5.4E-4
POCP	88.5	2.5	5.8	3.2	1.2E-5	1.9E-5

[1] Raw materials (apart from UVO) and auxiliaries used in transesterification process

[2] Transport includes the collection of used vegetable oil and fuel delivery

SPI (Sustainable Process Index) in $m^2 a$

GWP (Global Warming Potential) in kg CO_2 eq yr^{-1} MJ^{-1}

AP (Acidification Potential) in kg SO_2 eq yr^{-1} MJ^{-1}

EP (Eutriphication Potential) in kg PO_4 eq yr^{-1} MJ^{-1}

ADP (Abiotic Depletion Potential) in kg antimony eq yr^{-1} MJ^{-1}

POCP (Photooxidant Creation Potential) in kg ethylene eq yr^{-1} MJ^{-1}

The main impact to climate change is due to process chemicals. Methanol from fossil resources used at this stage accounts for 40 % of the overall global warming potential caused by biodiesel, whereas the impact of other chemicals, like KOH or H_2SO_4, can be neglected. The same holds true for the outcomes of the SPI. 22% of the total ecological footprint can be contributed to chemicals used during transesterification. The main share that makes up for 21% of the overall footprint is due to the depletion of fossil resources for methanol usage.

The largest factor of the overall footprint (42.5%) comes from the combustion of biodiesel itself causing high emissions of nitrogen oxides. This is reflected in the problem oriented approach with the high contribution of combustion emissions to the acidification, eutriphication and photooxidant creation potential of 85.5 %, 94.8 % and 88.5 %, respectively.

Both for the provision of process energy and for transport, the depletion of abiotic resources is the predominant impact category. Of course, the still prevalent use of fossil resources like diesel, charcoal, heavy and light fuel oil in energy provision systems is to be blamed, thus also contributing significantly to greenhouse gas emissions. It is noteworthy, that the SPI also takes into account radioactive substances emitted in the production of nuclear power. For example, according to the UK energy mix (37.3 % natural gas, 33.9 % mineral coal,

23.2 % nuclear power, 3.9 % hydro power, 1.6 % oil and 0.2 % brown coal) nuclear power would cause 41 % of the electricity's footprint.

As far as the absolute values of biodiesel compared to fossil diesel are concerned, both assessment methods show similar trends. Under the terms of the problem oriented approach biodiesel performs better in all impact categories except for the eutriphication potential (see Table III). The ecological footprint of fossil diesel is approximately the sixfold of the biodiesel's footprint.

For the comparison of the different scenarios, allocation methods and the system boundary expansion including substitution processes an aggregated evaluation method is very useful and therefore only the SPI of these different models will be presented at this point.

The distribution of the overall ecological footprint on the four subgroups of biodiesel production and usage in Scenario I (UVO and TME I) have already been discussed above. With the expansion of the system boundaries to scenario II (the life cycle of biodiesel from tallow starts at the gate of the slaughterhouse with a transportation step of render products to the rendering plant) a fifth life cycle stage accounting for the impact of the raw material used in the biodiesel process is introduced (see Figure 3). The footprint of the stages combustion (2.03 m²a/MJ), process energy (0.86 m²a/MJ) and process chemicals (1.06 m²a/MJ) is constant for all scenarios. This is also valid for the transport-stage (0.84 m²a/MJ), that comprises fuel delivery and the collection of raw material. For a better comparability the transport distances for UVO-collection and the distance from the rendering facility to the biodiesel plant are assumed to be each 100 km, which is close to realistic rates.

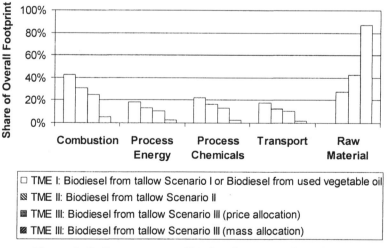

Figure 3. Different stages of biodiesel usage and production

In scenario II the raw material provision accounts for 28 % of the overall footprint of 6.6 m²a/MJ (for absolute values see also Figure 4). This is roughly the same share that combustion emissions have in scenario II. The major impact at the raw material stage has to be ascribed to the transport of render products to the rendering plant. Only about a fifth is due to the rendering process itself.

Including the whole agrarian chain that is necessary for cattle breeding leads us to Scenario III. Meat production is a very intensive process with high environmental impact. Therefore, the result of the footprint for this scenario is very much dependent on the allocation method applied at the slaughterhouse.

When allocating the render products that come out of the slaughterhouse by mass compared to meat, hides or other by-products, the agrarian chain has the most important environmental impact for TME. Price allocation leads to a reduction in the overall footprint per MJ energy resulting from combustion of TME of 37.79 m2a/MJ to 8.28 m2a/MJ. The absolute footprint (as shown in Figure 4) of biodiesel from tallow in scenario III with mass-allocation applied would be higher than fossil diesel.

Analysing scenario III the agrarian chain consists of the slaughtering process and cattle-breeding. Here the cattle-breeding plays an outstanding role in the production of biodiesel from tallow, where the main part of the cattle-breeding footprint can be attributed to fodder-production and emission of ammonia.

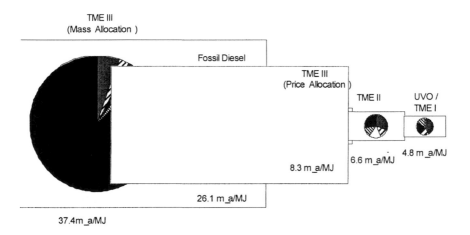

Figure 4. The Sustainable Process Index of different scenarios for Biodiesel compared to Fossil Diesel

In the case of system boundary expansion and introduction of substitution processes no allocation step in the transesterification process is applied. In order to be able to compare glycerol from the biodiesel process with synthetic glycerol it needs to be purified in a distillation step. This preparation accounts for 0.06 m²a/MJ and is added to the footprint of biodiesel. Whereas the footprint of the substitution processes is subtracted from the overall footprint. The equivalent process of synthetic glycerol production plays an outstanding role with a footprint of 6.36 m²a/MJ. Together with the substitution process potassium-fertilizer production (0.09 m²a/MJ) the overall footprint for biodiesel from used vegetable oil gets a negative value of -1.24 m²a/MJ. In the same extent the footprint for biodiesel from tallow is lowered to 0.85 m²a/MJ for scenario II and to 2.79 m²a/MJ and 36.24 m²a/MJ for scenario I according to price and mass allocation, respectively.

Conclusion

The outcomes of a LCA are very much dependent on the setting of system boundaries especially when considering side products from agriculture, as it was clearly shown in the example of biodiesel from tallow and waste cooking oil. For the environmental assessment of biodiesel from tallow it is crucial to decide whether tallow is a by-product of meat production or whether such very low-value products like render products should be considered as waste. It can be argued that the agricultural upstream process – with their costs and environmental impact – would not be undertaken alone for the production of by-products such as tallow. Mass allocation of the agricultural upstream processes yields an environmental impact that might be prohibitive to the production of biodiesel from tallow from an ecological point of view. Price allocation changes the environmental impacts of biodiesel from tallow considerably. In this case, prices reflect the interest of society to have meat and tallow produced and hence the driving force for this production.

A further large influence on the results of LCA can be found in the introduction of substitution process when expanding the system boundaries. In the case of biodiesel from used vegetable oil the ecological footprint is even shifted to a negative value from 4.8 m²a to -1.2 m²a per MJ combustion energy.

Regardless the setting of system boundaries, assuming that tallow should not be made responsible for the environmental impact of the agrarian chain to the same extent as meat, biodiesel from tallow and used vegetable oil have a lower ecological footprint than fossil diesel. Biodiesel from tallow and waste cooking oil perform better compared to other sorts of biodiesel that are not based on by-products or waste, but, for example, on rape seed oil. The ecological footprint for RME was calculated as 10.3 m²a/MJ based on data from Reinhardt (9). This LCA also includes substitution processes such as soy meal or glycerol production and therefore needs to be compared with -1.2 m²a/MJ UVO, 0.85 m²a/MJ TME II and 2.79 m²a/MJ TME III.

As far as the impact assessment with the two different methods CML and SPI are concerned the results show concordant trends. The CML method addresses with its impact categories the various environmental problems in detail and is capable of highlighting the trade-offs between them. In contrast to this the Sustainable Process Index is aggregated across different impact categories and allows comparison of these effects based on natural flows and natural qualities.

From the application point of view these two assessment methods both have their distinctive advantages. The CML method clearly focuses on well known environmental problems and valuates the contribution of a given life cycle to these problems. This view is especially important for the analysis of legal compliance of different steps within the life cycle in question as well as for the discussion between different stakeholders, namely environmental organisations, government agencies and business.

The SPI has the advantage to offer a clearer picture on the overall impact of a life cycle, allowing for direct comparison of different alternatives. On top of that the SPI allows valuation of trade-offs for different technological as well as organisational alternatives, thus supporting technology design and life cycle optimisation. As the data necessary for the SPI evaluation are generally known at a very early state of design, costly design failures may be avoided and the general set-up of a certain life cycle may be optimised.

An important feature of the SPI is that it focuses the attention of designers and decision makers on the most pressing environmental questions, which usually arise as a combination of different impacts emerging from a specific step in the life cycle. This guarantees on the one hand that problematic sections of a life cycle get the right scrutiny. On the other hand it saves crucial development resources as the main work can be dedicated to those sections of a life cycle that really matter.

Concluding we may say that not the absolute values of any assessment count, but Life Cycle Assessment should be a participatory process, as it is a good communication and learning tool. By assessing different scenarios a systematic discussion built on agreed upon assumptions with stakeholders and decision makers about optimal solutions for sustainable development becomes possible.

References

1. EN ISO 14040. *Life Cycle Assessment* 1997.
2. *Life Cycle Assessment, An operational guide to the ISO standards* Guinée, J. B.; Gorrée, M.; Heijungs, R.; et. al. 2001.
3. Heijungs, R. *CMLCA 4.0 Chain Management by Life Cycle Assessment Short Manual* Centre of Environmental Science; Leiden University 2003.

4. PreConsultants *SimaPro 5 User, Version 2.0, Manual* Amersfoort, 2002.
5. Krotscheck, C.; Narodoslawsky, M. *Ecol. Eng.* **1996**, 6, 241-258.
6. Ceuterick, D.; De Nocker, L.; Spirinckx, C. *Comparative Life Cycle Assessment of Biodiesel and Fossil Diesel Fuel* 2000.
7. Groves, A. P. *Well to Wheels assessment of Rapeseed Methyl Ester Biodiesel in the UK.* 2002.
8. Kaltschmitt, M.; Reinhardt, G. A.; Stelzer, T. *Biomass Bioener.* **1997**, 12, 121-134.
9. Reinhardt, G. A. *Gutachten: Ressourcen- und Emissionsbilanz: Rapsöl und RME im Vergleich zu Dieselkraftstoff* Institut für Energie- und Umweltforschung Heidelberg GmbH - ifeu 1997.
10. Gärtner, S. O.; Reinhardt, G. A. *Erweiterung der Ökobilanz für RME* 2003
11. Scharmer, K. *Biodiesel-Energie- und Umweltbilanz Rapsölmethylester* Union zur Förderung von Öl- und Proteinpflanzen E.V.: 2001.
12. Sheehan, J.; Camobreco, V.; Duffield, J.; et. al. *Life Cycle Inventory of Biodiesel and Petroleum Diesel for Use in an Urban Bus*, 1998.
13. Jungmeier, G.; Hausberger, S.; Canella, L., *Treibhausgas-Emissionen und Kosten von Transportsystemen, Vergleich von biogenen mit fossilen Treibstoffen* Graz, Austria, 2003.
14. Riedl, C. Master thesis Graz University of Technology, AT 2003.
15. Weidema, B. *J. Ind. Ecol.* **2001**, 4, 11-33.
16. ESU-ETHZ *Ökoinventare von Energiesystemen* Zürich, CH, 1996
17. Niederl, A.; Narodoslawsky, M. Life Cycle Assessment as an engineer's tool? PRES 04, 2004; pp. 1360.
18. Narodoslawsky, M.; Krotscheck, C. *J. Haz. Mat.* **1995**, 41, 383-397.

Chapter 19

Polyhydroxyalkanoate Production in Crops

Gregory M. Bohlmann

Process Economics Program, SRI Consulting, Menlo Park, CA 94025

Polyhydroxyalkanoates (PHAs) are biodegradable polyesters produced by microorganisms as intracellular energy reserves. The metabolic pathway from these microorganisms can be bioengineered into a variety of plants for making PHAs. While this new scheme is not yet commercial, it may have the potential for large scale manufacture at very low cost. Numerous challenges, both technical and non-technical, are associated with commercializing this technology. One is to achieve a high level of polymer production in the plant without a decrease in crop yield. Another is to economically recover the polymer from the plant biomass. There are also barriers associated with utilization of agricultural infrastructure for production of industrial products. This paper presents an analysis of the process economics for producing PHAs in agricultural crops such as soybean or switchgrass. The economics are compared to those for PHA production by *E. coli* fermentation.

Introduction

Biodegradable polyhydroxyalkanoates (PHAs) are synthesized by a wide variety of bacteria as a carbon reserve and electron sink. Bacterial fermentation has been employed as a commercial means of producing PHAs since their introduction in 1981. The metabolic pathways from these microorganisms can be bioengineered into a variety of plants for making PHAs. This new scheme for producing PHAs is not yet commercial. However, there has been significant research undertaken by government, academic and commercial organizations over the past decade to develop the technology. The main rationale for the synthesis of PHA in plants is the potential for producing the polymer on a large scale at a cost lower than bacterial fermentation. One company active in this field hopes to drive PHA cost down to under $1/lb using bioengineered plants (*1*).

Over 100 different monomers have been found to be included in bacterial PHAs and the metabolic pathways involved in the synthesis of a variety of these pathways have been elucidated by research scientists. Through bioengineering, it is possible to modify plant cells to incorporate PHA metabolic pathways from bacteria.

The chemical diversity of PHA translates into a wide spectrum of physical properties, ranging from stiff and brittle plastics to softer plastics, elastomers, rubbers and glues. The major diversity is in the length and presence of functional groups in the side chain of the polymer. Table 1 summarizes selected important PHAs.

Table I. Selected PHAs

PHA	Side Chain
Polyhydrobutyrate (PHB)	$-CH_3$
Polyhydoxyvalerate (PHV)	$-CH_2CH_3$
Polyhydroxybutyrate-valerate (PHBV, Biopol)	$-CH_3$ and $-CH_2CH_3$
PHBHx (Kaneka)	$-CH_3$ and $-CH_2CH_2CH_3$
PHBO (Nodax)	$-CH_3$ and $-(CH_2)_4CH_3$
PHBOd	$-CH_3$ and $-(CH_2)_{14}CH_3$

Source: Reproduced with permission from reference 2. Copyright 2003 Wiley.

PHB is a stiff and brittle polymer, so a variety of other copolymers have been developed that have physical properties more suitable for commercial applications. Copolymerization of 3-hydroxybutyrate with 3-hydroxyvalerate or with medium chain length hydroxyalkanoate units is effective for improving the brittleness.

One of the first pioneer companies to develop a biodegradable polymer that was completely biodegradable and also have good properties was ICI in the United Kingdom with its PHA product known as BIOPOL™, commercialized in 1981. Monsanto purchased the BIOPOL™ business from Zeneca Bio Products (formerly ICI) in April of 1996. Monsanto manufactured PHBV at its Knowsley, England fermentation facility until 1999. At that time Monsanto elected to exit the biodegradable polymers business after failing to make BIOPOL™ costs more competitive with petroleum based polymers. Monsanto indicated that it cost roughly $8.8 to make a kilogram ($4/lb) of PHBV through fermentation and that its goal of using bio-engineered crops to lower production costs was at least 5-7 years away (3).

In 2001 Metabolix purchased Monsanto's Biopol patent assets to add to its PHA technology portfolio. Metabolix is a research company that does not own or operate commercial facilities for PHA production. In 2002 the company's fermentation technology was demonstrated at the 50,000 liter scale to produce PHA. Metabolix indicates that potential production costs are well under $1/lb for its fermentation process (4). Metabolix also has R&D efforts towards bio-engineering plants to produce PHAs. The efforts are partly funded by the DOE. In 2003 BASF and Metabolix entered into a research collaboration agreement on PHAs. Metabolix will produce PHA from sugar using fermentation technology and supply BASF with pilot scale quantities to investigate the material properties. Metabolix indicates that planting of large field acreages for production from crops is likely to be at least five years away (5). Metabolix hopes to drive PHBV cost down to under $1/lb using bioengineered plants (1).

Procter & Gamble has developed a family of PHAs known as Nodax™. P&G plans to license different aspects of the production and application of Nodax. P&G and Kaneka Corp. in Japan are jointly developing PHAs, including poly(3-hydroxybutyrate-co-3-hydroxyhexanoate) (PHBH). The companies are producing PHBH at a rate of 50 kg/week at two locations in the United States and are considering sites in Europe and in Asia for a 30,000 metric tons per year plant (6).

Metabolic Pathways

Over 40 bacteria harbor the genes which are related to PHA biosynthesis. In addition, more than 200 microorganisms can use PHAs as an energy source, which makes PHA entirely biodegradable in a wide range of environmental conditions ranging from soil to sea water. The wealth of knowledge concerning the various bacteria which can biosynthesize PHA is a valuable starting point for bioengineering plants towards this end. Bacteria synthesizing PHAs have been broadly subdivided into two groups based on monomer chain length. Metabolic pathways have been identified for both major categories of PHAs:

- **Short chain length (SCL) PHAs** that contain 3-hydroxyacid monomers ranging from three to five carbons in length. PHB and PHBV are both included in this category. The most prominent bacteria with a SCL metabolic pathway is *Ralstonia eutropha.*
- **Medium chain length (MCL) PHAs** that contain 3-hydroxyacid monomers ranging from six to sixteen carbons in length. A number of Pseudomonads have MCL metabolic pathways.

The division between SCL and MCL PHA is mainly determined by the substrate specificity of the polymerization enzyme in the organism. The division is not strict since several bacteria have been found that can synthesize a "hybrid" PHA that can include monomers from four to eight carbons (7).

A number of enzymes and metabolic pathways have been identified by scientists. PHA synthases are the key enzymes of PHA biosynthesis in all of these pathways. These enzymes catalyze the covalent linkage between the hydroxyl group of one and the carboxyl group of another hydroxyalkanoic acid. The substrates of PHA synthases are the coenzyme A thioesters of hydroxyalkanoic acids. There is no evidence that PHA synthases can utilize either free hydroxyalkanoic acids or other derivatives. Most PHA synthases incorporate either short carbon chain length hydroxyalkanoic acids with 3-5 carbon atoms or medium carbon chain length hydroxyalkanoic acids with 6-16 carbon atoms. There are only a few PHA synthases which can incorporate both short chain and medium chain monomers. Examples are the PHA synthases of *T. pfennigii* and *Aeromonas caviae* (8).

PHB is the most widespread and thoroughly characterized PHA found in bacteria. A large part of the current knowledge on PHB biosynthesis has been obtained from *Ralstonia eutropha* (formerly *Alcaligenes eutrophus*) (7). In this bacterium, PHB is synthesized from acetyl-coenzyme A (CoA) by the sequential action of three enzymes (see Figure 1):

- 3-Ketothiolase catalyzes the reversible condensation of two acetyl-CoA moieties to form acetoacetyl-CoA. The acetyl-CoA metabolite is naturally available in many plants and bacteria.
- Acetoacetyl-CoA reductase subsequently reduces acetoacetyl-CoA to R-(-)-3-hydroxybutyryl-CoA.
- PHA synthase polymerizes R-(-)-3-hydroxybutyryl-CoA to form PHB.

Transgenic Plants

Advances in plant genetic engineering, combined with growing concerns about the environment and decreasing petroleum reserves have created new interest to use plants for the large scale production of renewable chemicals and polymers. Good examples of commodity chemicals produced efficiently in plants are starch and oils, which both can be harvested for food and non-food uses. Recent progress in molecular biology and plant transformation has enabled the creation of plants that have the capacity to produce new industrially useful products that are not naturally found in plants.

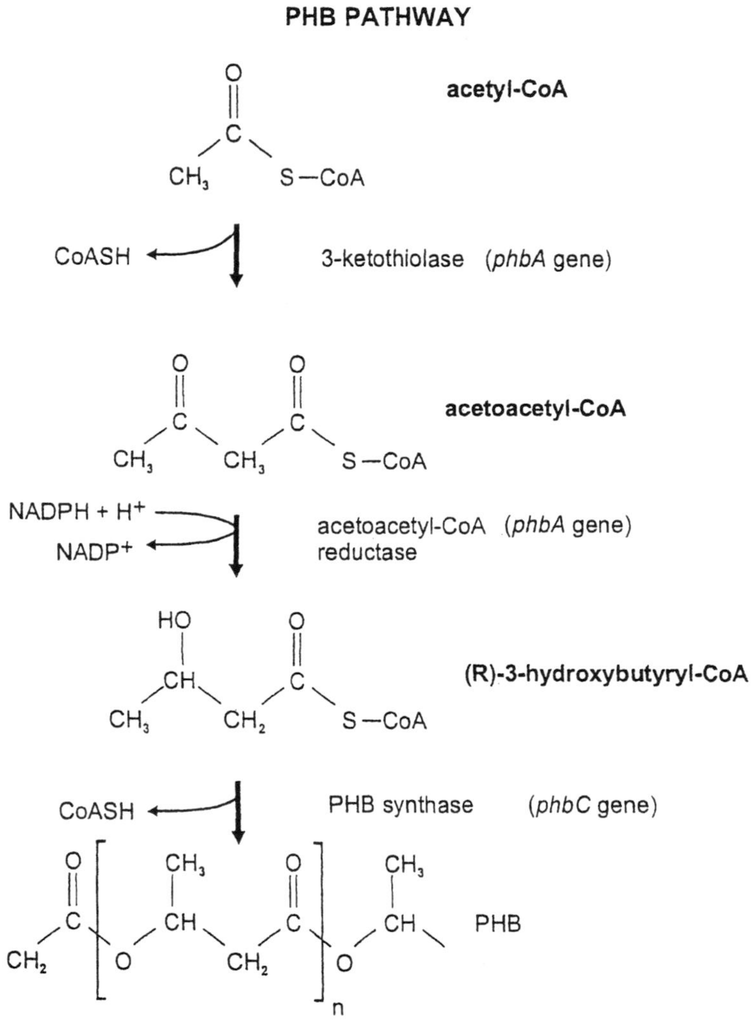

PHB PATHWAY

Figure 1. PHB Pathway

The challenge of using plants for bioengineered products such as PHAs is to achieve a high level of production within the plant without a decrease in crop yield. This must also be accompanied by the means for efficient, economical extraction of PHA from the plant. Companies in this field expect that polymer concentrations in plants will need to reach at least 15% of dry weight for economical production to be feasible (10). These objectives are accomplished through matching the appropriate plant with the desired metabolic pathway. A common substrate used in bacterial pathways for PHA is acetyl-CoA. This intermediate is also found in certain plant pathways. Acetyl-CoA is present in plant cells in several subcellular compartments including the cytosol, plastid, mitochrondrian and peroxisome. By incorporating a PHA metabolic pathway, the synthesis of PHA can, in theory, be achieved in any of these subcellular compartments.

In addition to the technical and economic challenges of commercializing PHA production in plants, there are also other large barriers. One of these barriers is the need for keeping PHA crops from spreading its genes to a compatible non-PHA plant. In assessing the risks posed by various PHA crops, the worst possible plant with regards to confining its genes would be one that:

- routinely breeds with related crop varieties
- produces large amounts of pollen and seed (and the seeds are particularly small)
- serves as an important food and feed crop
- spontaneously mates with wild relatives and
- is widely planted throughout the world.

One example of a crop like this is maize (corn). If this species were to be planted, great lengths would have to be taken to avoid contamination, especially in the U.S. corn belt. Modified corn intended for pharmaceutical use is greatly restricted on this basis. However, other specialty corn crops are commercially grown in the United States that require identity preservation management practices. This value-added market is small, but an increasing part of crop production in the U.S.

Even as the agbiotech industry embarks on innovations such as the use of crops to produce pharmaceuticals or chemicals, it continues to face concerns from the public about the technology. Several food industry groups in the U.S. have urged the government to halt genetically engineered crops until stricter regulations can be put in place to prevent contamination of food crops (11). In 2003, the Union of Concerned Scientists (UCS) convened an expert workshop on protecting the U.S. food and feed supply from contamination by crops genetically engineered to produce pharmaceuticals and industrial chemicals. One recommendation from the UCS workshop is for the USDA to spearhead a campaign to fund alternatives to the use of food and feed crops in pharma and industrial crop production, particularly the search for suitable non-food/feed crops.

During the initial research in this field, oilseed crops were hypothesized to be the ideal plant system for bioengineered PHA production because their fatty acid biosynthesis pathways are highly efficient. Synthesis of PHA in plants was first demonstrated in 1992 in the *Arabidopsis thaliana*. Although of no commercial agricultural value, this plant was chosen because it could be easily transformed through bioengineering and its short life cycle (approximately 2 months from seed to seed) enabled the rapid creation of hybrid plants. *A. thaliana* is a plant that accumulates approximately 40% lipids in its seeds and thus can be used as a good model for the synthesis of PHA in major oil crops. Over the 1990s, scientific research into PHA production in plants has expanded to include a number of species as shown in Table II.

Table II. Summary of Transgenic Plants Producing PHAs

Plant Species	Subcellular PHA compartment	Tissue	Type PHA	PHA %dwt
Arabidopsis thaliana	plastid	shoot	PHB	14-40
	plastid	shoot	PHBV	1.6
	peroxisome	whole plant	MCL PHA	0.6
Alfalfa	plastid	shoot	PHB	0.2
Corn	plastid	shoot	PHB	6
Cotton	cytoplasm	fiber	PHB	0.3
	plastid	fiber	PHB	0.05
Maize	peroxisome	cell suspension	PHB	2
Potato	plastid	shoot	PHB	0.02
Rapeseed	cytoplasm	shoot	PHB	0.1
	plastid	seed	PHB	8
	plastid	seed	PHBV	2.3
Tobacco	cytoplasm	shoot	PHB	0.01
	plastid	shoot	PHB	0.04

Source: Reproduced with permission from reference 12. Copyright 2002 Wiley.

Choice of the plant host and choice of the location within the plant for PHA synthesis can have a strong effect on yield as well as recovery and purification methods required. Also, many of the row crops listed in Table II suffer from a reduction in plant growth and fertility when bioengineered to produce PHAs. Grasses, which have particularly efficient mechanisms for the assimilation of carbon, have recently been studied as viable alternatives to row crops. For example, researchers in Australia have investigated transgenic sugar cane as a plant-based bioreactor for producing PHAs (*13*).

Determining the best crop system to use for commercially producing PHAs is a complicated matter. When Monsanto was exploring the use of an oilseed crop, there seemed to be apparent benefits such as efficiency of the plant metabolics. Concerns over using a food crop for such a purpose is leading to other options. Metabolix received a five year DOE award of $7.4 million to expand development of PHA production directly in switchgrass, a perennial plant that grows on marginal land. It is also being studied by the DOE as a potential energy crop. Table III compares soybean versus switchgrass as bioengineered crops for producing PHAs.

Table III. Soybean vs. Switchgrass as Transgenic Plants Producing PHAs

	Soybean	*Switchgrass*
Primary Commercial Use	Food crop	None
Existing U.S. Infrastructure	145,000 tons/day crushing capacity	None
PHA Recovery Process	Solvent extraction oil and meal recovery	Solvent extraction no coproducts
Estimated Delivered Feedstock Cost	$236/metric ton with growers premium	$55/metric ton

Barriers associated with the utilization of agricultural infrastructure for production are large, but not necessarily insurmountable. Identity preservation of PHA crops needs to be provided within the infrastructure and could potentially be quite costly. Transportation and storage infrastructure must be appropriate for identity preservation as well as polymer quality preservation. Depending on the crop, climate changes could also potentially impact yield and polymer structure. And finally there is the political barrier of managing public opinion on biotech crops. All of these issues are outside the scope of this analysis, but could be more important than the technical issues when considering overall production costs.

Some precedents have been set in the United States for commercial production of specialty corn and soybean varieties. While the vast majority of corn and soybean grown in the United States is produced and sold as a commodity, value enhanced varieties have been developed that provide a differentiated product to the market. These varieties require management practices that preserve grain identity from plating through storage. There is also a premium price associated with encouraging farmers to grow value enhanced varieties.

Patented Processes

More than a dozen organizations have recently patented technology relating to production of PHA in plants. Most of these patents describe bioengineering the metabolic pathways of plants to enable synthesis of PHAs. Of the approximately 50 recent patents in this field, ten describe separation processes for PHA recovery from plant biomass. About half of the separation patents rely on solvent extraction for PHA recovery. Currently, patents and other scientific literature on the purification of PHAs from plants is essentially limited to small laboratory scale experiments.

Solvent Extraction

PHA producing microorganisms produce PHA to greater than 60% total dry weight and are readily extractable by organic solvent. Much of the early effort directed to identifying methods for recovery of PHA have focused on recovery from bacterial sources using halogenated hydrocarbon solvents. Chloroform was commonly used and other solvents used include dichloromethane, dichloroethane and dichloropropane. The early purification of PHA from plants was based on these methods. In a typical laboratory process, plant material is harvested, lyophilized, and ground to a fine powder. The dry powder is first extracted with methanol to remove the lipids and chlorophylls, followed by extraction in chloroform to solubilize the PHA. The chloroform solution is cleared from debris either by water extraction or filtering. The volume is reduced by evaporation and PHA is precipitated by the addition of methanol (*14*).

PHA copolymers can be significantly more soluble in a range of non-chlorinated solvents than PHB homopolymer. PHA composition and morphology are determinants of polymer solubility characteristics. Generally, polymers with high crystallinity are more difficult to dissolve than those with low crystallinity. The thermal history of the polymer may also affect solubility. If the PHA has side chains, as the size of the chain increases the number of methylene groups in the polymer increases and therefore the polarity of the polymer changes. However, with a change in the size of the side chains, the crystallinity of the polymer is also affected which in turn affects the solubility characteristics. Such variables make it difficult to accurately predict PHA solubility from simple criteria such as similarities in chemical architecture.

The environmental implications and human toxicities associated with halogenated compounds create the need for PHA solvents with more benign properties. In 1999 Monsanto was issued a patent for PHA recovery using a solvent which is a lower ketone, dialkyl ether or a lower alcohol or ester (*15*). It is preferred that oil be extracted from the vegetable matter before PHA extraction. In an example, PHB is recovered from *Arabidopsis thaliana* that contains 15% polymer. Isopropanol is used to extract PHB and water is added and the volume reduced to recover the polymer. The patent indicates that solvents having solubility parameters between 15 and 30 $J^{1/2}/cm^{3/2}$ are preferred for a PHA such as PHB or PHBV.

In 2000 Monsanto was issued two additional patents for non-halogenated solvent extraction of PHAs (*16,17*). In US 6,043,063, PHA and oil may be co-dissolved from oil-bearing seeds using a non-halogenated solvent. This produces a PHA-enriched solvent/oil mixture. The separation of the meal from the PHA solution via filtration can be problematic because the meal often has a consistency that makes it difficult for the PHA solution to permeate through the meal. As a result, the filter can become plugged. It may be preferred to wash the meal with water or hexane prior to the PHA dissolution step to improve filtration characteristics of the meal. Color bodies from the meal may leach into the solvent resulting in a polymer which is brown/yellow in color. To minimize this effect and reduce color bodies, a bleaching/decolorization step may be undertaken such as activated carbon or clay treatment. In US 6,087,471, PHA is extracted using a non-halogenated PHA-poor solvent that dissolves less than 1% of the PHA at temperatures less than the solvent boiling point. Following extraction under pressure at a temperature above 80°C, PHA is precipitated by cooling the PHA-enriched solvent mixture.

Procter & Gamble has patented a solvent extraction process for PHA separation that employs a marginal nonsolvent (*18*). The process is intended to be adaptable for PHA recovery both from fermentation based production and oilseed production. Biomass is treated with a PHA solvent and marginal nonsolvent to permit removal of insoluble biomass. The PHA solvent is removed resulting in a suspension of precipitated PHA in the marginal nonsolvent. Figure 2 illustrates the process. In an example recovering poly(3-hydroxybutyrate-co-3-hydroxyheptanoate) from potato, acetone is used as the solvent and vegetable oil as the marginal nonsolvent.

Metabolix has patented a solvent extraction process that might be appropriate for processing PHAs from plant biomass on a large scale (*19*). Two methods are described in the patent, each with an option to recover PHA as the polymer or as a chemically transformed derivative:

- Figure 3 illustrates the first method whereby plant biomass is extracted with a solvent in which the oil is soluble and the PHA and meal are not highly soluble. This produces an oil fraction essentially free of PHA (less than 10 wt%). The PHA/meal mixture is then extracted with a second solvent in which the PHA is soluble. Alternatively, the PHA/meal mixture can be treated chemically or enzymatically to produce PHA derivatives which are then isolated from the meal.

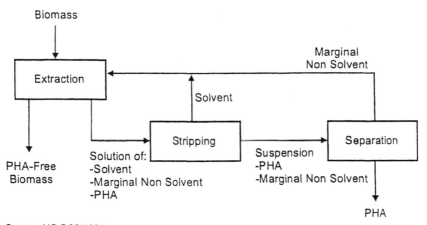

PHA RECOERY USING MARGINAL NON-SOLVENT

Biomass

Extraction

Marginal
Non Solvent

Solvent

PHA-Free
Biomass

Solution of:
-Solvent
-Marginal Non Solvent
-PHA

Stripping

Suspension
-PHA
-Marginal Non Solvent

Separation

PHA

Source: US 5,821,299

Figure 2. PHA Recovery Using Marginal Non-Solvent

• Figure 4 illustrates the second method whereby plant biomass is extracted with a solvent in which both the oil and PHA are soluble and the meal is not highly soluble. Less than 10 wt% of oil and PHA remain in the meal. PHA and oil are separated by a physical separation such as distillation. Alternatively, the PHA/oil product may be modified by chemical or biological treatment to provide a PHA derivative that is subsequently purified.

Non-Solvent Extraction Processes

A variety of processes have been developed to avoid the use of organic solvents in the recovery of PHAs from plant material. Enzymes and oxidizing agents can be used to digest the non-PHA components of the vegetable matter similar to the procedures used for the recovery of PHA from bacteria (20). Starting with an oilseed which accumulates PHA, the seeds would be crushed and extracted with hexane to recover the oil. The defatted meal would then be ground and wetted in order to make it more accessible to the enzymes and reagents involved in solubilizing non-PHA components. The carbohydrates, present mostly in the cell walls, would be stripped by a cocktail of carbohydratases containing arabanase, cellulase, beta-glucanase, hemi-cellulase, pectinase, and xylanase. Solubilization of glycoprotein would be done enzymatically with lysozyme whereas the solubilization of nucleic acids and proteins would be done either enzymatically with nuclease and protease, or chemically using surfactants and oxidizing agents such as peroxide or hypochlorite. The PHA granules would then have to be recovered from the partially digested plant material using either decantation, filtration and/or centrifugation (14). This process is not likely to be economic on an industrial scale. One must consider the high cost of using enzymes and the fact that no meals valuable for animal feed are recovered from the defatted cake. Monsanto's patent application describing this process experimentally demonstrates only the partial solubilization of the cell wall by a cocktail of enzymes and gives no data for the recovery of the PHA (21).

In the late 1990s Procter & Gamble patented processes for the isolation of PHA granules based on air classification (22) and centrifugal fractionation (23). These two processes are related to the well established wet and dry milling methods used in the corn industry for the fractionation of corn seeds into endosperm, germ and hull for the isolation of starch grains.

The air classification method separates dry solid particles according to weight and/or size by suspension in and settling from an air stream of appropriate velocity. Centrifugal fractionation separates particles suspended in a

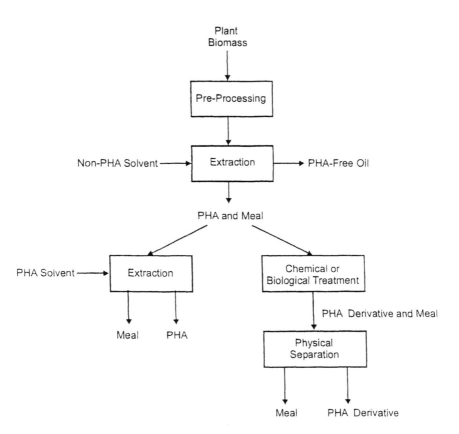

Figure 3. PHA Recovery by Extraction with Non-PHA Solvent

266

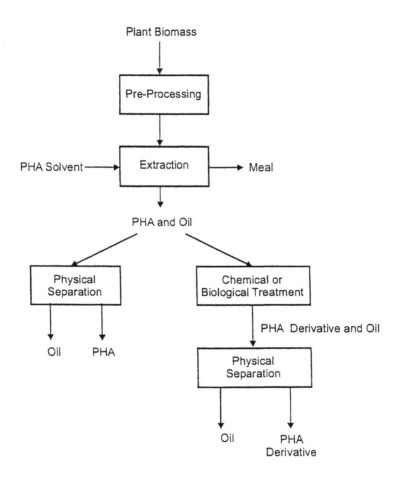

Figure 4. PHA Recovery by Extraction with PHA Solvent

solution based on size and/or density. PHA granules are typically 0.2-1 µm in diameter and thus expected to be one of the smallest particles present in plant cells in comparison to starch grains or protein bodies (*14*). This size difference can thus be used to separate and purify PHA granules from the other plant components. P&G's centrifugal fractionation patent provides examples of recovering PHA using this technique from soybeans, maize, tobacco, coconuts and potatoes (*23*). Typically, a poly(3-hydroxybutyrate-co-3-hydroxyoctanoate) granule purity greater than 95% is achieved and a yield of about 85%.

Process Economics

Process economics for the manufacture of PHA have been estimated as part of the Process Economics Program (PEP). Since PHA is currently not produced on a large commercial scale, the economics are based on engineering estimates developed from patents in the field. Process economics have been modeled for three scenarios that have been considered for large scale production (*2*):

- PHA is produced in a batch, aerobic fermentation process utilizing *Escherichia coli* as the fermentation microorganism. The primary substrate used is fatty acid, but other co-substrates such as glucose may also be used. PHA is recovered from fermentation broth by centrifugation and diafiltration. After solvent removal, PHA granules are dried, extruded and pelletized. A patent issued to Metabolix serves as the main basis for this technology (*24*).

- Soybeans genetically modified to contain PHA are prepared for solvent extraction by continuous dehulling and flaking. Soybean oil is recovered from the flaked soybeans by solvent extraction. The de-oiled meal is desolventized prior to further extraction to recover PHA. Acetone and soybean oil is used to extract PHA from the meal, which is desolventized and dried. PHA is recovered from the acetone/soybean oil solution by evaporating the acetone to form a PHA suspension. PHA is separated by centrifugation, washed with acetone, dried, extruded and pelletized. The polymer recovery process is based on a patent issued to Procter & Gamble (*18*).

- Switchgrass genetically modified to contain PHA is processed in an analogous manner to the soybean extraction process. However, PHA recovery from switchgrass is simplified because no oil is recovered from the plant. Also, the residual biomass after recovery probably has no value except as fuel.

Extraction efficiency of PHA from plants is likely to be a major factor affecting the production cost of PHA from crops. Complicating factors such as the presence of oil make extraction from plants more difficult than from bacteria. Economical polymer extraction will depend, among other factors, on polymer concentration in the seed. In contrast to production of PHA by bacterial fermentation, PHA will likely accumulate at low levels in plants. Companies in this field expect that polymer concentrations in plants will need to reach at least 15% of dry weight for economical production to be feasible (*10*). For the purpose of this analysis, we have developed process economics with the assumption that that the PHA content in genetically modified soybean or switchgrass is 15% after harvest.

A plant capacity of 185 million lb/yr (84,000 t/yr) PHA is selected as the primary basis for the PEP analysis. The corresponding soybean crush rate is 2,000 mt/day. The selected capacity reflects a typical soybean crushing facility in the United States. This capacity may be larger than initial commercial practice will be for production by fermentation. The next commercial scale PHA plant to be built will probably have a capacity of 66 million lb/yr (30,000 t/yr) (*6*). Table IV summarizes PHA process economics.

Table IV. Summary of PHA Process Economics[a]

	E. coli Fermentation	GMO Soybean	GMO Switchgrass
Total Fixed Capital, $ million	203	147	63
Raw materials, ¢/lb PHA	40[b]	84[c]	20[d]
By-products, ¢/lb PHA	0	-65	0
Utilities, ¢/lb PHA	9	8	2[e]
Other costs[f], ¢/lb PHA	12	6	6
Depreciation, ¢/lb PHA	9	7	5
G&A, Sales, Research, ¢/lb PHA	11	7	5
Total Production Cost, ¢/lb PHA	81	47	38

1. For a plant capacity of 185 million lb/yr (84,000 t/yr) and PEP Cost Index of 620
2. For fatty acid unit cost of 25¢/lb
3. For GMO soybean unit cost of $236/metric ton
4. For GMO switchgrass unit cost of $55/metric ton
5. Assumes all steam use generated by combustion of residue switchgrass
6. Other costs include labor, maintenance materials, operating supplies, plant overhead, taxes and insurance.

Source: Reproduced from reference 2. Copyright 2003 SRI.

Crucial to the economics of PHA recovery from soybeans is the recovery of meal and crude oil for sale. At the price levels used in this estimate, the sales credit for these coproducts (65¢/lb of PHA) represents 78% of the cost of the soybeans. The overall PHA economics are very sensitive to the level of coproducts recovered, and recovery at levels lower than assumed in this economic model would significantly impact the feasibility of this process.

The environmental and toxicity issues associated with halogenated solvents drive the need for cost effective PHA solvents with more benign properties. Hexane, which is widely used to extract soybean oil, does not solubilize most PHAs so it is not a good candidate (*14*). Acetone is a viable candidate for commercial PHA extraction as it is an affordable solvent (approximately 30-40¢/lb in 2004), solubilizes PHAs with medium chain lengths, and has low oral toxicity.

One of the potential advantages that switchgrass may have over soybean is its cost. Although switchgrass is currently not grown as a commercial crop in the United States, the Oak Ridge National Laboratory (DOE-ORNL) has studied the potential cost to grow it as a bioenergy crop. ORNL studies indicate a farmgate price (i.e. no transportation costs included) of $30-40 per dry U.S. ton for switchgrass. For the analysis presented in Table IV, we have assumed a switchgrass cost of $55 per dry metric ton delivered at the plant gate. On this basis as well as lower capital requirements, the total PHA production cost from GMO switchgrass is about 25% lower than from GMO soybean.

The process for producing PHAs in GMO plants is still in the early stages of development and there are opportunities to improve the economics through:

- Higher PHA content in plant, 15 wt% was assumed
- Higher PHA recovery, 85% was assumed
- Lower utility consumption, this estimate is based on a high solvent use for extraction.

Not included in the economics is a technology cost, such as licensing fee. Generally accepted accounting principles require immediate expensing of R&D costs. So the large costs to develop a biotech crop are not necessarily included in these economics, although there is some allowance in the general, administration, sales and research line item for on-going R&D expenses.

Conclusions

Technical progress has been made towards developing transgenic crops for commercial production of PHAs, but much work remains to be done (*25*). Technology improvements that are needed include higher product yields and more efficient recovery technologies. Agricultural infrastructure needs to be developed and concerns over genetically modified crops must be thoroughly addressed. Preliminary process economic comparisons of transgenic crops with fermentation processing suggest that the potential economic benefits of PHA production using transgenic crops merits further research.

References

1. Leaversuch, R. *Plastics Technol.* **2002**, *48*, 66.
2. Bohlmann, G. *Biodegradable Polymers from Plants*; Process Economics Program; SRI Consulting: Menlo Park, CA, 2003.
3. Van Arnum, P. *Chemical Market Reporter* **1999**, *255*, 7, 16.
4. Anon. *Chemical Engineering Progress* **2002**, *98*, 10, 15.
5. Baker, J. *European Chemical News* February 18-24, 2002, p 24.
6. Alperowicz, N. *Chemical Week* **2003**, *165*, 23, 26.
7. Poirier, Y. *Prog. Lipid Res.* **2002**, *41*, 131-155.
8. Steinbuchel, A.; Hein, S. *Adv. Biochem. Eng. Biotechnol.* **2001**, *71*, 81-123.
9. Somerville, C.; Nawrath, C.; Poirier, Y. U.S. Patent 5,610,041, 1997.
10. Slater, S.; Mitsky, T.; Houmiel, K.; Hao, M.; Reiser, S.; Taylor, N.; Tran, M.; Valentin H.; Rodriguez, D.; Stone, D.; Padgette, S.; Kishore, G.; Gruys, K. *Nature Biotechnol.* **1999**, *17*, 1011-1016.
11. Guzman, D. *Chemical Market Reporter* **2003**, *263*, 12, 16.
12. Poirier, Y.; Gruys, K. In *Biopolymers*; Doi, Y.; Steinbuchel, A., Ed.; Polyesters I; WILEY-VCH Verlag GmbH: Weinheim, Germany, 2002; pp 401-435.
13. Brumbley, S.; Purnel, M.; Chong, B.; Petrasovits, L.; Nielsen, L.; McQualter, R. W.O. Patent Application 2004/006657, 2004.
14. Poirier, Y. *Adv. Biochem. Eng. Biotechnol.* **2001**, *71*, 209-240.
15. Liddell, J. U.S. Patent 5,894,062, 1999.
16. Kurdikar, D.; Strauser, F.; Solodar, A.; Paster, M; Asrar, J. U.S. Patent 6,043,063, 2000.
17. Kurdikar, D.; Strauser, F.; Solodar, A.; Paster, M. U.S. Patent 6,087,471, 2000.
18. Noda, I. U.S. Patent 5,821,299, 1998.
19. Martin, D.; Peoples, O.; Williams, S. U.S. Patent 6,083,729, 2000.
20. Bohlmann, G. *Biodegradable Polymers*; Process Economics Program; SRI Consulting: Menlo Park, CA, 1998.
21. Liddell, J. W.O. Patent Application 97/17459, 1997.
22. Noda, I. European Patent 763125B1, 1999.
23. Noda, I. U.S. Patent 5,899,339, 1999.
24. Horowitz, D. U.S. Patent 6,340,580, 2002.
25. Snell, K.; Peoples, O. *Metabolic Engineering* **2002**, *4*, 29-40.

Chapter 20

The Biofine Technology: A "Bio-Refinery" Concept Based on Thermochemical Conversion of Cellulosic Biomass

Stephen W. Fitzpatrick

Biofine Technologies LLC, 300 Bear Hill Road, Waltham, MA 02451

The Biofine process is a high temperature, dilute acid-catalyzed rapid hydrolysis of lignocellulosic biomass. The process refines the biomass feed into four products: Levulinic acid, a versatile platform chemical, formic acid and furfural, commodity chemicals and a carbonaceous powder that can be burned or gasified to produce steam and electric power. The process is carried out in a reactor system that enhances the yield of the major products making it commercially viable. The process is flexible enough to utilize a wide range of lignocellulose. Derivative products of interest include automotive fuels, monomers, herbicides and general chemicals. A commercial scale process is now under construction. The process could potentially allow biomass to displace crude oil as the primary source of fuels and chemicals.

Introduction

In the light of unstable crude oil supplies and increasing environmental constraints, the use of abundant renewable cellulosic resources for energy, transportation and materials would appear to be an obvious strategy for industry to pursue. Provided that an efficient means of conversion and high volume markets for derivatives can be developed, use of plant-derived raw materials to fulfill markets presently based on crude oil holds the promise of a highly profitable, sustainable industrial enterprise. (7,26)

In addition to this primary financial driver the use of renewable natural resources to displace crude oil as the main source of commodities has been receiving much attention due to a number of secondary benefits including:

- *Domestic energy security* – The use of domestically produced renewable resources could significantly reduce U.S. and European dependence on crude oil imports from an increasingly uncertain global supply system. Even a partial displacement of imports would have the beneficial effect of buffering the market against increases in crude oil price; (22,23)

- *Reduction in global warming* – The use of renewable resources based on plant-derived matter is carbon dioxide neutral and would consequently eliminate increases in net greenhouse gas emissions due to fossil fuel use (32, 34, 35).

- *Stimulation of the rural economy* – The increased demand for crops grown specifically to supply energy would provide a renewed profit potential for the farming industry (22);

- *Environmental quality* – The derivative chemicals from cellulose are for the most part oxygenated and biodegradable. This generally leads to cleaner burning fuels and more environmentally friendly materials (6);

- *Stimulation of the recycling economy* – The cellulosic component is the largest fraction of municipal solid waste. It is also the most difficult to recycle due to contamination from other co-mingled wastes. An efficient cellulose conversion process capable of using non-recyclable paper and cardboard would be key to improving the economics of the recycling business (36).

- *Ecological risk reduction* - The use of domestically produced renewable resources will reduce or eliminate the ecological damage caused by drilling, transportation and refining of crude oil (13).

The following chapter describes a novel process developed by Biofine based on thermochemical conversion of cellulosic biomass to value-added fuels and chemicals. The technology has the potential to be economically competitive with existing crude oil based refining. Its commercial application offers the prospect of an economically viable fuels and chemicals industry based on domestically grown raw materials.

Cellulose is the most abundant polymer on the surface of the earth. It is the largest (non-aqueous) fraction of plant biomass. Annually, through photosynthesis, the solar energy equivalent to many times the world's annual use of energy is stored in plants. It is generated from atmospheric carbon dioxide using solar power. It therefore represents stored solar energy and it is the only renewable source of carbon. Cellulose is also the largest component of solid wastes produced by society comprising municipal, industrial and agricultural residuals (1, 25).

With the appropriate conversion technology, plant-derived biomass could become the primary source of energy, chemicals and materials over the next century. It is estimated that by dedicating a third of current forest and marginally arable land to production of short-rotation hybrid species or grassy energy crops it would be feasible to supply all transportation needs and a large fraction of petrochemical needs from biomass sources (2,7).

The Biofine process allows the conversion of cellulosic biomass to well-defined primary chemical products that can, in turn be converted to a wide range of derivative chemical products used in many sectors of the present day petrochemical-based industry. The process can utilize cellulosic from almost any source ranging from wood, and agricultural residues to paper sludge and municipal wastes. A refining process using crude biomass to produce a range of value-added chemical and fuel products could justifiably be termed, in concept, a "Biorefinery" (Figure 1) (37). The Biofine technology provides the initial key cracking step unlocking the potential that biomass holds as a renewable, domestic source of industrial chemicals and fuels (3,9).

The process works by "cracking" biomass under the influence of dilute mineral acid and moderate temperature. The cellulose fraction, consisting of hexosan (C-6 sugar polymers) is broken down to form a key intermediate chemical, levulinic acid in high yield and a co-product formic acid. Hemicellulose, consisting of pentosan (C5 sugar polymers), if present in the feed, is cracked to furfural. Lignin present in the feed goes through the process

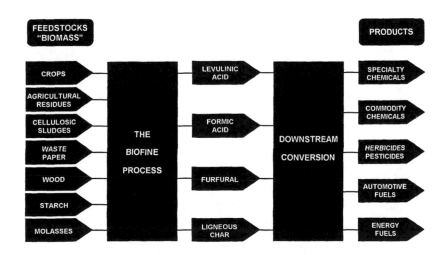

Figure 1. The "Biorefinery" (block diagram)

essentially unchanged although some depolymerization may occur. Both cellulose and hemicellulose degradation reactions produce and a carbon-rich char byproduct that with the lignin, is obtained as a dry solid mixture ("ligneous char"). These primary products become the platform chemicals from which other chemical products are produced.

The Primary Chemical Products

As mentioned in the previous section, the primary chemical products resulting from the hydrolysis process are levulinic acid, formic acid, furfural and a carbon-rich ligneous char.

Levulinic acid has been known as a versatile chemical for over fifty years but its high price ($5.00 per pound) has inhibited its large scale use. The levulinic acid molecule is a C-5 gamma keto acid. It has two functional groups which result in a wide spectrum of chemical reactivity (Figure 2). The Biofine process can produce levulinic acid for a cost in the range that allows it to be an economic starting material for production of fuel substitutes, monomers, novel pesticides and a wide range of commodity chemicals (*8, 33, 45, 46, 47*).

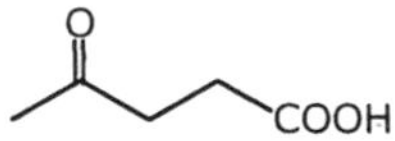

Figure 2. Structure of Levulinic Acid

Formic acid is a well-known commodity chemical with existing large volume uses in the manufacture of rubber, plasticizers, pharmaceuticals and textiles. The estimated world market for formic acid is presently over 450,000 tons per year (*44*). It is also used in large volume for agricultural silage in Europe and there are several very large volume uses such as formaldehyde and road de-icing agents that will become economically feasible with formic acid available at lower cost (*17*). A potentially very large future market for formic acid is in the domestic catalyst market. Formic acid is used in the manufacture of nickel, aluminum and other catalysts. It is also used in the regeneration of catalyst metals poisoned with sulfur. As exhaust emission limits for sulfur are tightened there will be increased demand for ultra-low sulfur fuels requiring increased use of metal catalysts used in their production. Esters of formic acid such as methyl and ethyl formate have also shown promise as automobile fuel components and as platform chemicals. Formic acid can also be gasified or bio-treated to recover its energy value in the event that there is no available market outlet.

Furfural is a commodity chemical with a present-day world market of over 250,000 tons per year. It is used primarily in the manufacture of furan resins, lubricating oils and textiles for leisure wear (*30, 42*). Furfural can be converted to the main product, levulinic acid, via hydrogenation to furfuryl alcohol in a well-known chemical conversion that has been carried out commercially for many years (*17,42*). Conversion of hemicellulose-derived furfural in this way, significantly increases the overall levulinic acid yield from the process.

The carbon rich ligneous char material is produced in the process as a dry powder. It is a mixture of lignin and char from the hydrolysis reactions and is composed of over 60% carbon. It has an energy content of around 27,000 KJ/Kg. It is potentially a suitable feedstock for carbon production or for a gasification reactor for conversion to a high energy "synthesis gas" that can either be used as a source of chemicals via a Fischer Tropsch conversion process or burned as fuel gas in a boiler or gas turbine for energy. Sufficient energy is concentrated in the char to provide both the steam and electric power needs of the process and, at larger scale (300 dry tons per day and over) has the potential to produce excess electric power for sale. This material is also being tested as a potential soil conditioner. The yields of primary products obtained in the process are shown in Table I.

Table I. Primary Products from the Biofine Hydrolysis Process

Primary product	Yield	Based on:
Levulinic acid	50%	Cellulose content
Formic acid	20%	Cellulose content
Furfural	50%	Hemicellulose content
Carbon-rich ligneous char (byproduct char from cellulose and hemicellulose with lignin)	30% 50% 100%	Cellulose Hemicellulose Lignin

Derivative Products

The derivatives that can be produced by conversion of the primary products apply to almost every sector of the industrial chemicals, fuels and energy market (Figure 3). By far the largest potential market for levulinic acid is in the production of oxygenated fuels for both transportation (gasoline and diesel) and energy generation (heating oil and gas turbine fuels). Levulinic acid can be directly converted by hydrogenation into the oxygenated gasoline fuel additive, methyltetrahydrofuran (MTHF) (*38*). MTHF has several attractive properties as a gasoline fuel additive: It has reasonable anti-knock properties (antiknock index (R+M)/2 = 80) and energy density and it has a relatively low volatility (R.V.P = 5.7 psi) (*30*). It also has the interesting property that it acts as a co-solvent for ethanol in gasoline mixtures i.e. MTHF significantly reduces the vapor pressure of ethanol when co-blended in gasoline and is now being used as a cosolvent for ethanol in P Series fuel (a mixture of ethanol, light hydrocarbons and MTHF) recently approved by the U.S. Department of Energy as an alternative gasoline, meeting the requirements of the Energy Policy Act for automobile fleet usage (*5,6,18*). MTHF is also an excellent general solvent being superior to tetrahydrofuran (THF) in many regards. Selected physical properties are shown in Table II (*28*).

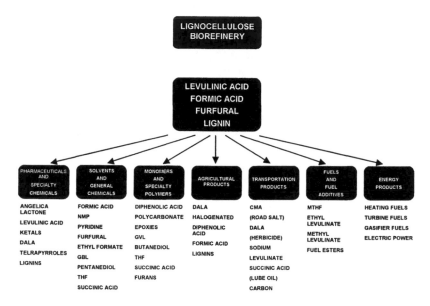

Figure 3. Range of Chemical and Fuel Products from the Biofine Biorefinery

Table II. Selected Properties of MTHF

Boiling pt. (102 mm Hg) Deg. C.	20
Boiling pt. (Atm.) Deg. C.	80
Flash pt. (TCC) Deg. C.	-11
RVP psi	5.7
LHV KJ/Kg	32,000
Specific gravity	0.813
Octane rating (R+M)/2	80

Esters of levulinic acid produced from either methanol or ethanol are under active development as blend components in diesel formulations. These esters are similar to the biodiesel fatty acid methyl esters (FAME) now used in European low emission diesel formulations. FAME has certain disadvantages as automotive fuel components in diesel due to cold flow properties (*24*) and gum formation. Addition of ethyl or methyl levulinate to FAME would be expected to alleviate both of these drawbacks. Levulinate esters also contribute greatly to the lubricity of the diesel mixture. Biofine has shown, in sponsored research, that the addition of 20% EL to a standard No. 2 base fuel was found to improve the lubricity from 410 to 275 using the HFRR index (43). This will be a very significant advantage in low sulfur diesels that have been subjected to high degrees of hydrodesufurization that reduces sulfur levels but also reduces fuel lubricity. Levulinate esters have also the potential to replace kerosene as home heating oil and as fuel for direct firing of gas turbines for electrical generation. (19). Production of levulinic acid esters has the added advantage that there is no co-production of glycerol for commercial disposal. Selected physical properties are shown in Table III.

A second very large potential market for levulinic acid is in the production of the photodynamic pesticide delta aminolevulinic acid (DALA). DALA is the active ingredient in a range of environmentally benign broad-spectrum herbicides and insecticides under development at the University of Illinois, Urbana Champaign (14). Biofine, in conjunction with the National Renewable Energy Laboratory (NREL), Golden, CO has developed and patented a process for conversion of levulinic acid to DALA in reasonable yield (15).

Table III. Selected Properties of Ethyl Levulinate

Boiling pt. (18 mm Hg) Deg. C.	93
Boiling pt. (atm) Deg C.	206.2
Flash pt. Deg C. (closed cup)	195
RVP psi	<0.01
LHV KJ/Kg	24,300
Specific gravity	1.016
Cetane Number (IQT)	<10
Lubricity (HFRR microns)	287

Another promising product produced from levulinic acid is diphenolic acid (DPA). DPA is a direct replacement for bisphenol A (BPA) in polycarbonates, epoxy resins, polyarylates and many other polymer formulations. DPA can copolymerize with BPA or can replace it in various formulations. In addition, DPA has an extra carboxyl group compared with BPA which confers additional polymer functionality allowing it the potential to be used for graft co-polymers, highly crosslinked polymers, charged polymeric materials (or ionomers) and a variety of polycarbonate polymers with novel properties. DPA is produced from levulinic acid and phenol. It was once used commercially in various resin formulations and decorative coatings before it was supplanted by lower cost petrochemically-derived BPA. Biofine has been working with NYSERDA and Rensselaer Polytechnic Institute on applications for DPA (16).

Other direct uses for levulinic acid include production of succinic acid by oxidation. Succinic acid allows production of tetrahydrofuran, 1,4-butanediol and gamma butyrolactone. Levulinic acid can also be hydrogenated easily to a range of interesting monomers and solvents with superior properties to existing products (*33*).

The overall Biorefinery concept is similar to the classical oil refinery model. The low unit cost benefits of economy of scale are obtained by processing of large quantities of feedstock for production of fuels and energy products. At the same time smaller quantities of levulinic acid and other intermediates are diverted into production of higher value added commodity and specialty chemicals making the venture highly attractive financially (*29*).

The Process Technology

The Acid Hydrolysis System

The Biofine process involves high temperature, dilute acid catalyzed hydrolytic breakdown of cellulose to form levulinic acid. The chemistry of conversion has been studied for many years and several reaction schemes have been proposed (*10,39*). Low yields due to uncontrolled char-forming side-reactions have inhibited commercial interest (Figure 4). Biofine has developed a novel reactor configuration that promotes production of levulinic acid while reducing char formation. This results in a sufficiently high yield of levulinic acid to be commercially attractive. Yields of levulinic acid from cellulose of over 70% of theoretical have been attained (*9*). This equates to a mass yield of 0.5 Kg per Kg of cellulose in the feed. In parallel, in the same reaction, hemicellulose in the feed is converted to furfural (*42*). Formic acid is co-produced with the levulinic acid. Lignin is believed to pass through the reaction essentially unchanged (although some depolymerization is suspected). Furfural can be converted to levulinic acid via furfuryl alcohol in a separate reaction using well-known hydrogenation technology increasing the levulinic acid yield proportionately. Biofine has also introduced the use of polymerization inhibitors to reduce undesirable char formation.

The reactor system consists of a plug flow reactor followed by a lower temperature completely mixed reactor. The conditions in the first stage favor the dominant fast first order, high temperature, acid catalyzed hydrolysis of cellulosic and hemicellulose to soluble intermediates. Based on published reaction kinetics (*20,48*), the first order reaction rate constant for the degradation at the first stage reaction conditions of cellulose from various feedstocks has been calculated to be in the range 3.0 to 12.0 reciprocal minutes. The completely mixed conditions in the second stage reactor favor the first order reaction sequence leading to levulinic acid at the expense of the higher order condensation reactions leading to tars (9,10). Based on published reaction kinetics the first order reaction rate constant for glucose degradation at the conditions in the second stage reactor has been calculated to be 0.146 reciprocal minutes (20,48) – at least an order of magnitude less than the cellulose degredation rate in the first stage. Additionally, the reaction conditions in the first stage followed by vapor separation in the lower pressure second stage favor high yields of furfural from the hemicellulose fraction of the feed (49). The reaction conditions are as Shown in Table IV.

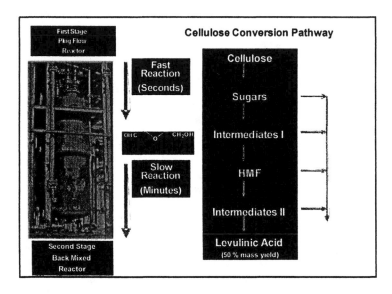

Figure 4. Simplified chemistry of acid hydrolysis of cellulose

The Overall Process

The complete process leading to a commercial grade levulinic acid consists of five steps carried out continuously (Figure 5).

1. **Feedstock Preparation and Mixing**, in which raw feedstock is ground to a size of 0.5 to 1 cm, and mixed with recycle dilute mineral acid,
2. **Hydrolysis**, where the main conversion reactions occur and ligneous char is separated from the reaction mixture,
3. **Product Concentration** where the water concentration is adjusted and formic acid and furfural, if present, is recovered.
4. **Recycle Acid Separation** where the product is removed from the acid to allow the mineral acid to be recycled

5. **Product Recovery** where the product is either converted to derivative products or further purified, if necessary.

Table IV. Biofine Hydrolysis Reaction Conditions (9)

Stage One:	
Mixing Configuration -	-Plug Flow
Acid Catalyst	-Sulfuric Acid
Acid concentration	-1.5% to 3%
Temperature	-210 to 220 Deg. C.
Pressure	-25 barg
Residence time	-12 seconds
Stage Two:	
Mixing Configuration	-CSTR (Backmix)
Acid Catalyst	-Sulfuric acid
Acid concentration	-1.5% to 3%
Temperature	-190 to 200 Deg C.
Pressure	-14 barg
Residence time	-20 minutes

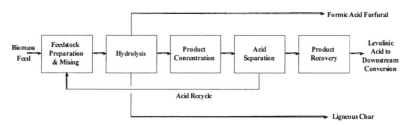

Figure 5. Process block flow diagram

Process Advantages

The process has several important advantages over any existing technology for production of levulinic acid. These are:

1. _Compactness:_ The reaction is fast resulting in a short residence time in the reactor system. This necessitates only a small reactor volume for a high throughput. The process is sufficiently compact that a self-contained 1000 ton per day facility can be accommodated on an ocean-

going "Panamax" barge according to a feasibility conducted by a marine architectural company (*40*).

2.   *Low Cost:* The only reaction catalyst is low cost dilute mineral acid that is recycled within the process eliminating the need for disposal of waste salts.

3.   *Feedstock flexibility:* The process is capable of treating a wide range of low grade, variable composition cellulosic feedstocks or on proposed dedicated energy crops such as willow, poplar, *miscanthus* Sp. or switchgrass (*41*).

4.   *Robustness:* The chemistry of conversion depends only on dilute mineral acid-catalysed hydrolysis of the carbohydrate polymers and is unaffected by contaminants typically found in waste feedstocks.

5.   *Ease of Operation:* The hydrolysis process is continuous. This affords the potential for significant process energy integration and reduces equipment size, labor requirements, maintenance and energy costs compared with comparable batch operations.

6.   *Byproduct flexibility:* Byproduct credits do not significantly affect the process economics. The furfural by-product produced from hemicellulose can be easily converted to levulinic acid (via hydrogenation to furfuryl alcohol) thereby significantly increasing the yield of the main product. Formic acid can be sold, converted to derivatives or waste-treated by gasification or anaerobic digestion to recover its energy value.

7.   *Energy self-sufficiency:* The process is energy self-sufficient. There is sufficient energy in the char byproduct to produce all the steam and electric power needs of the process. At larger scale (greater than 300 dt/d), the process produces significant excess power that can be exported to the grid.

8.   *Ease of scale-up:* The simplicity of the reactor makes scale-up to larger capacities more predictable for the following reasons:

- The process uses standard unit operations involving commercially available equipment.
- The need for expensive, corrosion-resistant materials used in the reactor is minimized due to the small reactor volume required. This allows several parallel reactors to be installed, if necessary, without significantly affecting plant capital cost.
- Even at high throughput capacities, the process is compact. The environmental impact of a process installation is thus minimized and the process can be fitted into limited space locations such as existing chemical sites or an ocean-going barge

Feedstocks

As mentioned previously, the process can use a very wide variety of cellulosic feed materials. As the technology is introduced and gains commercial acceptance, feedstock availability will be the most important determinant of site feasibility and it is likely that plants will be sited to take advantage of local feedstock opportunities. Longer term, once feedstock production becomes dedicated, plants will be located at strategic centers of feed supply. Cellulosic feedstock can therefore usefully be categorized as either "renewable" or "recurring".

Renewable feedstocks are those that are grown and harvested deliberately from dedicated "energy farms". These include short rotation hybrid tree species such as willow, poplar or acacia or crops such as miscanthus and switchgrass. The advantages of renewable feedstocks are that they are predictable in composition and quality and potentially available in huge quantities. A further advantage is that unprocessed biomass usually contains a high proportion of hemicellulose that the Biofine process converts directly to the valuable byproduct furfural. Using modern forestry techniques to grow short rotation species such as willow, poplar, switchgrass or miscanthus, sustainable biomass yields of over 12.5 dry metric tons per hectare are presently achieved and up to 25 dry metric tons per hectare per year have been predicted (1,12). There have been many studies on the long term sustainable cost of energy crops (25,31,41). From these it would seem that renewable biomass could be consistently available within the price range $40 to $75 per dry metric ton in the foreseeable future. This price range makes cellulosic biomass an economically viable farm crop competitive with crude oil as an energy feedstock. At a price for crude oil of $50 per barrel ($10 per million BTU) a renewable biomass price of $60 per dry ton ($3.5 per million BTU) should start to attract serious consideration as a starting point for a profitable and sustainable manufacturing industry

Recurring feedstocks are those that are produced as wastes from other activities. They are usually considered as nuisance wastes that require significant costs and resources for control and disposal. These include paper sludge, municipal solid waste (fiber fraction), conventional forestry waste wood, agricultural residues etc. (21,37). Estimates of the availability vary widely for industrial and agricultural residue, however this type of waste is estimated to amount to around 300 million dry metric tons of per year (26). The quantity of municipal waste is also very large. It is estimated that in the United States every member of the population disposes of over four pounds of solid waste containing over two pounds of cellulosic biomass each day.

The advantages of recurring feedstocks are that they are usually available in one place in large quantities as a result of an existing collection system, and their disposal is *usually* associated with a disposal fee (or tipping fee) for land-filling or incineration. For this reason, it is likely that recurring feedstocks will be available at a significant discount compared with energy crops. This will benefit early adopters of biomass utilizing technology.

Long term, the Biofine process has the potential to become a central technology in the establishment of a fuels and chemicals industry based on renewable biomass. It could thus offer an alternative (non-food) use for the output of farming activities. This would have the advantage of producing fuels and chemicals from non-fossil sources and preserving strategic farming capacity and infrastructure in the face of global economic challenges (*11, 41*).

A fascinating group of feedstocks that is in a somewhat unique category are the marine and aquacultural feedstocks such as kelp, algae and water hyacinth. Kelp could conceivably be harvested and processed at sea using a barge-mounted acid hydrolysis based biorefinery. Such an operation could be entirely carried out at sea in locations where kelp or other such crops proliferate (*25*). Products would be shipped by tanker to land-based processes for finishing and production of derivatives. Productivities for these feedstocks are estimated to be at least an order of magnitude higher than land-based plant crops.

Process Development and Commercialization

The Biofine process has been under development since 1988. Initial patent data was obtained processing ground-wood at semi-works scale at NREL. The process was then operated at bench scale at Dartmouth College, NH using a variety of feedstock, primarily paper sludge and agricultural residue. In 1998 Biofine completed the construction of a plant at South Glens Falls, NY under a grant from the U.S. Department of Energy and New York State Energy Research and Development Authority. This plant is capable of processing one dry ton per day of biomass raw material. It has been operated in a development and small scale production mode over the last five years (*3,9*). In 1999 Biofine was awarded the Presidential Green Chemistry Challenge Award for small business innovation (*27*).

The technology is presently being scaled up to commercial operation in the Republic of Italy. In a project jointly funded by the E.U. and private investment, a 50 dry ton per day facility is currently under construction. This facility is designed to operate on a mixture of paper sludge, agricultural residue and waste paper. The output products will be primarily levulinic acid and levulinate esters for fuel use. In this facility, the char residue will be gasified to produce a medium BTU auxiliary fuel gas for the boilers. Commercial scale plants are currently planned for Ireland, U.K. and U.S.

Process Economics

The estimated production cost of ethyl levulinate for a range of feedstock cost at three plant scales of operation are presented in Figure 6. The process economics calculations are based on the parameters shown in Table V. These pro-forma calculations show that for a large plant processing 1000 dry metric tons per day the production cost of ethyl levulinate, *net of capital related charges* varies from $187 to $385 per metric ton as feedstock price goes from $0 to $75 per dry metric ton. At this cost range it is competitive with base gasoline, diesel or established biodiesel additives such as FAME. In this case, the projected production cost for ethyl levulinate ($290 per metric ton) is equivalent to an energy cost of $12 per million BTU.

Figure 7 shows the variation of estimated capital cost with production capacity. Capital costs, in this case, are estimated for a self-sufficient, fully integrated, grass-roots plant and are factored estimates based on equipment prices. (Capital and operating costs are provided for conceptual study only. These costs vary considerably depending on location and feedstock type and are therefore only for conceptual purposes.)

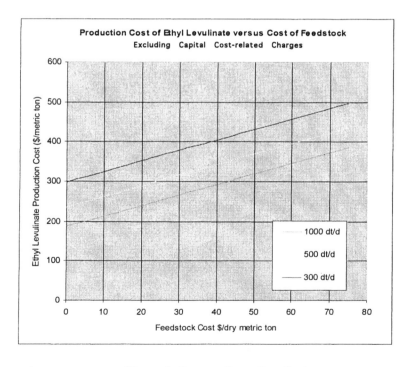

Figure 6. Process Operating Costs

Table V. Operating Cost Assumptions

Feedstock cost range	$0 to $75 per dry metric ton
Feedstock	Idealized typical mixture (50% cellulose, 20% hemicellulose, 25% lignin, 5% ash) available shredded at 50% moisture
Capital cost	All inclusive "grass roots" construction (U.S. Gulf Coast basis)
Capital related charges	Zero
Main product	Ethyl levulinate for fuel use (furfural converted to levulinic acid via hydrogenation to furfuryl alcohol followed by hydrolysis)
Byproducts	Formic acid (recovered from dilute aqueous solution using proprietary technology), electric power
Byproduct pricing	Formic acid $0.05/lb; Power $0.06/kwhr
On-stream factor	350 days per year (continuous operation)

Scale of Operations vs Capital Cost and Production Output for Ethyl Levulinate

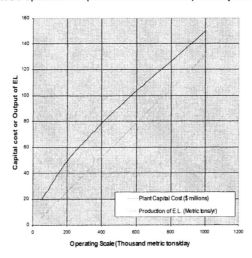

Figure 7. Process Capital Costs and Production Capacity

Acknowledgements

The author would like to acknowledge the following groups that have contributed significantly to the development of this technology: U.S. Department of Energy, Office of Science and Technology (Funding of the Biofine pilot plant and derivatives development); New York State Energy Research and Development Authority (NYSERDA) (Funding of development work and partial funding of the Biofine pilot plant); U.S. E.P.A. – Presidential Green Chemistry Award. – 1999 – Biofine Inc.; BioMetics Inc. Process development and engineering assistance.

References

1. Renewable Energy Resources: Opportunities and Constraints 1990 - 2020". World Energy Council, 1993, London, U.K.
2. Sterzinger, G. *Technol. Rev.*, **1995**, *98*, 34.
3. Jarnefeld, J. et al Proceedings of the Seventh National Bioenergy Conference, 1996, vol. 2, p. 1083. .
4. Rudolf, T.W. and Thomas, R.W. *Biomass*, **1988**, *16*, 33-49.
5. Lucas et al. SAE Technical Paper Series, # 932675 (1993).
6. Paul, Stephen F. American Chemical Society, July 1998,.
7. Plant/Crop-Based Renewable Resources 2020, DOE/GO-10098-385 January 1998.
8. Bozell, J. J.; Moens, L.; Elliott, D. C.; Wang, Y.; Neuenscwander, G. G.; Fitzpatrick, S. W.; Bilski, R. J.; and Jarnefeld, J. L. *Resources, Conserv. Recycl.*, **2000**, *28*, 227.
9. Fitzpatrick, S.W. U.S. Patents 4,897,497 and 5,608,105.
10. Timokhin, B.V. et al. *Russ. Chem. Rev.* **1999**, *68*, 73 – 84.
11. Spedding, C. "Biofuels and the Future,"– British Association for Bio Fuels and Oils. 12/12/2002
12. Borjesson, P. *Biomass Bioener.* **1999**, *16*, 291.
13. Gustavsson, L.; Borjesson, P. *Ener. Pol.*, **1998**, *26*, 699.
14. Rebeiz, C.A. *Enz. Micro. Tech.* **1984**, *6*, 390-401.
15. L. Moens, U.S. Patent 5,907,058.
16. Zhang R.; Moore, J. A. *Polym. Prep.* **2003**, *44*, 737.
17. Chemical Economics Handbook – SRI International 1996
18. Paul, S. U.S. Patent 5,697,987.
19. Erner, W.E. U.S. Patent 4,364,743.
20. Antal, M.J. et al. Carb. Res. **1990**, *199*, 91-109.
21. Glenn, J. *BioCycle* **1997**, no. 11, 30-36
22. "Before the Wells Run Dry" Douthwaite, Richard, Feasta/Lilliput Press (Ireland), 2003.

23. European Commission Communication on Alternative Fuels for Road Transportation, COM (2001) 547, November 7, 2001

24. Huang, C.; Wilson, D., 91st AOCS Annual Mtg., April 26th, 2000.

25. Towler, G. et al. AIChE. Annual Mtg., November 18, 2003

26. "Alternative Feedstocks Program – Thermal/Chemical and Bioprocessing Components" U.S. D.O.E., Bozell, J.J. et al eds., 1993.

27. Dugani, R. *Chem. Eng. News*, July 5, 1999.

28. Great Lakes Chemical Co., Methylterahydrofuran Information brochure, West Lafayette, IN.

29. "Marketing Plan for Levulinic acid Derivatives, Antares Group Inc., Landover, MD, November 16th, 2000

30. Bayan, S. and Beati, E. *Chemica e Industria,* **1941**, *23*, 432 – 434.

31. Lynd, L.R., Alternative Resources Workshop, U.S. D.O.E., Golden, CO Dec. 10th, 1998.

32. "Five Year Research Plan, Biofuels and Municipal Waste Technology Program (1998), U.S. D.O.E., DOE/CM 10093 – 25, DE 88001181

33. Mullin, R. *Chem. Eng. News*, Nov 8, 2004.

34. Pacula, S.; Socolow, R. *Science* **2004**, *305*, 968 – 969.

35. Hileman, B. *Chem. Eng. News,* Nov 17, 1997.

36. Andersen, S.L., *Tappi J.*, **1997**, *80 (4)*, 59.

37. Gravitis, J. and Suzuki, M. Proceedings of the Fourth International Conference Energy, Environment and Technical Innovation, Rome, Italy, September 20 – 24, 1999, 1, 695.

38. Elliott, D.C. et al. U.S. Patent 5,883,266.

39. Manzer, L. personal communication

40. Seaworthy Systems Inc., SSI Report #403-01-01, May 1997.

41. Hayes, D. "Biomass Fractionation in Ireland, University of Limerick Dept of Chemical and Environmental Sciences, 2003.

42. International Furan Chemicals web site: - www.furan.com/content_**physical**.htm

43. Texaco/NYSERDA/BIOFINE ethyl levulinate D-975 Diesel Additive Test Program, Glenham, NY, June, 2000.

44. European Chemical News, "Formic Acid" April, 2003

45. Leonard, R.H. *Ind. Eng. Chem.*, **1956**, *48*, 1331.

46. Shilling, W. L. *Tappi J.* **1965**, *48(10)*, 105.

47. Kitano, M. et al. *Chem. Econ. Eng. Rev.* (Japan) **1975**, *7*, 25.

48. Fagan, R.; Grethlein, H. E.; Converse, A.O.; Porteous, A. *Environ. Sci. Technol.* **1971**, *5*, 545.

49. Root, D.F.; Saeman, J.F.; Harris, J.F.; Neill, W.K., *Forest Prod. J.* **1959**, 158.

Chapter 21

The Development of Biopolymer-Based Nanostructured Materials: Plastics, Gels, IPNs, and Nanofoams

J. J. G. van Soest

Agrotechnology and Food Innovations (A&F) B.V., Box 17,
6700 AA Wageningen, The Netherlands (Jeroen.vanSoest@wur.nl)

The ability to design products with structural features on a nanometric scale is a major technology driver in materials research. Nanostructured materials are defined as materials with structural features on a sub-micron scale determining specific properties. They consist of materials such as metals, polymers, ceramics, composites and biomaterials. Future applications include ultra-precise drug-delivery, transparent nanofoams, nanoelectronics, coatings and ultraselective molecular sieves. The preparation of biopolymer-based nano-structured materials only recently gained attention. The main disadvantages of biopolymer materials like plastics and coatings are water sensitivity or low mechanical strength. New ways were developed, improving the properties of soft biopolymeric materials, such as the development of starch colloids, biopolymer interpenetrating networks and organic-inorganic hybrids. A description is given of several new classes of polysaccharide or protein based materials.

Introduction

During the last decades the development of inorganic-organic hybrid materials gained increasing attention because of the extraordinary properties originating from the synergistic effect between the components. Materials obtained by the incorporation of inorganic building blocks in organic polymers are of great interest because of positive influences on mechanical, thermal, electrical, optical and magnetic properties. The research area still lacks sufficient data on process-structure-property relationships. The use of unmodified biopolymers such as starch and cellulose, in high-quality materials is limited due to inferior properties (insufficient mechanical strength and thermal stability, undesired aging and water sensitivity).

The properties of biopolymer materials can be significantly improved by physically introducing inorganics on a sub-micron level. Examples are nanoclay starch plastics and the incorporation of nanosized silicon oxide in carbohydrate-based interpenetrating networks. Novel routes based on sol-gel processes are available opening the door to the (chemical) coupling of inorganics and organics on a nanometric scale. Progress was made in the development of metallo-organic precursors based on natural polymers.

The aim of the work at A&F is to develop a range of biopolymer-based nanostructured products or hybrids with a range of structural features suitable for various applications (with focus on aerogels and coatings). The mechanisms behind the formation of the structures need to be investigated. The research program covers the following aspects:

- influence of processing routes and synthesis conditions on material structure (uniformity, phase continuity, domain sizes, molecular mixing),
- formation of strong (covalent, co-ordination, ionic) or weak (van der Waals, hydrogen bonding, hydrophilic-hydrophobic balance) interactions at the interfaces,
- relationships between synthesis conditions and (mechanical, optical) properties,
- effects of drying procedures.

The A&F research program contributes to the development of novel high added value materials for the future and the necessary know-how and use of renewable resources. In this paper a short overview will be given of some of the research topics and materials.

Background

One of the technology drivers of this century is the ability to design materials with structural features in the nanometer or sub-micron scale.

Nanostructured materials can be developed from almost every type of material, including polymers, metals, biomaterials and composites (1). Some of the many assorted applications that are likely to become effective include ultra-precise controlled delivery systems, nano-electronics, ultraselective molecular sieves, coatings, transparent aerogels or heat-isolating nanofoams. Using biopolymers for making nanostructured materials only recently gained attention of the scientific community. The incorporation of inorganics such as fillers or porous layered structures in polymers is scientifically well explored. The study of inorganic-organic hybrids, nanocomposite or nanostructured materials is a rapidly expanding research area. Extraordinary properties can be obtained by decreasing the size of inorganic components down to nanometric dimensions.

The main disadvantages of (bio-)polymer-based materials, such as plastics, foams and coatings, are:
- high sensitivity for solvents such as water
- low mechanical strength
- low thermal stability
- aging or low stability over time

Advantages of biopolymers are:
- renewability and degradability
- availability of a wide range of derivatives and structures
- compatibility with biological systems
- well-known chemistry

The most significant disadvantage of pure inorganic materials is brittleness.

The combination of structurally different materials is often used to adapt the properties of a material to meet specific needs, like water and chemical resistance, impact strength, flexibility and weatherability. The method, most commonly used, is blending. As a rule complete miscibility is not attained so that the desired thermal and mechanical properties are not realized.

The properties of soft polymeric materials can be improved by integrating an organic polymer with inorganic building blocks (Hybrid) (2-4). Especially, combinations of nanosized inorganic moieties with organic polymers have a high potential in future applications. Hybrid inorganic-organic materials were shown to be promising systems arising from the synergistic effect of the components. The role of the inner interfaces is predominant. In hybrids, the inorganic moieties give good thermal and optical properties, while the organic parts provide flexibility, toughness, hydrophobicity or wetting. A major problem for the synthesis of hybrid materials is due to the stability of the components: thermally stable inorganics are often formed at high temperatures, whereas biopolymers decompose above 250 °C. Hence, hybrids need to be prepared and processed at low temperatures. An ideal route is the sol-gel process.

Inorganic-organic hybrids are divided into two classes based on the nature of the interface. In Class-I materials the organic and inorganic phases are embedded. Weak interactions (hydrogen bonding, van der Waals and hydrophilic-hydrophobic) are responsible for the cohesion in the structure. Examples of Class-I materials are:

- Organic polymerized monomers or crosslinked polymers embedded in sol-gel matrices
- Inorganic nanoparticles embedded in a polymer
- Polymers filled with *in situ* generated inorganics
- Simultaneous formed interpenetrating networks or IPN's

In Class-II materials components are chemically linked by strong covalent or iono-covalent bonds. The molecules used as starting building blocks possess two distinct functionalities: alkoxy groups (R-O-M bonds) and metal-carbon (M-C) links. The alkoxy groups can be formed into an oxo-polymer network by hydrolysis-polycondensation reactions in a sol-gel. Hybrids can be obtained from organically modified silicon alkoxides such as polyfunctional or polymer functionalised alkoxysilanes. The network-forming functionalities can be covalently connected in a sol-gel in several ways:

- functional groups for the organic polymerization and inorganic polycondensation are present in a single molecule (multifunctionality),
- functional groups are present in separate (end-capped) molecules,
- a preformed organic polymer contains reactive inorganic groups or a preformed inorganic polymer contains reactive organic groups (pendant functionality).

Several nanostructured hybrids based on synthetic polymers are already described. It is also possible to make biopolymer-clay hybrids by exfoliation, blending or on the basis of organic-organic IPNs. Almost no information is available on the preparation of biopolymer-inorganic hybrid nanocomposites and the relationships between synthesis conditions and final morphology and properties of materials consisting of biopolymers and inorganics. The purpose of this project is to design several inorganic-organic hybrid gels or networks based on carbohydrates and inorganic building blocks. The influence of drying procedures will be studied to be able to make dry materials with specific and tunable structural features from the wet gels. Process-structure-property relationships will be investigated. Future applications, resulting from this research, are coatings and lightweight optically transparent insulating aerogels or nanofoams. Coatings could benefit from increased resistance to water, organic solvents and mechanical abrasion, weatherability, and from better adhesion to metal or glass surfaces. Hybrid aerogels would benefit from lower brittleness and better insulating properties.

State of the art

Chemistry - Polysaccharide modification is a well-known research area (*5-7*). A range of physical, enzymatic and chemical modifications are already described and easily applicable. Interesting modifications are cross-linking, esterification and introduction of reactive epoxy and acrylic groups as shown in Figure 1 (*8*).

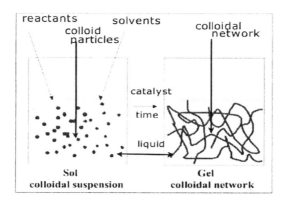

Figure 1. Starch chemistry: Some examples of the chemical modifications used during the preparation of nanostructured materials based on carbohydrates

Extensive knowledge is available on the use of acrylic-grafted carbohydrates and hydrogels prepared from cross-linked starches. Recently, it was shown at A&F that it is relatively easy to make IPNs based on starch or cellulose. IPNs form a class of materials where different polymers are intertwined on a nanometric level with domain sizes generally in the range of 20-200 nm (*9-12*). Sperling defined IPNs as combinations of two or more network polymers synthesized in juxtaposition (*9*). This means that there should be some kind of interpenetrating action on a molecular level. However, this requirement is hardly ever met and therefore the term is used more broadly. Usually finely divided phases of tens of nanometers in size are formed. In some IPNs the phases are both continuous on a macroscopic scale. Mixing polymers results in multicomponent polymer materials. IPNs are usually thermosets. IPNs show high chemical and mechanical stability and creep and flow are highly suppressed compared to polymer blends. Until now most IPNs were based on synthetic polymers. IPNs based on polysaccharides are still relatively unknown and unused.

Recently polysaccharide-based nanofoams or aerogels were made (*13*). No literature reports were found on the preparation of aerogels or nanofoams using starch. Applications of biopolymer-based IPNs are drug-delivery, non-wovens, absorbents and diapers.

Sol-gel (Figure 2) is the term for the process for the synthesis of solids performed in a liquid at low temperatures (below 100 °C) (*3-4,14-16*). Sol-gel processes are used to form inorganics in solvents, through the growth of metal-oxo polymers. The chemistry is based on inorganic polymerization reactions. The first step is the hydroxylation of a metal alkoxide, involving M-OH-M or M-O-M bridges, followed by the polycondensation of reactive hydroxy groups. Metal-oxo-based oligomers and polymers containing residual hydroxy and alkoxy groups result. The structure and morphology of the macromolecular networks depend on the reaction conditions.

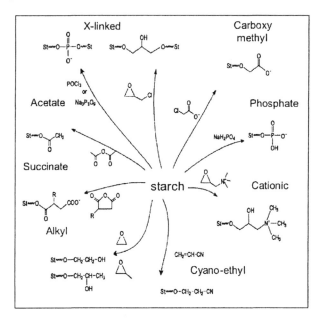

Figure 2. Schematic representation of sol-gel chemistry

Incorporation of inorganic building blocks - Physical mixing of organic polymers and preformed inorganics (by exfoliation) usually results in phase separation and poorer properties. The incorporation of unmodified colloidal particles promotes the aggregation and phase separation of the polymers. The following strategies are used to avoid these drawbacks (*3-4,14-20*):

- encapsulation of particles in a polymer (core- shell)
- crosslinking or functionalization of the organic polymer matrix
- surface modification of inorganic particles with stabilizing, polymerizable or initiating groups

- in situ sol-gel generation of inorganic particles
- simultaneously preparation of hybrid IPNs using the sol-gel process
- use of dual network precursors or (end-capped) polymers with pendant functionalities

Inorganic carbohydrate hybrid systems - Limited information is available on the use of saccharides in nanostructured inorganic hybrids. Some papers describe the incorporation of nanosized clays in thermoplastic starch by exfoliation (*21-22*). Recently the synthesis of starch/clay superabsorbents was reported (*23*). Several studies concern the preparation and properties of organo-silicon derivatives and cellulose-based hybrids (*24-25*). Inorganic starch hybrids with electrorheological activity were prepared using *in situ* sol-gel synthesis (*26*). Several carbohydrate-based metallo-organic precursors were developed (*27*). Functional oligomers and polymers having hydroxy groups were formed, which can react with metal-oxo polymers. Reactive alkoxysilane groups were grafted onto a variety of oligomers and polymers (*3-4*).

Drying procedures - Evaporation of a liquid from a wet gel is complex and different stages are encountered (*28-29*). Most important is the control of shrinkage, collapse and cracking of the network. Several ways were developed to influence the capillary forces involved, such as aging, addition of tensides, organic templates, or drying-control chemicals. However, this was not sufficient to completely control shrinkage. Methods were developed to reduce the surface tension during drying:

- Supercritical-(sc)-drying
- Freeze-drying
- Ambient-pressure-drying

Expensive methods, such as freeze- or sc-drying, must be avoided to make large-scale application possible. Finding improved routes for the exchange of solvent by air is vital. Ambient-pressure drying was introduced to prepare silicon-based aerogels and involves solvent-exchange processes and modifications of the inner surfaces of the gels. Companies such as Hoechst (*30-31*) and Nanopore developed cheap alternatives of this route (*32*).

Recently developed nanostructured biopolymer products

A short description and some product examples will be given below of some of the recent developments at A&F.

Starch-based sub-micron particles or microgels - At A&F research is being performed in the area of colloidal particle technology based on starches, i.e. sub-micron particles or microgels (*33-36*). Starch colloids form stable dispersions in an aqueous environment, and exhibit interesting properties which can be used in both food and non-food applications (rheology modifier, thickener, latex or binder polymer, protective and functional coating, adhesive, hydrogel and superabsorbents, controlled release of active ingredients and delivery of medicine). A variety of starch particles, with adjustable particle sizes ranging from 40 nm to 40 μm and differences in crosslinking density, have been made using methods such as extrusion and emulsion processing (Figure 3). Colloidal behavior is found for sub-micron particles. The interactions between the small particles lead to the formation of stable dispersions in water and interesting rheological behavior (*33-34*). Especially, extrusion gives novel and unique starch particles with a very narrow size distribution, which are referred to as latex particles (Figure 4) (*36*).

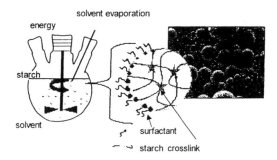

Figure 3: Schematics of the preparation of sub-micron particles via emulsion processing

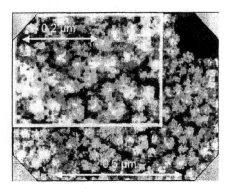

Figure 4: Scanning electron microscopy photograph of latex or sub-micron particles prepared via extrusion processing

Biopolymer-based interpenetrating networks (IPNs) - Crosslinking of single polymer networks can lead to brittle materials. Biopolymer-based IPNs can result in materials with good strength, less brittleness and increased flexibility, low shrinking, lower sensitivity to solvents and heat, less wearing, and improved

296

compatibility with substrates. Some of the presently already commercialized applications of IPNs are contact lenses, perm-selective and permeable membranes, dental fillings, sound and vibration damping, tough rubber and plastic materials, pressure-sensitive adhesives and coatings, nail polish, medical devices and plastic surgery materials, and textiles. Amongst the possible new applications are superabsorbent materials, diapers to retain body fluids, agriculture and packaging, hydrogels and controlled release, transparent plastics and fire retardants.

A range of protein, carboxymethyl cellulose (CMC), starch and acrylate based IPN materials were prepared (unpublished results) (37). In particular the sequential route, as schematically shown in Figure 5, was successful in obtaining transparent or translucent (hydro-)gel biopolymer-based networks.

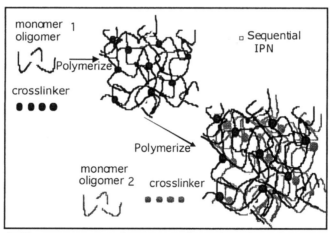

Figure 5: The preparation of a sequential IPN based on oligomers and-/or monomers as the two different building blocks of the two polymeric network

An example of a strong carbohydrate-polyacrylate-silicon IPN network is given in Figure 6 (left). This typical material is translucent and gives a strong gel network in water. The material shrinks only slightly in a 30% (v/v) ethanol-water mixture. Figure 6 (right) also shows two examples of gelatin-acrylate (semi-)IPN foams. It is possible to make translucent or transparent films based on protein-acrylates as well with good mechanical strength.

Carbohydrate-based nanofoams - Aerogels or nanofoams can be made by replacing the liquid solvent in a gel by air with the objective to minimize the substantial alteration of the gel network structure and volume. Nanofoams have unique physical and thermal properties such as low density, high specific surface area, low thermal conductivity, and low dielectric permittivity. The replacement of the liquid in a gel with air can be done with supercritical drying, but also ambient-pressure drying is an option (23,28-32). There are three classes of

Figure 6: Several examples of IPNs based on protein (right) or carbohydrate (left)

nanofoams: fully inorganic, inorganic/organic hybrids and fully organic or polymer-based. Inorganic nanofoams can be made with very low densities (< 0.005 g/cm3). They can be made with a high compressive strength, but they are usually brittle. Organic nanofoams based on (natural) polymers or derivatives are less fragile than inorganic aerogels. When the structural network of an inorganic network is combined with organic functional groups or polymers, hybrid inorganic-organic nanofoams can be formed, having an improved elasticity over inorganic nanofoams.

A&F research is focused on the development of polysaccharide-based and hybrid nanofoams for application as highly transparent, low weight and insulating polymer glass and high quality coatings (*unpublished result*) (*2,38*). In the research program, in particular starch and cellulose esters and carboxymethyl ethers are used as starting polymers. In this way a range of hydrophilic and hydrophobic gel networks were made in water or organic solvents (*13,37-38*). The hydroxyl, carboxyl and ester groups can be crosslinked with a range of crosslinkers. Crosslinking is a well-known method for increasing the strength of polysaccharide and cellulosic materials such as paper. In Figure 7 the chemical structure is shown of a carboxymethyl cellulose (CMC) crosslinked with epichlorohydrin in water.

It was shown that there is a clear relationship between the crosslinking degree and the compressive strength of the gels. Translucent and transparent microfoams were prepared using ambient pressure-temperature air or freeze-drying based on gels prepared in aqueous environment. Nanofoam were obtained based on hydrophobic materials, which were prepared in organic solvents and subsequent sc-CO_2 drying. An example is given in Figure 8. The materials had a porosity of circa 30-40% and were translucent or transparent.

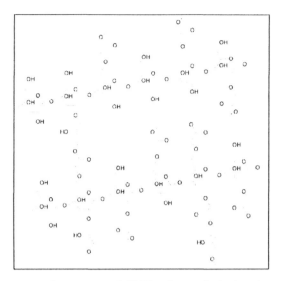

Figure 7: The chemical structure of CMC gel crosslinked with epichlorohydrin.

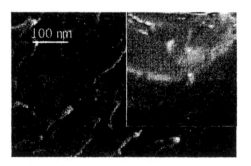

Figure 8: A polysaccharide based nanofoam, based on a crosslinked cellulose ester made controlled sc-CO$_2$ drying of the gel in acetone. A field-emission-gun-SEM photograph shows the nanoporous structure. The inset is a photograph of the dried product (0.5 cm large piece).

Carbohydrate-based inorganic or water-absorbing clay-hybrid gels - It was shown by numerous authors that it is possible to make carbohydrate-based superabsorbing polymers (SAP's) (*39-41*). Materials were prepared with water binding capacity of more than 400 g water per g dry material. However, the most striking shortcoming of these materials is the swelling capacity under elevated pressures. Research is being performed in preparing carbohydrate based hydrogel or water binding polymer gels with increased strength by incorporation of inorganics (*23*). It was shown for example that it is possible to incorporate clays and nanoclays by means of a combination of crosslinking and exfoliation in aqueous environment. In Figure 9 an example is given of a water binding starch-based material containing inorganic clay (based on bentonite). The porous structure can be seen in the SEM photograph. The presence of the clay was confirmed with X-ray diffractometry showing the characteristic crystalline structure of the clay. Possible applications are diapers and hygiene products, rheology modifiers and thickeners and controlled release systems.

Biopolymer-nanoclay hybrid plastics - Polymer nanocomposites are used as packaging materials. Application of nanocomposites in the biopolymer packaging industry (e.g. films and bottles) potentially yields materials with low water vapor permeability (WVP) accompanied by a high transparency of the

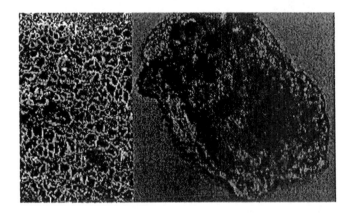

Figure 9: SEM of starch-clay hydrogel (left: x500 magnification – 80 µm piece). Right part shows the polarised microscope picture of a 0.3 mm piece of a clay-hybrid water absorbent polymer with the birefringence as a result of the clay.

final product. Proper processing of nanocomposites is essential in order to obtain complete dispersion, exfoliation and orientation of the nanoparticles and therewith enhanced barrier properties. Exfoliation is the process of reducing micron size clay particles to nanosize flakes or platelets and dispersing the thin platelets throughout the biopolymer matrix. The improved properties are acquired by means of a variety in processing methods such as melt compounding (blending of molten polymers at high shear), extrusion or *in situ* polymerization. Complete dispersion and exfoliation of the nanoparticles will only occur if the nano-material is compatible with the biopolymer. Therefore often some additional chemical modification of the clay is essential. In addition to complete dispersion and exfoliation, orientation of the nanoparticles is critical to enhance the barrier properties of the nano-composite (*42*). It was shown that preparation of thermoplastic starch (TPS)-hybrid films via melt intercalation results in a decrease of the water vapor transmission rate by nearly 40% at clay contents of 10 wt% (*43*).

300

TPS-hybrid-films were prepared via sheet-extrusion. It is shown that the relative WVP (hybrid WVP / pristine WVP) decreases to nearly 50% of the pristine TPS-film by adding solely 3 wt% of nanoclay (Figure 10). It suggests that the layered structure of the clay platelets decreases the migration of water through the film matrix.

The decrease in WVP is mainly attributed to the presence of oriented and exfoliated, large aspect ratio silicate layers in the TPS-matrix (*44*). Due to the oriented nanoclay-platelets the water is forced to traverse the film by a tortuous path through the TPS-matrix, as shown schematically in Figure 11. The decrease in WVP is important in evaluating TPS for utilization in for instance food packaging where efficient barrier properties are required. In addition, reduction in WVP can decrease the film thickness with equal barrier properties. This will

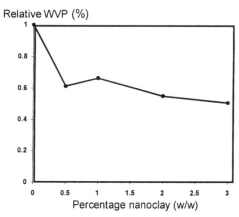

Figure 10: The effect of nanoclay on the WVP of starch plastics

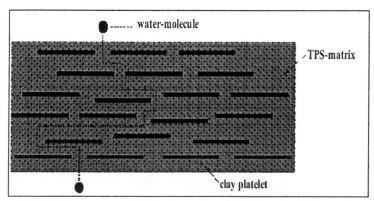

Figure 11: Explanatory model for the reduced WVP in starch-based nanoclay hybrid plastics.

decrease material usage (costs) and increase the clarity (transparency) of the film. In addition charged clay platelets can bind water molecules and lower water migration rate.

Conclusions

An interesting start has been made for the preparation of biopolymer-based nanostructured materials. The materials show an interesting range of properties with high potential for various applications. Starch microparticles are interesting because of their colloidal and rheological properties. Applications are thickeners, dispersion adhesives, paper coatings and controlled release systems. IPNs show improvements in strength and solvent resistance, making them interesting for coating, paints, hydrogels, and controlled release materials. The use of sc-CO_2 drying in combination with the modification of polysaccharides makes it possible to make lightweight or low-density transparent materials (nanofoams) with good mechanical properties. Inorganic hybrids were shown to result in materials with improved properties (mechanical, water-binding and water vapor barrier) compared to original starch and protein-based plastics or films and water-binding polymeric materials (superabsorbents).

More detailed studies on the relationships between synthesis, structure and properties would be helpful in future research programs in the area of product development.

Acknowledgements

The authors wishes to thank the following persons for their contributions to this paper: F. Giezen, R. Jongboom, R. van Tuil, G. Schennink, A. Burylo, K. Ralla, A. Kunze, J. Laufer, F. Kappen, Y. Dziechciarek, V. Blanchard., M. Andre and J. Timmermans (all A&F or former A&F).

References

1. Rühle, M.; Dosch, H.; Mittemeijer, E. J.; van de Voorde, M. J. *European White Book on fundamental research in materials science*, URL http://www.mpg.de/english/illustrationsDocumentation/documentation/euro pWhiteBook/index.html, Max-Planck-Institut für Metallforschung Stuttgart
2. Sonneveld, P. J.; van Hulzen, M., *PT-Industrie* **2000**, 8, 10.
3. Kickelbick, G. *Prog. Polym. Sci.* **2003**, 28, 83-114.
4. Judeinstein, P.; Sanchez, C. *J. Mat. Sci.* **1996**, 6, 511-525.

302

5. Zhang, L. M. *Macromol. Mat. Engin.* **2001**, 286, 267-275.
6. Edgar, K. J. *Progr. Pol. Sci.* **2001**, 26, 1605-1688.
7. *Starch: Chemistry and Technology;* Whistler, R. L.; Bemiller, J. N.; Paschall, E. F., Eds.; Academic Press: New York, NY, 1984; pp 1-201.
8. Gotlieb, K. F.; Timmermans, J. W.; Slaghek, T. M. PCT Patent Application WO187986, 2001.
9. Klempner, D.; Sperling, L. H.; Utracki, L. A. *Adv. Chem. Ser.* **1994**, 239, 1-28.
10. Lipatov, Y. S. *Progr. Polym. Sci.* **2002**, 27, 1721-1801.
11. Kim, S. J.; Park, S. J.; Kim, S. I. *React. Funct. Polym.* **2003**, 55, 53-59.
12. Jordhamo, G. M.; Manson J. A.; Sperling, L. H. *Polym. Eng. Sci.* **1986**, 26, 525-530.
13. Tan, C.; Fung, B. M.; Newman, J.K.; Vu, C. *Adv. Mat.* **2001**, 13, 644-645.
14. Arkles, B. *MRS Bull.* May 2001, 402–407.
15. Livage, J.; Gangulib, D. *Solar Energy Mat. Solar Cells*, **2001**, 68, 365-381.
16. *Sol-Gel Materials, Chemistry and Applications*; Wright, J.D.; Sommerdijk, N. A. J. M., Eds.; Adv. Chem. Texts; Gordon & Breach Science Publishers: Cambridge, UK, 2001, Vol. 4, pp. 1-136.
17. Bounor-Legaré, V.: Angelloz, C.; Blanc, P.; Cassagnau, P.; Michel, A. *Polymer* **2004**, 45, 1485-1493.
18. Yano, S.; Iwata, K.; Kurita, K. *Mat. Sci. Eng.* **1998** , C6, 75–90.
19. Sanchez, C.; Soler-Illia, G. J. D. A. A.; Ribot, F.; Grosso, D. *Comptes Rendus Chimie* **2003**, 6, 1131-1151 .
20. Mutin, P.H.; Guerrero, G.; Vioux, A. *Comptes Rendus Chimie* **2003**, 6, 1153-1164 .
21. Wilhelm, H.-M.; Sierakowski, M.-R.; Souza, G. P.; Wypych, F. *Carb. Polym.* **2003**, 52, 101-110.
22. Berger, W.; Lutz, J.; Opitz, G.; de Vlieger, J.; Fischer, S. E.P. Patent 1,229,075, 2002.
23. Wu, J. H.; Lin, J. M.; Zhou, M.; Wei, C. R. *Macromol. Rapid Comm.* **2000**, 21, 1032-1034.
24. Zoppi, R. A.; Gonçalves, M. C. *J. Appl. Polym. Sci.* **2002**, 84, 2196-2205.
25. Hunter, M. J. G.B. Patent 671,721, 1952.
26. Vieira, S. L.; de Arruda, A. C. F. *Int. J. Modern Phys. B* **1999**, 3, 1908-1916.
27. Wagner, R.; Richter, L.; Wersig, R.; Schmaucks, G.; Weiland, B.; Weissmuller, J.; Reiners, J. *Appl. Organometallic Chem.* **1996**, 10, 421-435.
28. Hüsing, N.; Schubert, U. *Angew. Chem. Int. Ed.* **1998**, 37, 22-38.
29. Pierre, A. C.; Pajonk, G. M. *Chem. Rev.* **2002**, 102, 4243-4244.
30. Dierk, F.; Zimmermann, A. U.S. Patent 5,789,075, 1998.
31. Dierk, F.; Zimmermann, A. U.S. Patent 6,316,092, 2001.
32. Smith , D. M.; Johnston, G. P.; Ackerman, W.C.; Stoltz, R. A.; Maskara, A.; Ramos, T.; Jeng; S.-P.; Gnade, B. E.; U.S. Patent 6,380,105, 2002.

33. Dziechciarek, Y.; van Soest, J. J. G.; Philipse, A. P. *J. Coll. Interfac. Sci.* **2002**, 246, 48-59.

34. van Soest, J. J. G.; Dziechciarek, Y.; Philipse, A. P. In *Starch and Starch containing Origins – Structure, Properties and new Technologies;* Yuryev, V.; Cesaro, A.; Bergthaller, W., Eds.; NOVA Publishers: Hauppauge, NY, 2002; pp. 199-213.

35. van Soest, J .J. G.; van Schijndel, R. J. G.; Stappers, F. J. M.; Gotlieb, K. F.; Feil, H. PCT Patent Application WO00/40617, 2000.

36. Giezen, F. E.; Jongboom, R. O. J.; Feil, H.; Gotlieb, K. F.; Boersma, A. PCT Patent Application WO00/69916, 2000.

37. van Soest, J. J. G., Development of IPNs period 2000-2004, 2004, A&F, Wageningen, The Netherlands.

38. van Soest, J. J. G., Timmermans, J. W., Development of nanofoam period 2000-2002, 2002, A&F, Wageningen, The Netherlands.

39. Feil, H.; van Soest, J.J.G.; Schijndel, R. J. G. PCT Patent Application WO9912976, 1999.

40. Besemer, A. C.; Thornton, J. W. PCT Patent Application WO9827117, 1998.

41. Bergeron, D.; Couture, C.; Picard, F. PCT Patent Application WO 02/38614, 2003.

42. Yano, K.; Usuku, A.; Okada, A. *J. Polym. Sci., Part A: Polym. Chem.* **1997**, 35, 2289-2294.

43. Park, H. M.; Lee, W. K.; Park, C. Y.; Cho, W. J.; Ha, C. S. *J. Mat. Sci.* **2003**, 38, 905-915.

44. Ward, W. J.; Gaines, G. L. G.; Ward, W. J.; Alger, M. M.; Stanley, T. J. *J. Membr. Sci.* **1991**, 55, 173-180.

Chapter 22

Polyols: An Alternative Sugar Platform for Conversion of Biomass to Fuels and Chemicals

J. Michael Robinson[1], Caroline E. Burgess[1,7], Melissa A. Bently[1], Chris D. Brasher[1], Bruce O. Horne[2], Danny M. Lillard[2], José M. Macias[2], Laura D. Marrufo[3], Hari D. Mandal[1], Samuel C. Mills[2], Kevin D. O'Hara[1], Justin T. Pon[4], Annette F. Raigoza[5], Ernesto H. Sanchez[1], José S. Villarreal[1], and Qian Xiang[6]

[1]Chemistry Department, The University of Texas of the Permian Basin, 4901 East University Boulevard, Odessa, TX 79762
[2]Champion Technologies, 115 Proctor, Odessa, TX 79762
[3]Pfizer Global Research and Discovery, St Louis, MO 63017
[4]Chemistry Department, University of Texas at Dallas, Richardson, TX 75083
[5]Chemistry Department, University of Notre Dame, South Bend, IN 46556
[6]Abenogoa, 1400 Elbridge Payne Road, Suite 212, Chesterfield, MO 63017
[7]Current address: Energy Institute, Pennsylvania State University, University Park, PA 16802

Polyols provide an alternative "sugar platform" for biomass conversion to hydrocarbon fuels, alcohols, chemicals and hydrogen. Polyols are obtained directly from biomass by an "intercepted dilute acid hydrolysis and hydrogenation" (**IDAHH**), wherein the incipient unstable aldoses are intercepted by catalytic hydrogenation to produce a solution of stable polyols with no detectable phenols. Granular catalyst is retained by a screen and insoluble lignin is simply filtered from the product slurry. Complete conversions are obtained for several biomass types within 3-6 hr at ~185 °C with 0.8% H_3PO_4. Minimum cost for polyols ranges from $0.12-0.16/lb.

Introduction

Two classical biomass processing platforms (Figure 1) are described by the US Dept of Energy's Office of Energy Efficiency and Renewable Energy as the "Sugar Platform" and the "Thermochemical Platform". These platforms generally follow lines of considerable efforts in fermentation of *aldose* sugars and direct pyrolysis of biomass. Neither platform has been overly successful because of the inability to obtain the desired monomeric sugars (e.g., glucose) in suitable yields and thermal transformations are nonselective and give low quality products. New technologies are emerging along chemical pathways that are neither fermentations nor pyrolyses. Rather than only seeking a myriad of slight improvements to each of these classical platforms, feedstocks for the future must embrace new alternative pathways where developments may provide more attractive economic solutions to the current inefficient biomass conversions and may afford high quality fuels.

1. Sugar Platform (Fermentation)

    Biomass ➡ $C_6H_{12}O_6$ ➡ $2\ CH_3CH_2OH\ +\ 2\ CO_2$

                   "aldose sugar"   < 67% Carbon efficiency

2. Thermochemical Platform (Pyrolysis)

    Biomass ➡ (gas) ➡ oil + char + CO_2

                                       < 80% Carbon efficiency

Figure 1. Classical Biomass Platforms

Our research has established a clean fractionation of biomass carbohydrate polymers into monomeric polyols which serve as a new alternative "sugar" platform. Polyols are obtained from biomass by *chemical* means, rather than by enzymatic degradation, and are further transformed by *chemical* means into quality fuels and other chemicals.

Strategy to Solve the Biomass Fractionation Problem

Obtaining a clean fractionation (*1*) of biomass polysaccharides has been a difficult task. Ultimately, utilization of biomass to produce liquid fuels economically must solve this challenge. Many government agencies (*2*), national labs (*3*), industry and academia (*4,5,6,7,8*) have attempted to solve this problem by almost as many methods and improvements. The reactive nature of the monomeric carbohydrates generated during acidic hydrolysis unfortunately provides for continued reactions (degradation). Currently, the digestion of polysaccharides is limited by the simultaneous undesirable side reactions of aldoses. Hydrolysis conditions and rates of hydrolysis and degradation reactions (k_1 & k_2) have to be delicately balanced in order to achieve a reasonable yield of desired sugars as shown in Figure 2, which is a pictorial statement of the problem. High glucose yields are not available even at high temperatures for very short reactions in dilute acid. In a special case, high conversion (including a secondary lower temperature digestion) was achieved by a fast flow shear factor (*9*). However, the shrinking bed reactor that was employed is not practical for industrial use and a clean separation from lignin was not achieved.

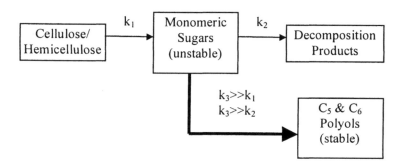

Figure 2. Intercepted Hydrolysis Strategy

Herein, a strategy is defined which provides for trapping the reactive carbonyl carbohydrates immediately upon hydrolysis from biomass polysaccharides. The lower section of Figure 2 depicts how incipient aldoses are rapidly reacted (e.g., catalytic hydrogenation) to ~100% polyols, which are much more stable to further reaction. This reductive mode defines the "intercepted dilute acid hydrolysis and hydrogenation" (**IDAHH**) strategy wherein the rate limiting step is the rate of hydrolysis. Figure 3 displays tangent lines to the dissolution curve of cellulose and the production curve of glucose calculated using Saeman's kinetics for 1% H_2SO_4 at the relatively low temperature of 170 °C. These tangents represent the continued initial reaction rates for both curves with such an interception reaction strategy, whereupon cellulose might be completely digested and high yields of a stable derivative, sorbitol, might be produced in less than 2 hr, rather than the limited amounts of glucose that would otherwise continue to decompose under these conditions.

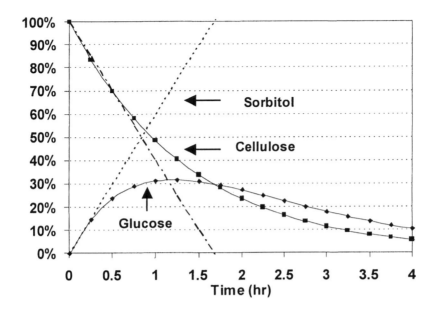

Figure 3. Calculated Cellulose Hydrolysis Profile in 1% H₂SO₄ at 170 °C.

In fact, this preparation of inexpensive polyols is the first step in the proposed new platform for a bio-refinery (Figure 4) that is capable of producing several types of products from these intermediate polyols ranging from hydrocarbons to alcohols and hydrogen.

In the massive literature of biomass hydrolysis efforts, almost all have sought to produce glucose for fermentation. Few have attempted to *intercept* the hydrolysis at the aldose stage *before* other undesirable degradation reactions occur. Choosing to use an interception strategy provides a distinct advantage. For example, if simultaneous hydrogenation is employed during the polysaccharide hydrolysis, each *incipient* aldehyde group is reduced irreversibly. If the speed of the reduction is greater than the hydrolysis, then aldose concentration is kept very small. Consequently, raw biomass is converted to a solution of C_6/C_5 polyols, predominantly sorbitol and xylitol, which reflects the hemicellulose and cellulose content of the biomass resource utilized. Lignin is not appreciably affected and can be simply filtered from solution. A remarkably facile biomass fractionation results.

Several factors must be considered for this strategy to deliver an *inexpensive* polyol mixture and ultimately, inexpensive final products. These include: 1) single step reaction, 2) aqueous solvent system, 3) total carbohydrate perspective, 4) the stability of polyol products to reaction conditions, and 5) a highly active and stable catalyst. Acidic conditions are the best choice for hydrolysis because both the acetal group forming the carbohydrate polymer and the ring opening of the cyclic hemiacetal group to the reducible acyclic aldose form are both acid catalyzed.

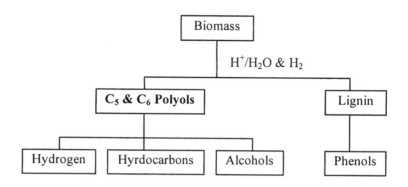

Figure 4. Biomass Refinery via Polyols Platform.

Indeed, a simultaneous acid hydrolysis/hydrogenation was found in the Soviet literature (*10,11,12,13*). One report briefly discussed the use of dilute H_3PO_4 (0.7%) with a Ru/C catalyst at an initial hydrogen pressure of 440-735 psi and at 155-160 °C for 1 hr to generate sorbitol from pine sawdust. However, the available (English) papers were brief, particularly with respect to raw biomass conversion. Therefore, our efforts to develop an "intercepted dilute acid hydrolysis and hydrogenation" (**IDAHH**) process for biomass fractionation began by re-examining these conditions for a variety of biomass resources (*14*).

Results and Discussion

Biomass Resource Survey for IDAHH

Table I shows the biomass resource characterization by C_5 and C_6 aldose as well as lignin content. FTIR analyses of our biomass samples agreed closely with these literature values. However, Table 1 simplifies the complex mix of carbohydrates that differ considerably between resources. In addition to mainly glucose and xylose, the hemicellulose fractions contain aldoses such as mannose and galactose, but also have small amounts of acid forms such as glucuronic acid. Nevertheless, these diverse resources are all hydrogenated to mixed polyols (sorbitol, mannitol, xylitiol, etc.) which contain small amounts of polyhydroxy acids, such as gluconic acid. While the stereochemical differences between the starting sugars (e. g., mannose and glucose) are retained in the polyol products, they have no impact in later *chemical* reactions where they are reduced into fuels with no stereochemistry.

Since −35 mesh oak sawdust was available locally, it was selected to study the initial variables (temperature, time, biomass load, intrinsic acidity, and acid concentration) needed for **IDAHH**. Seven raw biomass resources as well as C_6 standards, cellulose and starch, were surveyed. These feedstocks have 59.0-73.2% holocellulose and include a hardwood, softwood, perennial energy crop, and an agricultural waste. Switchgrass and corn stover have been deemed as the "best quantity" biomass resources with the advantage of fast growth, cyclical harvest, and ready availability. However, it is important to note that such "crops" have a significant protein content which may contribute to catalyst poisoning in our reactions.

Table I. Major Components of Selected Biomass Resources

Biomass	%C_6	%C_5	C_6/C_5	%C_6+C_5	%Lignin	%Total
Woods						
Aspen	57.6	15.6	3.7	73.2	15.5	88.7
Cedar	61.4	8.1	7.6	69.5	32.5	102
Pine	58.1	11.2	5.2	69.3	29.4	98.7
Oak	46.9	20.4	2.3	67.3	23.2	90.5
Crops						
Switchgrass	38.5	27.4	1.4	65.9	17.6	83.5
Baggasse	41.0	23.0	1.8	64.0	25.2	89.2
Corn Fiber	36.9	22.1	1.7	59.0	16.7	75.7

SOURCE: Adapted from Reference 14. Copyright 2004 Elsevier.

Most biomass resources gave essentially complete conversion of their carbohydrates to polyols within 3-5 hr at 165-190 °C with 5-15% (w/v) biomass loading (Table II). A negative Benedict's test (Cu) for reducing sugar (≤100 ppm) in many reactions represents essentially quantitative reduction. Acetone extracted cedar sawdust was the most difficult to digest and perhaps is the result of high lignin content, greater crystallinity of cellulose, or the prior processing.

The other woody resources also gave moderately positive copper tests due to a slow cellulose hydrolysis. However, cane, switchgrass and corn fiber all gave strongly positive Benedict's tests. These three biomass resources also did not wet easily and their protein content may have interfered with the hydrogenation. Other experimental details are given in reference 14.

Table II. Selected Biomass Reaction Data

Rxn	Biomass	°C	hr	% Conv	% Solid	%Mat. Bal.	Test
1	Oak	160	3.0	49.8	54.1	103.9	
2	Oak	187	3.0	71.8	35.8	106.2	
3	Oak	.182	17.0	71.5	20.4	90.5	
4	Oak	193	3.2	64.4	19.4	82.6	+ Fe
5A	Oak	190	4.3	64.8	25.9	90.7	
5B		190	4.7	55.7	41.0	96.8	
5C		189	5.2	47.6	na	na	
6	Pine	186	4.3	68.1	36.5	104.7	+ Cu
7	Aspen	186	4.3	58.8	42.8	101.6	+ Cu
8	Corn Fiber	167	4.0	76.5	23.4	99.9	++ Cu
9	Cellulose	182	4.8		39.0		
10	Cellulose	189	4.1		9.0		
11	Cellulose	170	12.0				

SOURCE: Adapted from Reference 14. Copyright 2004 Elsevier.

Crystalline cellulose is known to hydrolyze slowly and also showed a wetting problem. The higher temperature cellulose reactions shown in Table II afforded only 91% conversion at 189 °C compared to the slower digestion at 182 °C. These reactions show that the hydrolysis rate can be almost doubled, albeit with faster stirring, with an ~10 °C increase in temperature. Both reactions illustrated a continuing slow hydrolysis. An overnight reaction at 170 °C gave complete cellulose conversion.

Conversion Estimation

Upon hydrolysis of C_6 polysaccharides shown in eq **1** and the subsequent hydrogenation, the molecular weight increases by the values of water and hydrogen to afford sorbitol. Therefore, the effective formula weight for the monomeric unit is 162 g/mol.

$$
\begin{array}{cccc}
& H+ & & H_2 \\
(C_6H_{10}O_5)_n + H_2O & \rightarrow & (C_6H_{10}O_5)_{n-1} + C_6H_{14}O_6 & \rightarrow & C_6H_{14}O_6 \quad\quad (1) \\
(EW = 162)_n \quad MW = 18 & & MW = 180 & MW = 182
\end{array}
$$

Similarly, a C_5 polysaccharide anhydrous monomeric unit (132 g/mol) gives xylitol, (MW = 152 g/mol). Since hemicellulose is a mixture of different size (C_5 & C_6) and type (aldehyde and acid) of sugars that vary with each biomass resource, exact calculations are complex. However, the amount of total C_5 and C_6 contents are known from the literature. Such data allow calculation of the theoretical amount of mixed polyols that should result from each biomass resource. The conversion of carbohydrates shown in Table II is a comparison of the actual conversion to the theoretical values for each biomass resource as listed in Table I. Adding this conversion of solubilized biomass (polyols, gravimetric) together with the insoluble solids (lignin) gives a material balance *estimate*. A balance of this type provide an internal quality control and error checks for drying, sampling, etc, whereas a gross conversion views only unreacted biomass. For example, a low conversion (64.4%) to soluble polyols and a low solids recovery (19.4%) totals to a low material balance for Reaction 4 (82.6%) which indicates that some further reaction must be occurring (discussed later). Further, the 111.6% material balance for the switchgrass reaction, points to a sampling or drying error. Even though a few such imbalances are shown in Table II, these data demonstrate the importance of the conversion comparison to effectively present this survey of the **IDAHH** process.

Reactor #1 used a loose glass jar liner that was open at the top. While the liner was convenient for contents removal, it was also required to preclude corrosion in long-term reactions. Reactions therein could not be vigorously stirred because they would splash over the top and not continue to react properly. As a result, small amounts of biomass were deposited above the solvent line and only partially reacted. These unreacted solids were typically 0.5-9% of the total reaction. Such variation is reflected in results shown in Table II which are therefore considered as close estimates.

Acid Catalyst Survey

Initial reactions patterned after Soviet reports were conducted, with and without the glass liner, in Reactor #1. Reaction 1 (Table II) was for 3 hours at 160 °C whereupon only half of the material had reacted. Therefore, longer time and higher temperature than in the literature was investigated. A reaction for >20 hr with 0.7% H_3PO_4 and with no liner present led to a green colored solution containing Fe and Ni. Similar reactions with 0.7% H_2SO_4 showed even more evidence (color) of corrosion. One experiment with twice the H_3PO_4 concentration (1.5%) and a 20 hr reaction afforded a solution with sizeable amounts of leached metals: 260 ppm Fe, 67 ppm Cr, 46 ppm Ni, and 3.9 ppm Ru. A 0.7% H_3PO_4 solution obtained in the same manner averaged about 6 times less metal ions for all except Ru, which was below the detection limits. Presumably, the Ru might be 0.6 ppm and which would represent about 1% of the Ru used. Local leaching and re-reduction on remaining metal surfaces might be the cause of increased aggregation of spent Ru catalysts. Less leaching occurs in shorter reaction times but will still lessen catalyst life.

The glass liner greatly reduced corrosion of the stainless steel reactor. Removing the cooling coil then left only the stirring shaft and propeller exposed to acid but lengthened the cooling time after each reaction and before the system could be opened. In this mode Reaction 3 (17 hr, 182 °C) showed little corrosion. In some contrast to the Soviet reports, lower acid content could perhaps be used successfully. For example, Reaction 2 used 0.35% H_3PO_4 solution to give almost complete reaction in 3 hr at 187 °C with 10% biomass load, indicating that a slightly longer reaction time would be necessary. Since even slower digestions were experienced with crystalline cellulose, the use of 0.7-0.8% H_3PO_4 to speed the rate of hydrolysis was continued.

Hydrogen Consumption

IDAHH reactions were monitored by data acquisition and control software. Hydrogen consumption can be visualized by a differential hydrogen pressure (actual versus theoretical). A typical plot of the relative hydrogen consumption versus time and temperature is shown for Reaction 6 (Figure 5). Reduction clearly began during the rise to operating temperature, at ~100 °C. Easily hydrolyzed hemicellulose provides aldoses for the initial fast portion of the reduction. The change in slope of the hydrogen consumption shows that the easily hydrolyzed components were completely reacted after 2 hr while the continuing gradual slope shows the same rate as the more difficultly hydrolyzed microcrystalline cellulose. Since the 5% cellulose reaction at 189 °C was only 91% complete in 4.1 hr, it was not too surprising that the cellulose portion of pine reaction shown in Figure 5 was not quite complete using similar conditions, as evidenced by the hydrogen consumption curve not reaching a constant plateau. Additional evidence for incomplete reaction was the high proportion of solids remaining (36.5%) compared to the theoretical 29.4% (Table I) as well as a positive test for reducing sugars. Furthermore, batch reactions of this type follow an infinite dilution pattern for the remaining substrate. Therefore reaction times, even for starch, were often extended past the initial plateau point of the hydrogen consumption plot for ~1 hr so that only traces (<100 ppm) of aldoses remained.

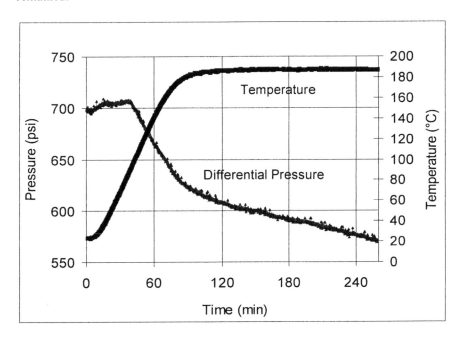

Figure 5. Hydrogen Consumption Curve for Reaction 6: Pine.

Selective Reaction of Carbohydrates versus Lignin

Lignin does not react to liberate any detectable phenols under conditions where carbohydrates are converted virtually quantitatively to polyols. Indeed, this proves the *cleanness* of the **IDAHH** process since lignin separation and removal has historically been the most difficult task. However, a clean separation from lignin does not infer that the polyols are pure; other small contaminants (salts, acids, amino acids, etc.) may be removed by other means.

Multiple reactions were conducted with 1% oak at 165 °C for periods up to 35 hr and all gave negative tests for phenols with the FeCl$_3$ method (*15*). Reactions 2-4 (Table II) show phenols were not present at ~180°C for 17 hr as well as in 3 hr reactions at ~185 °C, but began to show a slightly positive test above 190 °C for 3 hr. Four reactions on oak sawdust at ~185 °C had material balances of ≤100% (no drying error) and averaged 21.2% unreacted solids (lignin). Such lignin values are in good agreement with values also obtained by the H$_2$SO$_4$ method (*16*) used on our oak (21.1%) and pine sawdust (29.0%). This remarkable selectivity for digestion of carbohydrates versus lignin below 190 °C is due to both dilute solutions of H$_3$PO$_4$ along with its lower intrinsic acidity (pK$_a$ = -2) versus H$_2$SO$_4$ (pK$_a$ = -10). Furthermore, any "acid soluble lignin" dissolved under these conditions did not afford phenols and perhaps reprecipitated during the cooling phase after the reaction.

Catalyst Study

Most reactions were conducted with 5% ruthenium supported on carbon powder at 10% total weight versus the amount of biomass. Powdered catalyst with high surface area and high catalyst ratio was initially chosen to allow a fast hydrogenation reaction to intercept aldoses before the usual degradation occurred. Therefore, it also became important to evaluate the reuse of catalyst in subsequent reactions. Powdered catalyst was recovered together with unreacted lignin upon filtration from the polyol solution. Powered catalyst reuse was briefly studied using catalyst cleaned of lignin by an ambient pressure hot base extraction. Lignin extraction efficiency was ≥73%, limited in part by prior incomplete hydrogenolysis of biomass. Reactions 6 and 7 (Table II) sequentially reused catalyst that was initially used in Reaction 5. However, the conversion to polyols dropped significantly with each recycle, which was also mirrored by progressively slower consumption of hydrogen. This prompted a separate study of starch to both avoid many potential catalyst poisons as well as the concomitant need to strip lignin from the catalyst for each sequential reuse. In this manner a series of starch reactions with catalyst recycle was conducted in a Hastelloy C reactor at where insignificant corrosion occurs. Catalyst from the first reaction was reused sequentially three times, wherein polyol yields and rate of hydrogen consumption decreased steadily with each reaction. Catalyst leaching studies were inconclusive due to weight gain.

Catalyst deactivation may be due to 1) dramatic change in catalyst surface area, 2) poisoning due to leached reactor metals, albeit less likely with Hastelloy C, 3) poisoning by sulfur containing amino acids from protein hydrolysis, 4) oxidation by incidental air of active sites to give ruthenium oxides, and 5) absorption of gluconic acid which adds weight and blinds active sites. Arena's finding (*17*) supports the latter two possibilities; ruthenium catalyst life can be significantly extended if the feedstock streams (glucose and hydrogen) are reduced to <0.5 ppm of oxygen. Accordingly, the flushing cycles for these batch reactors were improved to remove almost all traces of oxygen, which did prove to extend catalyst activity. Ruthenium oxides might aid the oxidation of glucose to gluconic acid, or might be easily leached into acid media under these conditions. The most active and dispersed catalyst would leach first which would greatly diminish the effective surface area and overall catalyst activity. Arena also found a synergistic poisoning effect between iron, oxygen, and gluconic acid on Ru catalyst even under neutral conditions for glucose (18). We are developing a continuous flow reactor system to preclude opening to remove products and curtail catalyst exposure to ambient oxygen.

Ideally, an **IDAHH** process using raw biomass resources would necessarily require some facile removal of catalyst from both the polyol solution and the unreacted lignin solids. The Hastelloy C reactor was therefore fitted with a basket so that after a reaction the pelletized catalyst could be screened from the polyol/lignin slurry. Lignin is then simply filtered from solution. The pelletized catalyst and the improved oxygen flushing protocol were both tested during four sequential starch reactions reusing the same 0.5% Ru/C pellets. With the improved flushing protocol, little catalyst deactivation was observed. Hydrogen consumption shows a fast, complete reaction in ~45 min at 165-170 °C.

Arena (19) devised more active and more hydrothermally stable mixed titania supports for Ru hydrogenation of glucose at ~120 °C. Elliot has recently developed an even longer activity Ru/TiO_2 (rutile) catalyst (20). Operation at lower temperature (100 °C) and with pH neutral conditions (for glucose) obviously helps to extend catalyst life compared to the conditions required for the raw biomass reactions herein. However, the rutile-supported ruthenium might be less prone to absorption and is scheduled for **IDAHH** tests.

Higher Temperature and Time Effects on Secondary Reactions

Several starch reactions point to a secondary reaction occurrence. A 2 hr reaction at 163 °C was complete (97%). However, two 2.8 hr reactions at 170 °C both gave, gravimetrically, an apparent "loss" in polyol (74 and 81%). HPLC analyses indicated 95% sorbitol and a few minor polyols for the 2 hr reaction whereas the 2.8 hr reactions were almost identical with 70% and 71% sorbitol. More importantly, a peak at shorter retention time than any polyol standard appeared due to a continuing reaction of acyclic polyols to the anhydro form produced by dehydration. Similarly, Soviet research gave only sorbitan (90%) in stronger acid (2% H_2SO_4) at 180 °C for 2 hr. Dehydration explains the low yields of polyols not only with starch at 170 °C but also in other more severe reactions in this report, e.g., Reactions 4 and 5 (Table II) at ~190 °C. Obviously, such dehydrations to anhydro polyols exacerbate gravimetric material balances for hydrogenolyses even with the less acidic H_3PO_4. Isomerization is acid catalysed and results in small amounts of other polyols at higher temperatures. While mixed polyols, including sorbitan, may not be desirable for other uses, they are suitable for a "polyol platform" for conversion to other chemicals and fuels.

Estimated Cost of the IDAHH Polyols Process

Modeling this process with chemical engineering software (Superpro Designer) with default settings for plant costs, and using \$40/t for biomass, \$0.54/lb for H_2, \$0.35/lb for H_3PO_4, only a 20% slurry of biomass, and no recovery cost for acid, allows a gross estimation of costs (albeit not including catalyst costs). Costs then calculate as >\$0.06/lb for mixed polyols from pine. While not easily comparable without some costs for catalyst, acid recovery, etc., these *minimum* costs are considerably less than present costs of sorbitol (\$0.24-\$0.36/lb). A completely integrated scheme includes acid recovery when fuels are produced and the prorated cost of lignin content in the biomass is factored into the cost of phenols. The superb yields and relatively inexpensive raw biomass are the main factors contributing to such reduced processing costs.

Chemicals and Fuels from the Polyol Platform

Polyols produced with this strategy may not be suitable for fermentation into ethanol, but many other routes to chemicals and fuels now become available as Figure 4 shows. Products range from hydrogen to hydrocarbons to alcohols. Phenols can also be produced from lignin by-product. Any partial hydrogenation of lignin during the polyols fractionation is included in the subsequent complete hydrogenation to phenols and thus is not a lost cost. All routes and products shown herein are known to occur, but each is in a different stage of development.

Our earlier research established a *chemical* process (Figure 6) for the production of various hydrocarbon fuels and chemicals from biomass polysaccharides (*21,22,23*). Two steps are required to complete the process from polyols. Polyols are converted into mainly liquid hydrocarbons by reduction with boiling hydriodic acid. Hydrocarbons phase separate and the aqueous acid is recycled. Step 2 converts remaining halocarbons into alkenes. An electrochemical regeneration of the primary reducing solution provides an economically improved process and is the subject of another patent application.

In this process, the C_6 component is more desirable for two reasons: 1) less reduction cost and 2) C_6 polyols form mainly lower oligomers (C_{12}) while C_5 polyols tend to form higher oligomers. For the biomass resources tested, pine proved to be the most practical resource with high total carbohydrates (69%) and a significantly higher C_6/C_5 ratio (5.2). The C_{12} dimers consist of ~70 isomers within a boiling point range of 180-210 °C which is more like kerosene but with a much higher (78-83) blending octane number due to the preponderance of cyclic and branched structures. Similarly, the C_{11}-C_{12} cross dimers from mixed polyols provides a fuel comparable to commercial gasoline.

Presently, a very high purity value-added hydrocarbon, solvent grade hexane, is achieved from starch via sorbitol. This is a high volume solvent used for polymer reactions as well as for the extraction of oils from many types of seeds. With starch, prior processing has already achieved removal of insoluble

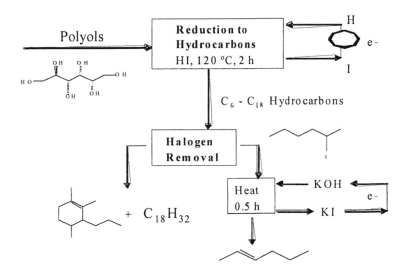

Figure 6. Polyols to Hydrocarbons Process

lignin and cellulose, as well as hemicellulose and ash, but to our surprise, traces (~0.5%) of protein remain but which can be dealt with. As expected, the more highly refined starch is more costly than raw biomass and complete reduction to hexane is slightly more expensive than the dimer hydrocarbons. However, this process appears economic for even hexane, especially at today's fossil oil derived costs. Again quite sophisticated software models calculate costs for hexane to be about $0.90/gal since about half the mass is lost as the oxygens are removed. Since the fuel alcohols from the polyols platform retain oxygen, a greater volume of these much lower heating value fuels is obtained for about half the hexane price.

Dumesic's process can produce hydrogen from monomeric sugars and the preferred substrate is indeed polyols (24). Such hydrogen can perhaps best be used internally in the biorefinery to initially fractionate biomass and in subsequent reactions as well. Alkenes produced in the hydrocarbon process discussed above can be hydrogenated to preferred products such as hexane. Additionally, mixtures of methanol and ethanol can be obtained directly from starch/glucose by catalytic hydrogenolysis using 5% H_2SO_4 and Ru/C at 200 $^{\circ}$C in 4 hr (25). Currently, we are developing a similar alcohol path from the polyols platform where certain raw biomass byproducts have already been removed. We have also confirmed Soviet reports to further react byproduct lignin by more severe hydrogenations in base to afford >35% liquid phenols.

Conclusion

Conversion of biomass carbohydrate polymers, hemicellulose and cellulose, to polyols can be complete (>99%) with 15% load at <190 $^{\circ}$C for 3-5 hr with simultaneous hydrolysis by ~0.8% H_3PO_4, efficient stirring and hydrogenation with Ru catalyst. Pine is the most practical biomass resource screened because of a high total carbohydrate content with high C_6/C_5 ratio, and only traces of protein. Reaction temperature of $\leq$190 $^{\circ}$C, with this acid, assures polyol solutions with no phenols; a very "clean fractionation" of biomass. Catalyst activity must be maintained at least by 1) use of acid resistant construction materials and 2) removal of ambient oxygen in the reactor. Further catalyst protections are currently under study. Both pelletized catalyst and unreacted lignin are recovered and separated respectively by simple mechanical means. Thus, there appears to be considerable potential in the further development of this strategy of "intercepted dilute acid hydrolysis" (**IDAHH**) to provide low cost polyols from biomass. Polyols provide an alternative sugar platform from which a variety of fuels and chemicals can be produced.

Acknowledgements

The Robert A. Welch Foundation Chemistry Department Grant, a National Renewable Energy Lab contract, a U.S. Department of Energy Basic Energy Sciences - Advanced Research Projects grant, and the Texas Higher Education Coordinating Board - Advanced Technology Program supported this research.

References

1. Bozell, J. J.; Black, S. K.; Myers, M. *Proceedings of the 8th International Symposium on Wood and Pulping Chemistry;* Vol. 1: Oral Presentations; Helsinki, Finland; 6-9 June 1995. pp 697-704.
2. Bulls, M. M.; Watson, J. R.; Lambert, R. O.; Barrier, J. W. In *Energy From Biomass and Wastes XIV.* Klass DL, Ed.; IGT: Chicago, IL, 1991; pp 1167-1179.
3. McLaughlin, S. B.; Samson, R.; Bransby, D.; Wiselogel, A. *Proceedings of the 7th National Bioenergy Conference. Southeastern Regional Biomass Energy Program;* Vol. I, Nashville, TN, 15-20 September 1996; p. 1-8.
4. Goldstein, I. S. In: *Organic Chemicals from Biomass;* Goldstein, I.S., Ed.; CRC Press: Boca Raton, FL, 1981; pp 101-124.
5. Gould, J. M. U.S. Patent 4,806,475, 1989.
6. Ladish, M. R.; Svarczkopf, J. A. *Bioresource Technology* **1991,** *36,* 83-95.
7. Xiang, Q.; Kim, J. S.; Lee, Y. Y. *Appl. Biochem. Biotech.,* **2003,** *105-108,* 337-352.
8. Liu, C.; Wyman, C. E. *Ind. Eng. Chem. Res.* **2003,** *42,* 5409-5416.
9. Torget, R. W.; Padukone, N.; Hatzis, C.; Wyman, C. E. U.S. Patent 6,022,419, 2000.
10. Balandin, A. A.; Vasyunina, N. A.; Barysheva, G. S.; Chepigo, S. V. *Bull. Acad. Sciences of USSR (Chem)* **1957,** 403.
11. Balandin, A. A.; Vasyunina, N. A.; Chepigo, S. V.; Barysheva, G. S. *Bull. Acad. Sciences of USSR (Chem.)* **1959,** 839-842.
12. Balandin, A. A.; Vasyunina, N. A.; Chepigo, S. V.; Barysheva, G. S. *Bull. Acad. Sciences of USSR (Chem.)* **1960,** 1419.
13. Sharkov, V. I. *Angew. Chem. Int. Ed. Eng.* **1963,** *2,* 405-409.
14. Robinson, J. M.; Burgess, C. E.; Bently, M. A.; Brasher, C. D.; Horne, B. O.; Lillard, D. M.; Macias, J. M.; Mandal, H. M.; Mills, S. C.; O'Hara, K. D.; Pon, J. T.; Raigoza, A. F.; Sanchez, E. H.; Villarreal, J. S. *Biomass and Bioenergy,* **2004,** *26(5),* 473-483.
15. Pavia, D. L.; Lampman, G. M.; Kriz, G. S. *Introduction to Organic Laboratory Techniques,* 3rd Edition; Saunders College Pub.: Ft. Worth, TX: 1988, p. 452.
16. TAPPI method T 222 om-88. In: TAPPI Test Methods; TAPPI Press: Atlanta, GA, 1994.
17. Arena, B. J. U.S. Patent 4,510,339, 1985.
18. Arena, B. J. *Appl. Catal. A: Gen.* **1992,** *87,* 219-229.
19. Arena, B. J. U.S. Patent 4,487,980, 1984.
20. Elliot, D. C.; Werpy, T. A.; Wang, Y.; Frye, J. G., Jr. U.S. Patent 6,235,797, 2001.

21. Robinson, J. M.; Burgess, C. E.; Mandal, H. D.; Brasher, C. D.; O'Hara, K. O.; Holland, P. L. *Amer. Chem. Soc. Fuel Chem. Div. Preprints* **1996**, *41(3)*, 1090-1094.
22. Robinson, J. M. U.S. Patent 5,516,960, 1996.
23. Robinson, J. M.; Banuelos, E. B.; Barber, W. C.; Burgess, C. E.; Chau, C.; Chesser, A. A.; Garrett, M. H.; Goodwin, C. H., Holland, P. L., Horne, B. O., Marrufo, L. D., Mechalke, E. J., Rashidi, J. R., Reynolds, B. D., Rogers, T. E., Sanchez, E. H.; Villarreal, J. S. *Amer. Chem. Soc. Fuel Chem. Div. Preprints* **1999**, *44(2)*, 224-228.
24. Cortright, R. D. .Davda, R. R.; Dumesic, J. A. *Nature* **2002,** *418*, 964-967,
25. Abreu, C. A. M.; Lima, N. M.; Zoulalian, A. *Biomass Bioenergy* **1995**, *9*, 487-492.

Chapter 23

Optimization of the Process Conditions for the Extraction of Heteropolysaccharides from Birch (*Betula pendula*)

Peter Karlsson[1], Johannes P. Roubroeks[1], Wolfgang G. Glasser[2], and Paul Gatenholm[1]

[1]Biopolymer Technology, Department of Materials and Surface Chemistry, Chalmers University of Technology, SE–412 96, Göteborg, Sweden
[2]Department of Wood Science and Forest Products, Virginia Polytechnic Institute and State University, Blacksburg, VA 24061

The extraction of a heteropolysaccharide ("hemicellulose") fraction from birch wood chips by mild acid hydrolysis was optimised using process conditions involving treatments with hot water, dilute acetic acid and dilute sulphurous acid. Variable parameters included time temperature and acid concentration. Following prehydrolysis (i.e., treatment prior to delignification for pulp production by the kraft process), the wood chips were extracted with alkali, neutralized, and purified by dialysis using a membrane with a nominal molecular weight cutoff of 12-14 thousand Daltons. The freeze-dried heteropolysaccharides were analyzed for yield, chemical composition, and molecular weight. Best yields were obtained with low acid concentration, high temperatures and long reaction times. Total yields reached levels of ca. 30% of available non-cellulosic polysaccharides. The composition of the highest yield fractions had xylose and lignin contents of 53-62 and 15-26%, respectively. Molecular weight measurements by GPC in DMSO suggested DP_w values ranging from about 30 to 150. The results suggest that it is possible to extract a polymeric heteropolysaccharide fraction from birch wood chips by mild acid-catalyzed prehydrolysis prior to kraft pulping.

Introduction

Wood is a complex hierarchical composite with excellent material properties. The secondary cell wall is a multiphase structure, consisting mainly of rigid cellulose microfibrils (40%) embedded in an amorphous matrix of non-cellulosic heteropolysaccharides (30%) associated with lignin (25%) (*1*). The most abundant heteropolysaccharide component in the secondary wall of hardwood is glucuronoxylan (GX), or its 4-O-methyl ether derivative, which constitutes 25-35% of the woody mass (*2*). GX is constituted of a backbone of (1→4)-β-D-xylopyranose residues substituted with α-glucuronic acid at the O-2 position. The substitution with glucuronic acid along the chain is one per ca. ten xylose residues (*3*). In the native state, GX is partially acetylated (~70%) at the C-2 or C-3 position (*4*). While GX is a major fraction of the non-cellulosic polysaccharides, other fractions include a variety of polymers containing rhamnose, arabinose, mannose, galactose, and glucose. The non-cellulosic wood fraction is therefore variably referred to as "hemicelluloses", "polyoses", or heteropolysaccharides (*5,6,7*). Hemicelluloses function as a compatibilizer or (anti-) plasticizer in the molecular organisation of cellulose with lignin (*8*). Moreover, it has been suggested that hemicelluloses contribute to the twisting of cellulose microfibrils during cell wall formation (*9*). The physiological role of lignin in the plant cell wall is mainly as a stiffening, sealing and complexing agent, and as an antioxidant (*10*). Lignin is a phenolic compound often linked to hemicelluloses through non-covalent or covalent bonds. The suggested ester linkages are degraded by alkaline conditions while some ether linkages are more stable (*11*).

It is suggested that hemicelluloses play an important role in pulp and papermaking (*12,13*). In kraft pulping, which involves severe alkaline treatment conditions at high temperatures, the heteropolysaccharides are degraded by molecular weight and by "endwise peeling" (*7,14*). This reduces the positive effects of high molecular weight and ionic charge. Due to value-added awareness many studies have been performed concerning the stabilisation of the carbohydrates under alkaline conditions. The use of sodium borohydride inhibits primary peeling and increases yield considerably, but it is quite expensive in commercial practice (*14*). Other methods for hemicellulose stabilisation include polysulphite (*7,14*). In contrast to stabilisation, attempts were also made to preserve hemicelluloses in pulps via organosolv pulping (*15*), or to recover solubilized non-cellulosic sugars by pre-treatment with high pressure steam ("steam explosion"). However, none of these processes has resulted in commercial practice (*16*). Other methods for the extraction of high molecular weight heteropolysaccharides from wood and agricultural harvesting residues have been demonstrated on pilot scale using a variety of approaches (*17,18,19,20*). One such extraction procedure involved delignification by either hypochlorite or organic solvents followed by extraction with alkali. This approach yielded a heteropolysaccharide fraction comprising between ½ and 2/3 of all non-cellulosic polysaccharides in biomass (*18,21*).

Chemical wood pulping is an important industrial process, with a relatively low yield. However, this process is attractive, because of the low requirements of wood quality and wood species. Moreover, it includes very short cooking times due to the easy penetration of alkaline solutions into wood. A well established processing of the spent liquor, generation of process heat, recovery of the pulping chemicals and production of valuable by-products such as turpentine and tall oil are also advantages of the kraft cooking. Major drawbacks are however, the odour and the dark colour of the unbleached pulp.

The central operation in a chemical pulp mill is the digester. This step separates the fibres from each other and simultaneously solubilizes the lignin and the hemicelluloses. Prior to cooking, the wood chips are impregnated by steam in order to facilitate the penetration of cooking chemicals.

In this study, we aimed at obtaining a heteropolysaccharide fraction in high yield by performing a hydrolytic pre-treatment step on wood chips followed by a room temperature extraction. Pre-treatment conditions involved acetic acid (AcOH), sulphur dioxide (SO_2) and high temperature water. The isolated fractions were analysed for yield, chemical composition (sugars, uronic acids, and lignin) and molecular mass.

Experimental

Material

The raw material used consisted of birch chips (*Betula pendula*) from the southern parts of Sweden. The wood chips were cut in a pulp mill (Södra Cell, Mörrum, Sweden). However, to obtain a more workable dimension, a knife was used to manually cut the chips down to an approximate size of 0.5-1.5 cm length, 0.5 cm wide and 0.3 cm thick.

Methods

Experimental Design

An experimental design (MODDE 4.0 Graphical software, UMETRI AB, Umeå, Sweden) was used to describe the variable parameters used in this study (Figure 1). The parameters time, temperature and acid (acetic acid, AcOH, and SO_2) concentration were varied to investigate the effects on the extracted hemicelluloses. The parameters selected for the study represented a practical range and aimed for solubilization of the hemicelluloses in high yield, high purity, and with preservation of molecular weight.

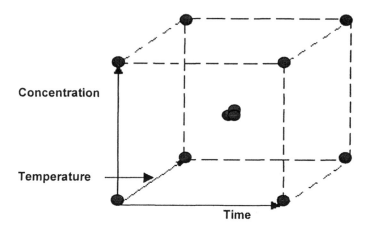

Figure 1. Graphic representation of the experimental design.

Treatment of wood chips

The wood chips were treated according to the experimental design and the flow diagram depicted in Figure 2. This flow scheme shows all the operations performed on the wood chips and all the fractions collected during the trials. After each treatment the wood chips were weighed and the moisture content was determined. All fractions were kept at 6°C until further analysis.

In the impregnation step, 100 g of wood chips were treated with water or acetic acid in a ratio of 5:1. In the case of SO_2, water was also used for the impregnation. It was made certain that all the wood chips were covered with impregnation liquid. In a first step, pressure was reduced for 30 min to remove air trapped in the wood chips. N_2-gas pressure (5 bar) was then applied for 30 min. Both treatments were performed at room temperature. After impregnation, the free (unbound) liquids were decanted and collected. After analysis for solids content, the samples were discarded. Prehydrolysis conditions were set according to the experimental design, and the experiments were performed using a pre-heated, rotating polyglycol bath. Samples treated with water were kept at 100°C (30 and 60 min), 120°C (45 min) and 140°C (30 and 60 min). AcOH-treatment employed three different concentrations: 0.01 M (30 and 60 min at 100°C and 140°C), 0.05 M (45 min at 120°C) and 0.1 M (30 and 60 min at 100°C and 140°C). The digestion with SO_2 also involved three different concentrations: 2 w/w% (30 and 60 min at 100°C and 140°C), 4 w/w% (45 min at 120°C) and 6 w/w% (30 and 60 min at 100°C and 140°C). The samples at 45 min and 120°C were performed in triplicate to estimate the variation of the experiment (Figure 1). After the treatment, the tubes were kept in a cooling water bath for at least 12 minutes. After cooling, the liquids were filtered with the aid of a Büchner funnel and collected. The wood chips were extracted with 1 M NaOH at room temperature making sure that all the wood chips were in full contact with the alkali. After 16 h of extraction, the alkaline liquid was collected and the wood chips were washed with 0.5 L water. The two fractions were combined. All fractions were neutralised (pH 7) to prevent further hydrolysis. During this neutralisation some precipitate ("pellet") formed, which was removed by centrifugation (4000 rpm, 10 min). The supernatant was decanted and the pellets were collected, air dried and stored in a freezer until further analysis.

All the fractions were subjected to dialysis to remove salts and low molecular weight hydrolysis products. This procedure was performed for 3 days against running deionised water. The nominal molecular weight cut-off of the tubing was given as 12-14 thousand Daltons by the manufacturer (Spectra/Por 5). All fractions were freeze dried. Yield is defined as the amount of total extracted material based on 100g of treated wood chips. The amount of extracted material comprised both solubles in the extraction liquids and solids in pellets.

Gas Chromatography

All the freeze dried samples, including the pellets, were subjected to gas chromatography (GC). The method is based on hydrolysis and derivatisation to alditol acetates of the sugar constituents and performed according to STFI method AH 23-17 and AH 23-18 (*22, 23*).

The equipment Varian 3380 with autosampler 8200 (Varian, Palo Alto, CA,USA) consisted of an injector (Varian 1079 split/splitless), a column (Supelco SP-2380 30 m × 0.25 mm) and a FID detector (Varian). The temperature starts at 150°C and increases to 250°C at 15°C/min. The nitrogen flow is 1 mL/min. The injection volume of the alditol acetates is 1 µL and 2-deoxy-galactose is used as an internal standard.

High-Performance Size-Exclusion Chromatography combined with Multi Angle Laser Light Scattering detection

For the separation of hemicellulose of the leaching and washing fractions a high-performance size-exclusion chromatography system combined with refractive index, UV and multi-angle laser light scattering detection (HPSEC-UV-MALLS-RI) was used. The system consists of a Waters 2690 instrument with online degasser, autosampler and column oven (Waters Milford, MA, USA) and three serially connected columns (OHpak SB-806M HQ, OHpak SB 804 HQ and OHpak SB-803 HQ, Shodex, Showa Denko KK, Miniato, Japan) were used. The eluent was 0.1 M NaNO$_3$ containing 0.01% NaN$_3$ at 0.5 mL/min. Detectors were refractive index (Optilab DSP, Wyatt Technology Corp., Santa-Barbara, CA, USA), UV monitor (Shimadzu SPD-10A/vp, Shimadzu Corp. Kyoto, Japan) and multi-angle laser light scattering (MALLS; Dawn DSP equipped with a He-Ne laser at 632.8 nm, Wyatt Technology Corp., Santa-Barbara, CA, USA). Columns and RI detector were controlled at 50°C.

Samples were ground in a vibrating mill (Perkin Elmer, Germany) for 1 min. and concentrations (typically 2.0-2.5 mg/mL) were filtered (0.45 µm) before injection of 100 µL. Data for molecular weight determination were analysed using ASTRA software (version 4.73.03, Wyatt Technology Corp., Santa-Barbara, CA, USA) based on a dn/dc of 0.110 (24).

In the DMSO-water system (90:10) with 0.05 M LiBr, three serially connected columns were used: Gram 30, 100 and 3000 (PSS). The detector system consisted of a UV detector (PLLC 1200), and a RI detector (Shodex RI-71). The flow rate was 0.4 ml/min at 60°C. Molar masses were calculated from the viscosity and RI-signals by universal calibration using pullulan standards.

Results and Discussion

Experimental Design

The experiments were performed according to the experimental design described in the experimental section. Three parameters, time, temperature and acid concentration (pH) were varied and the results were evaluated by the graphical analysis software programme MODDE 4.0. The experimental data were subjected to modelling using multiple linear regression software. The optimisation involved a function of all parameters measured and the interaction terms between the parameters:

$$F_{(x,y,z)} = ax + by + cz + dxy + exz + fyz$$

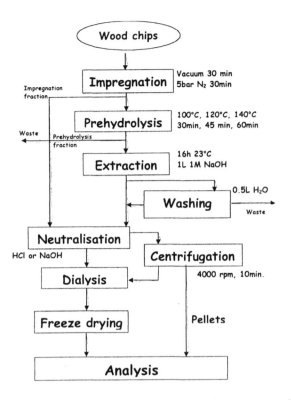

Figure 2. Flow scheme summarizing the isolation protocol of heteropolysaccharides by mild hydrolysis of wood chips.

where x = time, y = temperature, z = acid concentration (M for AcOH and w/w% for SO_2); and a, b, c, d, e, f are curve-fitting constants.

The typical contour-plots shown in Figure 3 reflect the parameters involved in the pre-hydrolysis process. The parameter effect on the yield of extracted material after alkaline extraction is revealed.

When water was used as a prehydrolysis liquid, it was observed that time and temperature had only a minor effect on yield of the extracted wood components. Yield was highest when temperature was high and pre-hydrolysis time was long. However, this sample also produced the highest weight loss. For this reason the sample selected for compositional analysis was that generated at low temperature and long pre-hydrolysis time with resulting modest yield (3.3% w/w). In general, elevated temperatures and longer hydrolysis times resulted in

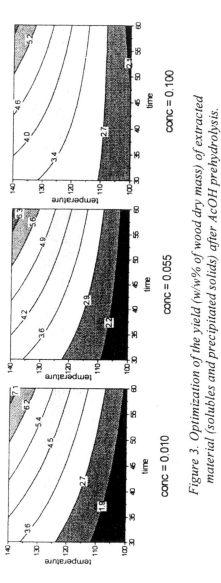

*Figure 3. Optimization of the yield (w/w% of wood dry mass) of extracted
material (solubles and precipitated solids) after AcOH prehydrolysis.*

higher yields and greater weight losses. At lower temperatures, time appeared to be more important for weight loss.

The experiments with AcOH produced the results illustrated in Figure 3. At 0.01 M AcOH (140°C, 60min) the highest yield was 8% w/w (upper right-hand corner). The yield was found to decrease with increasing AcOH concentration. This can be attributed to excessive hydrolysis of heteropolysaccharides and removal of mono- and oligosaccharides by dialysis. It can be observed (Figure 3) that time and pH are the most important parameters while temperature and the interaction between these parameters are less important. Weight loss of the remaining wood mass is mainly influenced by time and temperature and not by pH.

When SO_2 was used, it was observed that low temperature and long hydrolysis times had the greatest influence on product yield. Weight loss was dictated primarily by temperature. The lowest concentration of sulphur dioxide (2 w/w%), lowest temperatures and longest hydrolysis time resulted in the highest yield (8.3 w/w%). Higher concentrations of SO_2 resulted in extensive degradation of hemicelluloses.

In conclusion, a selection of samples was made for each treatment (water, AcOH and SO_2) based on the highest yield of extracted wood components in combination with lowest weight loss (i. e., preservation of wood mass for subsequent pulp production).

The composition of the selected samples is depicted in Figure 4. It can be observed that the analysis results obtained with the parent wood sample corresponded to previously reported values in the literature (*24*). The samples selected all exhibited similar xylose yields. The content of heteropolysaccharides was highest in the AcOH sample (7 w/w% based on extracted material), whereas the lignin content was highest in the sample treated with SO_2. This may be attributed to the conversion of lignin into water-soluble lignin sulfonates retained by the dialysis tubing as water-soluble substrates. Treatment at low temperature (100°C) resulted in little water-insoluble mass (ie., pellet fraction) in comparison with higher temperatures (more pellet mass). The samples with the highest content of isolated heteropolysaccharides and lowest lignin content are, of course, of greatest practical interest.

Molecular mass

The size exclusion chromatograms revealed the polymeric character of the heteropolysaccharide fractions (Figure 5). The light scattering (LS) intensities for each fraction were rather low which indicates that little mass is represented in an aggregated state. The refractive index (RI) responses indicated that the highest molecular mass in the extracted material is obtained after treatment with water. This is in contrast to the treatment with SO_2, which produced the lowest

330

molecular weight fractions. The treatment with AcOH resulted in fractions with broad molecular mass distributions and average molecular weights intermediate to the molecular masses of the fractions generated by the treatments with H_2O and SO_2. The calculated average molecular weights are summarized in Table I.

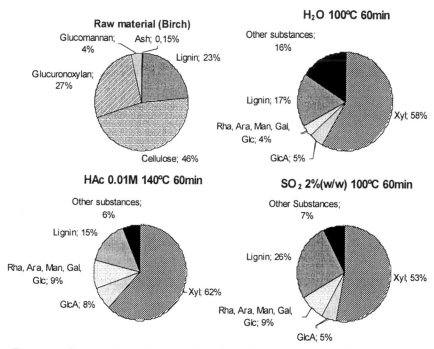

Figure 4. Composition of isolated hemicellulose fractions (in %w/w based on both pellets and soluble fraction).

In general, it can be seen that the molecular weights of the water-insoluble fractions (pellets), which are soluble in the DMSO-bases mobile phase, are higher than those of the soluble materials. The contribution of lignin in the AcOH and the H_2O samples is greater than in the SO_2 treated sample due to lignin solubilisation after sulfonation during pre-hydrolysis. In the DMSO-water system the average molecular mass is lower in comparison to the average molecular mass in the water system. This is most likely due to aggregated material, which is more easily produced in the water system. However, a similar trend can be observed. The highest molecular mass was obtained in the sample after water treatment while SO_2 treatment resulted in lower molecular mass.

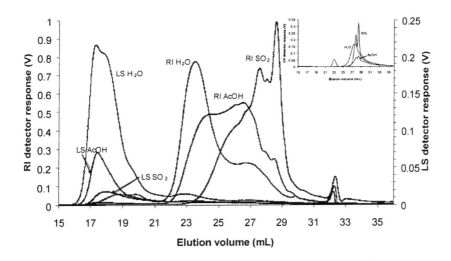

Figure 5. Size exclusion chromatograms showing molecular mass distributions of H₂O, AcOH and SO₂- treated samples. The inset shows the UV-signal for each sample based on the same time axis.

Table I. Comparison between average molecular masses of samples treated with H₂O, AcOH and SO₂ and analysed using a DMSO and a water-based system.

Sample	$(\overline{M_w})$ DMSO (g/mol)[a]	$(\overline{M_w})$ Water (g/mol)[b]
Water *Soluble*	11,330 *DP ≈ 76*	38,860 *DP ≈ 259*
Water *Pellets*	20,650 *DP ≈ 138*	---
HOAc *Soluble*	6,160 *DP ≈ 41*	23,010 *DP ≈ 153*
HOAc *Pellets*	16,800 *DP ≈ 112*	---
SO₂ *soluble*	4,490 *DP ≈ 30*	16,460 *DP ≈ 110*
SO₂ *pellets*	8,720 *DP ≈ 58*	---

[a]DMSO-water system 90:10   [b] 0.1 M NaNO₃

Conclusions

The isolation of a heteropolysaccharide fraction in high yield, high purity, and high molecular weight by acid-catalyzed pre-hydrolysis (prior to kraft delignification for pulp production) of birch wood chips has been optimised. The yield of total extracted material was found to be affected by the composition of the pre-hydrolysis solution (H_2O, dilute AcOH or dilute SO_2) and the processing conditions (pH, temperature and time). The amount of extracted material for samples with high heteropolysaccharide yield and low weight loss of wood was 3.2, 8.0 and 8.3 w/w% for treatment with H_2O, AcOH and SO_2, respectively. The yields based on available hemicelluloses in the parent wood were 7.5, 20.1 amd 17.8 w/w% for H_2O, AcOH and SO_2, respectively. The molecular weights of all samples were determined by GPC with LS, and they were found to vary between DP_w 34 and 150. They decreased from water treatment to SO_2 treatment. The SO_2 treated sample contained the largest amount of lignin. The sample with the highest heteropolysaccharide yield, lowest weight loss of wood, lowest lignin content and highest average molecular mass (AcOH treated sample) was chosen for future studies.

Acknowledgements

The authors would like to thank Dr. Per-Olof Larsson for valuable discussions concerning future application possibilities in pulp and paper industry. Dr. Harald Bredlid from the Department of Chemical Engineering and Environmental Science, Chalmers University of Technology, Gothenburg, Sweden for practical considerations in extraction experiments. Dr. Bodo Saake, Institute of Wood Chemistry and Chemical Technology of Wood, Federal Research Centre of Forestry and Forest Products, Hamburg, Germany for measurements of average molecular mass in a DMSO system and Södra Cell for financial support.

References

1. Betts, W. B.; Dart, R. K.; Ball, A.S.; Pedlar, S.L. In *Biodegradation of Natural and Synthetic Materials*; Betts, W. B., Ed.; Springer, Berlin, Germany. 1991; pp.139-155.
2. Carpita, N.; McCann, M. *Biochem. Mol. Biol. Plants* **2000**, 52-108.
3. Jacobs, A.; Larsson, P. T.; Dahlman, O. *Biomacromol.* **2001**, *2*, 979-990.
4. Aspinall, G. O. In *The Biochemistry of Plants: A Comprehensive Treatise*, Preiss: J., Ed.; Academic Press, New York, NY, 1980; Vol. 3, pp 473-500.

5. Timell, T. E., in *Cellular Ultrastructure of Woody Plants,* W. A. Cote, Jr., ed., Syracuse University Press, New York, 1965, p. 127.
6. Bolker, H. I., *Natural and Synthetic Polymers,* Marcel Dekker, Inc., New York, 1974; p. 688.
7. Fengel, D.; Wegener, G. *Wood: Chemistry, Ultrastructure, Reactions.* de Gruyter, Berlin, Germany, 1984; p. 613.
8. Popa, V. I. In *Polysaccharides in Medicinal Applications.* Dumitriu, S. Ed.; Dekker, New York, N.Y., 1996, pp. 107-124.
9. Reis, D.; Vian, B.; Roland, J.C. *Micron.* **1994,** *25,* 171-187.
10. Glasser, W. G., Kelley, S. S. *Lignin,* Section in *Enc. Polym. Sci. Eng.,* Vol. 8, John Wiley & Sons, New York, **1987,** p. 795-852.
11. Imamura, T.; Watanabe, T.; Kuwahara, M.; Koshijima, T. *Phytochem.* **1994,** *37,* 1165-1173.
12. Bhaduri, S. K.; Ghosh, I. N.; Deb Sarkar, N. L. *Ind. Crops Prod.* **1995,** *4,* 79-84.
13. Molin, U.; Teder, A. *Nord. Pulp Pap. Res. J.* **2002,** *17,* 14-19.
14. Sjöström, E. *Wood Chemistry: Fundamentals and Applications, 2nd edition;* Academic Press, New York, NY, 1993; pp 293.
15. Goyal, G. C., Lora, J. H., Pye, E. K. *Tappi,* **1992,** *75(2),* 110-116.
16. Puls, J.; Saake, B. In *Hemicelluloses: Science and Technology.* Gatenholm, P.; Tenkanen, M., Eds.; ACS Symposium Series 864; American Chemical Society, Washington, DC, 2004; pp 24-37.
17. Sun, R. C., Lawther, J. M., Banks, W. B. *Carbohydr. Polym.* **1996,** *29,* 325-331.
18. Glasser, W. G., Kaar, W. E., Jain, R. K., Sealey, J. E. *Cellulose,* **2000,** *7,* 299-317.
19. Glasser, W.G.; Jain, R.K.; Sjöstedt, M.A. *Thermoplastic pentosan-rich polysaccharides from biomass,* US Patent #5,430,142, **1995.**
20. Gustavsson, M.; Bengtsson, M.; Gatenholm, P.; Glasser, W. G.; Teleman, A.; Dahlman, O. In *Biorelated Polymers- Sustainable Polymer Science and Technology;* Chiellini, E.; Gil, M. H. M.; Buchert, J.; Braunegg, G.; Gatenholm, P.; van der Zee, M. Eds.; Kluwer Academic-Plenum Publishers, London, UK, 2001, pp 41-51.
21. Ibrahim, M., Glasser, W. G. *Bioresource Technology,* **1999,** *70,* 1881-92.
22. Theander, O.; Westerlund, E. A. *J. Agric. Food Chem.* **1986,** *34,* 330-336.
23. Schöning, A. G.; Johansson, G. *Sv. Papperstid.* **1965,** *68,* 607-613.
24. Roubroeks, J. P.; Saake, B.; Glasser, W. G.; Gatenholm, P. In *Hemicelluloses: Science and Technology.* Gatenholm, P.; Tenkanen, M., Eds.; ACS Symposium Series 864; American Chemical Society, Washington, DC, 2004; pp 167-183.
25. Zinbo, M.; Timell, T. E. *Sv. Papperstid.* **1965,** *68,* 647-662.

Chapter 24

Protein-Based Plastics and Composites as Smart Green Materials

Stéphane Guilbert, Marie-Hélène Morel, Nathalie Gontard, and Bernard Cuq

UMR 1208 Agropolymer Engineering and Emerging Technologies Joint Research Unit, INRA/Agro.M/UM2/CIRAD, 2 place P. Viala, 34060, Montpellier Cedex, France

Introduction

Many raw materials with high plant protein content (e.g. wheat and corn gluten, as well as protein concentrates and isolates from soybean, peanut, rapeseed, sunflower seed or cotton seed) or animal protein extracts (e.g. collagen or gelatin, fish or meat myofibrillar proteins, seric albumin or egg, casein) can be used to manufacture different materials (*1-4*). These materials are made with proteins that have unique functional properties (thermoplastic, elastomeric, thermoset, adhesive), which often differ markedly from the properties of standard "synthetic' materials, or even from materials developed using other hydrocolloids (e.g. starch or cellulose derivatives). There is considerable potential for modulating the properties of protein-based materials because of the variability in the amino-acid composition of proteins, the broad range of available formulations and material formation techniques, including casting or thermoplastic processes. Materials made from proteins are usually biodegradable, and sometimes even edible when food-grade components and manufacturing processes are used.

Plant proteins are generally not very expensive (around €0.5-1/kg⁻¹ for corn and wheat gluten, respectively, for protein contents ranging from 55 to 75%). High quantities of these proteins are often present in coproducts of processed field crops (e.g. wheat and corn gluten are coproducts of starch processing; protein concentrates or isolates are obtained from oilcake after extracting oil from oil crops). Plant proteins are thus widely available and relatively easy to handle. Animal proteins are more expensive (€2-10/kg⁻¹) than plant proteins.

They are also hampered by their poor reputation with respect to many applications (e.g. in cosmetics or as materials designed for contact with food products). However, their properties are essential for some applications (e.g. gelatin). Note that many proteins are allergens or their use sometimes gives rise to intolerance in the consumer, which could limit their applications.

At the outset of the 20th century, proteins were considered interesting raw materials for making plastics to eventually replace cellulose. Formaldehyde crosslinking of milk casein (galalith) is a process that was invented as early as 1897 to make molded objects such as buttons, broaches and various other items. The first patents were taken out in the 1920s on the use of zein to formulate different materials (coatings, resins, textile fibers). At that time, formaldehyde was widely used in blends with soybean proteins and slaughterhouse blood to make automotive parts, especially distributor caps. In addition, gelatin was used to produce films for foods, drug capsules and photography films. In the 1960s, synthetic products replaced proteins for many of these applications, and this protein boycott (except for gelatin and collagen) lasted for the next 30 years.

Since the 1980s, R&D on non-food uses of proteins has boomed as a result of the fresh interest in renewable and biodegradable raw materials, in using some field crops for non-food purposes, and in the unique specific properties of proteins. This research has led to very many patents, but there are still very few commercial applications.

Structure of materials, and physical and chemical modifications

It is important to compare the structural features of proteins with those of synthetic polymers or of polysaccharides in order to determine the specific properties of the derivative materials. Contrary to homopolymers or copolymers in which one or two monomers are repeated, proteins are heteropolymers consisting of amino acids. Proteins have a specific amino acid sequence and spatial conformation which determine their chemical reactivity and thus their potential for the formation of linkages that differ with respect to their position, nature and/or energy. This heterogenic structure provides many opportunities for potential crosslinking or chemical grafting—it even facilitates modification of the film-forming properties and end-product properties.

When only considering low-energy interactions, there is simultaneous involvement of Van der Waals interactions, hydrogen bonding, hydrophobic and ionic interactions in the structuring of protein arrangements. The respective participation of these different interactions depends on the proteins involved and can be adjusted according to the physicochemical conditions and treatments used. Protein-based materials are always polar to various extents (proteins with high apolar amino acid contents are known to be completely or partially water insoluble). They are closer to polysaccharide-based materials than to polyolefin- or polyester-based materials in this respect.

Protein treatment in oxidizing medium promotes the formation of covalent bonds, mainly between cystein residues or between lysine residues. These linkages, which are respectively called disulfide bonds and lysine bonds, induce very high molecular weight aggregate protein structuring when different polypeptide chains are involved. Depending on the density of covalent bonds, the behavior of the end products resembles that of elastomers (rubber-type) or of thermoset materials (e.g. highly resistant natural materials such as collagen-rich tissues with lysine-lysine bonds or hair with disulfide bonds).

Only very pure low molecular weight proteins are crystallizable. Most protein-based materials thus have an amorphic structure. However, α-helix and β-sheet type secondary structures which are highly stabilized by hydrogen bonds could be related to crystalline zones. Some studies have shown that the presence of β-sheets determines a cereal's potential use as an adhesive, coating or textile fiber (5). X-ray studies revealed that the stretching of protein fibers can lead to the formation of crystalline structures, thus enhancing their distortion resistance. The semi-crystalline structure of animal bristles—a material with irregular mechanical properties—would thus be the result of protein stretching when they are removed from the animal.

Protein molecular weights also have a substantial effect on the protein network structure. They also determine the presence of molecular overlapping, leading to the formation of physical nodes. As is the case for other polymers, overlapping could occur beyond a critical molecular weight (generally around 10^4 g.mol^{-1}), overlapping is possible and the material properties are stable.

Table I shows examples of protein raw materials that are commonly used in non-food applications, along with their composition, mean molecular weights and main constituent subunits.

The use of thermal and mechanical treatments and, to a lesser extent, UV or γ-ray treatments prompts crosslinking of protein networks (6). These treatments also have a slight anti-plasticizing effect, i.e. they lead to an increase in Tg for constant water contents. Protein crosslinking is thus triggered by a heat treatment carried out at temperatures around 20-80°C above the glass transition temperature (Figure 1). The crosslinking activation energy can thus be sharply reduced by the addition of mechanical energy (by a ratio of 1-3). Very high crosslinking density can thus be achieved in the presence of mechanical energy (shear) for reasonable time:temperature ratios and much lower than thermal decomposition conditions (7,8).

Table 1: Main physico-chemical characteristics of the proteins used as polymeric materials to form material (9).

Proteins	Amino acid ratios [a]			Main sub-units			
	A	B	C	Name	W_R	MW	S [b]
Corn zein	36	10	47	α-zein	80	21-25	IV
Wheat gluten	39	14	40	Gliadin	40	30-80	IV
				Glutenin	46	200-2000	III
Soy proteins	31	25	36	β-Conglycinin	35	185	II
				Glycinin	40	363	II
Peanut prot.	30	27	32	Arachin	75	330	II
Cottonseed	41	23	32	Albumin	30	10-25	I
Proteins				Globulin	60	113-180	II
Keratin	34	11	42	--	--	10	III
Collagen	13	13	40	Tropocollagen	--	300	III
Gelatin (A)	12	14	41	--	--	3-200	III
Caseins	31	20	44	α_{S1}, α_{S2}, β, κ, γ	--	19-25	(c)
Whey prot.	30	26	40	β-lactoglobulin	60	18	I
				α-lactalbumin	20	14	I
Myofibril							
- sardine	27	31	35	Myosin	50	16-200	II
- beef meat	27	27	39	Actin	20	42	II

Where MW is the molecular weight (kD) and W_R is the rate of sub-unit weight (%) in raw materials. 50% amidation rates of aspartic and glutamic acids are supposed for peanut proteins, keratin, collagen, and gelatin. [a] Amino acid ratios (mol/100 mol): -A- non-ionized polar (Asn, Cys, Gln, His, Ser, Thr, Tyr), -B- ionized polar (Arg, Asp, Glu, Lys) and -C- non-polar (Ala, Ile, Leu, Met, Phe, Pro, Trp, Val). [b] S is the solubility of proteins according to Osborne classification: I- in water, II- in diluted salt solutions, III- in diluted acidic or basic solutions, and IV- in ethanol (80%) solutions [c] Miscellaneous associations.

A very broad range of crosslinking rates can be obtained by controlling the thermomechanical treatment of gluten (15-95% of insoluble proteins in SDS, depending on the harshness of the thermomechanical treatment). Thermomechanical treatments can thus be used to structure protein materials without any chemical additive and without thermal decomposition, with a crosslinking rate similar to that obtained via the addition of formol or another crosslinking agent (Figure 2). When altered by heat treatments, crosslinking agents or radiation (UV, gamma, etc.), crosslinked proteins form insoluble networks with elastomer-type thermomechanical behavior, and different extents of puncture strength depending on the covalent crosslinking density (Figure 2).

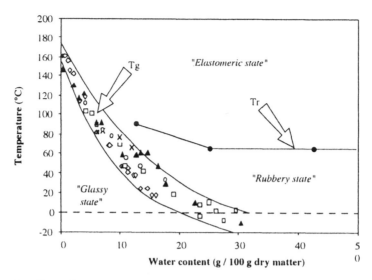

Figure 1: Effect of water content on the glass transition temperature (T_g) and on minimum thermosetting temperature (T_r) for wheat gluten proteins (10).

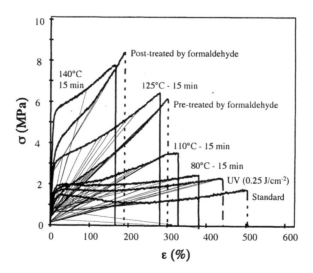

Figure 2: Influence of cross-linking physical (thermal or UV treatments) or chemical (pre- or post-treatments by formaldehyde) treatments on the mechanical properties (i.e. strength and deformation at break) of wheat gluten films (6).

Physical or chemical modifications of protein-based materials (films, materials, resins, adhesives, etc.) have also been proposed to enhance their water resistance and limit the impact of relative humidity on their properties or to modify their functionality (e.g. strengthen the surfactant effect) (*6,11-13*).

Technologies for manufacturing and forming protein-based materials

The following steps are required to form a material from a protein concentrate or extract: *(i)* scission of low-energy intermolecular bonds stabilizing systems in the native state, *(ii)* protein rearrangement, and *(iii)* the formation of a 3-dimensional network stabilized by new interactions or linkages, after removal of the intermolecular bond scission agent.

Two different technological strategies could be used to make protein-based materials: the "solvent process" (commonly called "casting" or "continuous flow") involving a protein solution or dispersion, and the "dry process" using the thermoplastic properties of the proteins under low hydration conditions.

The casting process is mainly controlled by the composition and concentration of the solvent systems used, the temperature, as well as the desolventization conditions (drying conditions).

Thermal or thermomechanical processes (e.g. extrusion or hot press molding) are used to form materials under low moisture conditions. The glass transition temperature (Tg)—because of their hydrophilic nature (which varies between proteins)—is highly affected by moisture (160-200°C decrease in Tg in the dry state and around 60-100°C for material with 10% moisture content). In practice, when protein materials contain about 15% water (i.e. which generally occurs when they are at equilibrium with 85% relative humidity at ambient temperature), their Tg is close to the ambient temperature. This effect is even more obvious in the presence of plasticizers.

Plasticizers are generally required for the formation of protein-based materials (*11,14-18*). These agents modify the raw material formation conditions and the functional properties of these protein-based materials (i.e. a decrease in resistance, rigidity and barrier properties and an increase in flexibility and maximal elongation of the materials). Polyols (e.g. glycerol and sorbitol), amines (e.g. tri-ethanolamine) and organic acids (e.g. lactic acid) are the most common plasticizers for such applications. Completely or partially water insoluble amphipolar plasticizers such as short-chain fatty acids (e.g. octanoic acid) can be used since some protein chain domains are markedly apolar.

The depressive effect of plasticizers has been simulated via different molecular models based on the molar fraction or molar volume according to the number of potential hydrogen bonds or the percentage of hydrophilic groups of the plasticizer (*15,17*).

It is essential to determine the phase equilibrium patterns (Figure 1) of protein-based materials according to the moisture (or plasticizer) contents in order to be able to control the material formation conditions and predict variations in the properties of the end products under different usage conditions (temperature and relative humidity) (6,10,19,20).

Protein-based materials can be obtained using "standard" systems for extrusion, calendering, extrusion blow molding, injection and thermoforming processes (9,21-25). These setups are generally tailored to synthetic material production processes for which the material formation properties are known and the processing parameters have been optimized. For efficient protein transformation, the relative thermal, mechanical and chemical sensitivities and viscosities in the malleable phases at temperatures above the Tg of each protein should be considered. In many cases and contrary to thermoplastic or polyolefin starches, thermal or thermomechanical treatments lead to covalent crosslinking of proteins and thus to a substantial increase in viscosity. Hence, it is very complicated to optimize the material formation treatment (temperature, plasticizer content, residence time, etc.) in order to stabilize the processing temperature between the Tg and the thermocrosslinking temperature of the material (Figure 1).

When using wheat gluten (or corn gluten) for instance, the possibility of obtaining a homogeneous malleable phase with thermal (at temperatures above the Tg), mechanical (shear) and chemical (additives and degradation) treatments was investigated on the basis of the viscoelastic and flow properties of the malleable matter relative to temperature, water content and time (7,8,22,24,26-29). "Plasticized gluten" resembles a structured viscoelastic solid with pseudo-plastic behavior. A study of rheological functions (G' and G'' moduli, complex viscosity) of plasticized gluten revealed that the time:temperature ratio could be applied (22,29). For given mixing conditions, the complex viscosity of plasticized gluten can be characterized by a power law with temperature and glycerol content as variables. Redl et al. (25,29) carried out studies on the extrusion of gluten-based materials in a corotating twin screw extruder, with simulation of flow properties and extrusion conditions. They show that gluten material is very difficult to form trough extrusion. This is due to the difficulty to control the processing conditions in order to plasticize the material without thermosetting. Then the operating window is very narrow.

Properties of protein-based materials

The macroscopic properties of protein-based materials partially depend on protein-network stabilizing interactions. The presence of "physical bonds" (i.e. folded chains), covalent intermolecular bonds and/or a high interaction density is sufficient to produce films that are completely or partially water insoluble and that will swell in water by 30-150% (*30*).

The mechanical properties of protein-based materials have been studied (Table II) and modeled (*9,31-33*).

Table II. Mechanical properties of various protein-based films and comparison with some synthetic films (*3*).

Films	Tensile Strength (MPa)	Elongati- -on (%)	X	T	RH (%)
Polyesters	178	85	--	--	--
High density polyethylene	25.9	300	--	--	--
Low density polyethylene	12.9	500	--	--	--
Methylcellulose	56.1	18.5	--	25	50
Myofibrillar proteins	17.1	22.7	34	25	57
Hydroxypropylcellulose	14.8	32.8	--	25	50
Whey protein isolate	13.9	30.8	--	23	50
Soy proteins	1.9	35.6	88	25	50
Wheat gluten + 10% glycerol	23.2	14.0	280	20	50
Wheat gluten + 30% glycerol	3.5	214	270	20	50
Wheat gluten + 30% glycerol + 15 min at 130°C	8.5	268	290	20	50
Corn zein proteins	1.4	--	81	26	50

Where X is the film thickness (μm); T is temperature (°C); RH is relative humidity (%).

The mechanical properties of protein-based materials closely depend on the plasticizer content, temperature and ambient relative humidity (*16,34,35*). At constant temperature and composition, an increase in relative humidity leads to a major change in the material properties, with a sharp drop in mechanical strength and a concomitant sharp rise in distortion. These modifications occur when the Tg of the material is surpassed (Figure 1). These variations can be reduced by implementing crosslinking treatments (physical or chemical) or using high cellulose or mineral loads (*22*).

Protein-based materials generally have high water vapor permeability (Table III). Water vapor permeation through protein films is facilitated by the steady presence of hydrophilic plasticizers which promote water molecule adsorption. Protein-based materials have much higher water permeability (around 5×10^{-12} mol m^{-1} s^{-1} Pa^{-1}) than synthetic materials (0.05×10^{-12} mol m^{-1} s^{-1} Pa^{-1} for low-density polyethylene). This feature could still be interesting for coatings on materials that need to "breathe", e.g. for fresh produce packaging, and films for agricultural or cosmetic applications.

These properties can be modified to resemble those of polyethylene films by adding lipid compounds (beeswax, paraffin, etc.) to the film formulation (36,37). As already noted for the mechanical properties, water barrier properties are highly dependent on the temperature and relative humidity of the protein material and decline suddenly during the glass transition phase (22).

Table III. Water vapor permeability (10^{-12} mol.m^{-1}.s^{-1}.Pa^{-1}) of various protein-based films and comparison with some synthetic films and biodegradable films (3).

Film	WVP	T	X	RH
Starch	142	38	1190	100 - 30
Sodium caseinate	24.7	25	--	100 - 0
Soy proteins - pH 3	23.0	25	83	100 - 50
Corn zein	6.45	21	200	85 - 0
HPMC	5.96	27	19	85 - 0
Glycerol monostearate	5.85	21	1750	100 - 75
Wheat gluten - Glycerol	5.08	30	50	100 - 0
Wheat gluten - Oleic acid	4.15	30	50	100 - 0
Fish myofibrillar proteins	3.91	25	60	100 - 0
Wheat gluten - Carnauba wax	3.90	30	50	11 - 0
Hydroxypropyl cellulose	2.89	30	25	97 - 0
Low density polyethylene	0.0482	38	25	95 - 0
Wheat gluten - Beeswax	0.0230	30	90	100 - 0
High density polyethylene	0.0122	38	25	97 - 0
Beeswax	0.0122	25	120	87 - 0
Aluminium foil	0.00029	38	25	95 - 0

Where WVP is water vapor permeability HPMC is Hydroxy propyl methylcellulose; T is the temperature (°C); X is the film thickness (μm); RH is the difference of relative humidity (%) across the film.

The gas barrier properties (O_2, CO_2 and ethylene) of protein-based materials are very attractive since they are minimal under low relative humidity conditions (38). O_2 permeability (around 1 amol.m^{-1}.s^{-1}.Pa^{-1}) is comparable to the EVOH properties (0.2 amol.m^{-1}.s^{-1}.Pa^{-1}) and much lower than the properties of low-density polyethylene (1000 amol.m^{-1}.s^{-1}.Pa^{-1}, Table IV). The O_2 permeability of protein films is about 10-fold higher that of EVOH-based films, mainly due to the high plasticizer content.

While the barrier properties of synthetic materials remain quite stable at high relative humidity, the gas-barrier properties of material proteins (as for all properties of hydrocolloid-based materials) are highly relative humidity- and temperature-dependent. The O_2 and CO_2 permeabilities are about 1000-fold higher for moist films than for films stored at 0% relative humidity. For protein-based films, this effect is much greater for "hydrophilic" gases (CO_2) than for "hydrophobic" gases (O_2). This modifies the CO_2/O_2 selectivity coefficient, which rises from 3 to more than 50 when the relative humidity rises from 0 to 100% and the temperature from 5 to 45°C as compared to constant values of around 3-5 for standard synthetic films. Gas permeability differences in protein materials are partly due to gas solubility differences in the film matrix. This could be mainly explained by the high affinity between CO_2, the polypeptide chain and many lateral amino acid groups (*39*).

Table IV. Oxygen permeability $P(O_2)$ (10^{-15} mol.m^{-1}.s^{-1}.Pa^{-1}) and carbon dioxide permeability $P(CO_2)$ (10^{-15} mol.m^{-1}.s^{-1}.Pa^{-1}) of various protein-based films and comparison with some synthetic films and biodegradable films (*3*).

Film	$P(O_2)$	$P(CO_2)$	T	a_w
Low density polyethylene	16.050	185.68	23	0
Polyester	0.192	1.67	23	0
Ethylene-vinyl alcohol	0.003	--	23	0
Methylcellulose	8.352	1315.60	30	0
Beeswax	7.680	--	25	0
Carnauba wax	1.296	--	25	0
Corn zein	0.560	9.50	38	0
Wheat gluten protein	0.016	0.31	25	0
Chitosan	0.009	--	25	0
High density polyethylene	3.580	--	23	1
Polyester	0.192	--	23	1
Ethylene-vinyl alcohol	0.096	--	23	0.95
Pectin	21.440	937.20	25	0.96
Wheat Gluten	20.640	1614.80	25	0.95
Starch	17.360	--	25	1
Chitosan	7.552	352.44	25	0.93

Where T is the temperature (°C) and a_w the equilibrium water activity.

A film with good O_2 barrier properties is interesting for the protection of oxidizable foods (rancidification, loss of oxidizable vitamins, etc.). However, some extent of permeability to O_2 and especially to CO_2 is required to decrease the metabolic activity of many fresh fruits and vegetables. The development of protein films with selective gas permeability features could thus be highly promising, especially for controlling respiratory exchange and improving the shelflife of fresh or slightly processed fruit and vegetables. Wheat gluten-based films were tested with the aim of creating atmospheric conditions suitable for preserving fresh vegetables. Measurements of changes in the gas composition of modified atmosphere packaged mushrooms under wheat gluten films confirmed the high selectivity of such materials, i.e. the CO_2 and O_2 composition ranged from 1-2% despite product respiration (40).

The aroma barrier properties of protein-based materials seem especially interesting for blocking apolar compound permeation. However, it is hard to determine the relationship between the physicochemical properties of aroma compounds and their retention by protein films (41).

Solute retention properties (especially antimicrobial and antioxidant agents) were investigated and modeled for wheat gluten (42-45). These properties indicate potential applications for the controlled release of functional agents. The antimicrobial efficacy of edible wheat gluten-based films containing antimicrobial agents has been extensively documented (46). The use of these films on high moisture model foods extended their shelflife by more than 15 days at 4 and 30°C. Many patents and publications recommend adding antioxidant agents to protein films and coatings, as already done in some commercial edible films. Guilbert (47) measured α-tocopherol retention in gelatin films applied to the surface of margarine blocks. No migration was noted after 50 days storage when the film was pretreated with a crosslinking agent (tannic acid), whereas tocopherol diffusivity was around 10 to $30 \times 10^{-11} m^2 s^{-1}$ without the film.

Environmental aspects: biodegradability and life cycle analysis

The "environmental" properties of protein-based materials are of special interest. For instance, an analysis of the life cycle of gluten-based materials (composed of gluten and glycerol) was conducted using the standard ISO 14040 method (Figure 3). The preliminary results obtained in the two environmental impact categories analyzed, i.e. energy use (10 MJ/kg gluten) and greenhouse gas emission (0.72 kg CO_2-eq/kg gluten), are shown in Table V. These values are very low in comparison to the results obtained with commercial bioplastics such as PLA or biodegradable polyester starch blends (e.g. Materbi®) (48).

Wheat protein-based materials are biodegradable and environment-friendly (27). A biodegradation study was performed in liquid medium using the standard ISO 14852 method. The wheat protein-based materials were completely degraded after 36 days (Figure 3) and no microbial inhibition due to toxic metabolite excretion was noted.

These results apply for all gluten crosslinking rates (up to 90% insoluble proteins in SDS) if crosslinking is achieved using thermal or thermomechanical treatments (Figure 4). The construction of protein networks and their chemical crosslinking could induce major changes in the conformation and resistance to

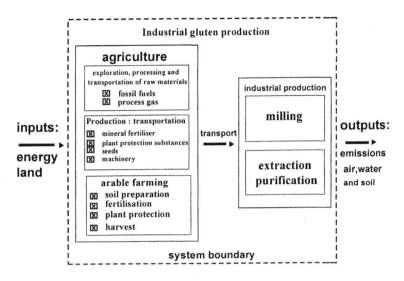

Figure 3: Life cycle assessment of wheat gluten materials (27)

Table V. Energy use and green house gas (GHG) emissions of some biodegradable and conventional materials (27,49).

Polymer	Energy use [MJ/kg]	GHG emissions [kg CO_2-eq]	Analysis scope
Low density polyethylene	80.6	5.04	Incineration
Starch	18.5	1.08	Cradle to gate
Thermoplastic starch	18.9	1.10	Cradle to gate
Mater-Bi foam grade	32.4	0.89	Composting
Polylactic acid	57	3.84	Incineration
Polyhydroxyalcanoates by fermentation	81	n.a.	n.a.
Polyhydroxyalcanoates, various processes	66-573	n.a.	n.a.
Wheat gluten (preliminary data)	10 (– 17)	0.72	Cradle to gate

346

enzymatic hydrolysis and chemical attacks of proteins. However, Garcia-Rodenas *et al* (*50*). showed that the susceptibility of casein and wheat gluten-based films to *in vitro* proteolysis did not significantly differ from that noted for native proteins.

The results of *in vitro* enzymatic digestion studies using proteases also confirmed that the protein network hydrolysis rate remained constant irrespective of the harshness of the thermomechanical crosslinking treatment applied, except for the harshest treatments where a slight reduction was noted (*51*).

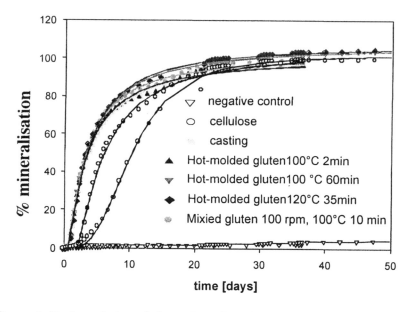

Figure 4. Biodegradation of wheat gluten based materials, comparison with cellulose; STURM method (52).

Conclusion

Plant proteins are generally inexpensive, widely available and relatively easy to process. Animal proteins are more expensive, but sometimes have no functional substitutes. Protein-based materials are generally homogeneous, transparent, grease and water resistant and water insoluble.

Solvent processes are tailored for fiber extrusion, production of coatings for seeds, drug pills and foods, for making cosmetic masks or varnishes and pharmaceutical capsules.

Heat casting of protein-based materials by techniques usually applied for synthetic thermoplastic polymers (extrusion, injection, molding, etc.) is more cost-effective. This process is often applied for making fibers, resins and flexible films (e.g. films for agricultural applications, packaging films, and cardboard coatings) or objects (e.g. biodegradable materials) that are sometimes reinforced with fibers (composite bioplastics for construction, automobile parts, etc.). Material protein properties can generally be modified in a wide range of ways via raw material choices and combinations, the proper use of fractioning techniques and rheological modifying additives, and also by adjusting the product formation process variables. Proteins are thus especially sensitive to thermomechanical treatments, so carefully controlling these treatments will in turn facilitate control of the material forming process and especially the properties—to even obtain highly crosslinked materials that are very strong and relatively resistant to water (swelling rate of less than 30% after 24 h).

The functional properties (especially optical, barrier and mechanical) of these protein-based materials are often specific and unique, and they could thus be used as raw materials for bioplastics with a wide range of agricultural, agrifood, pharmaceutical and medical industry applications. Their high moisture permeability is especially attractive for cheese, fruit and vegetable packaging, and for agricultural material and cosmetic applications. Protein-based materials have slightly lower mechanical properties than reference materials such as low-density polyethylene or plasticized PVC, but the addition of fibers (composite materials) can considerably improve them. The thermoplastic properties of proteins and their water resistance (for insoluble proteins) are especially interesting for natural resin uses to produce particleboard type materials.

Gas barrier properties (O_2, CO_2 and ethylene) of protein-based materials can be utilized in designing selective or active materials for modified atmosphere packaging of fresh products such as fruit, vegetables, cheese, etc. Solute retention properties (especially antimicrobial and antioxidant agents) are attractive for designing controlled-release systems for useful agents in food (e.g. active coatings, encapsulation), agriculture (e.g. coated seed), pharmacy (drug delivery) and cosmetic industries. Since protein based materials have very unique properties in this concern, they can be used for the design of "smart" materials with properties (mechanical, barrier and degradation) controlled by environmental conditions (temperature and relative humidity)

References

1. Cuq, B.; Gontard, N.; Guilbert, S. *Cereal Chem.* **1998**, *75*, 1-9.
2. Gennadios, A. *Protein Based Films and Coatings*; CRC Press: Boca Raton, **2002**.
3. Guilbert, S., Cuq, B. *Les protéines matériaux*; Édition Groupe Français d'Étude et d'Application des Polymères, **2001**; Vol. 13.
4. Torres, J. A. *Protein functionality in food systems*; Marcel Dekker: New York, **1994**.
5. Rothfus, J. A. *J. Agric. Food Chem.* **1996**, *44*, 3143-3152.
6. Micard, V.; Belamri, R.; Morel, M. H.; Guilbert, S. *J. Agric. Food Chem.* **2000**, *48*, 2948-2953.
7. Morel, M. H.; Redl, A.; Guilbert, S. *Biomacromolecules* **2002**, *3*, 488-497.
8. Redl, A.; Guilbert, S.; Morel, M. H. *J. Cereal. Sci.* **2003**, *38*, 105-114.
9. Di Gioia, L., ENSA, 1998.
10. Cuq, B.; Abecassis, J.; Guilbert, S. *Int. J. Food Sci. Technol.* **2003**, *38*, 759-766.
11. Galietta, G.; Di Gioia, L.; Guilbert, S.; Cuq, B. *J. Dairy Sci.* **1998**, *81*, 3123-3130.
12. Marquie, C.; Aymard, C.; Cuq, J. L.; Guilbert, S. *J. Agric. Food Chem.* **1995**, *43*, 2762-2767.
13. Pommet, M., Redl, A., Morel, M. H., Guilbert, S. *A way to improve the water resistance of gluten-based biomaterials: plasticization with fatty acids*; The Royal Society of Chemistry: Cambridge, **2004**.
14. Cuq, B.; Gontard, N.; Cuq, J. L.; Guilbert, S. *J. Agric. Food Chem.* **1997**, *45*, 622-626.
15. Di Gioia, L., Guilbert, S. *J. Agric. Food Chem.* **1999**, *47*, 1254-1261.
16. Gontard, N.; Guilbert, S.; Cuq, J. L. *J. Food Sci.* **1993**, *58*, 206-211.
17. Pommet, M.; Redl, A.; Morel, M. H.; Guilbert, S. *Polymer* **2003**, *44*, 115-122.
18. Pouplin, M.; Redl, A.; Gontard, N. *J. Agric. Food Chem.* **1999**, *47*, 538-543.
19. Di Gioia, L.; Cuq, B.; Guilbert, S. *Int. J. Biol. Macromol.* **1999**, *24*, 341-350.
20. Kokini, J. L.; Cocero, A. M.; Madeka, H. *Food Technology* **1995**, *49*, 74-&.
21. Cuq, B.; Gontard, N.; Guilbert, S. *Polymer* **1997**, *38*, 4071-4078.
22. Guilbert, S., Gontard, N., Morel, M. H., Chalier, P., Micard V., Redl, A. *Formation and properties of wheat gluten films and coatings*; CRC Press: Boca Raton, **2002**.

23. Mungara P.; Zhang, J. Z., S.; Jane, J. L. *Soy protein utilization in compression-molded, extruded, and injection-molded degradable plastics*; CRC Press: Boca Raton, **2002**.

24. Pommet, M.; Redl, A.; Morel, M. H.; Domenek, S.; Guilbert, S. *Macromol. Symp.* **2003**, *197*, 207-217.

25. Redl, A.; Morel, M. H.; Bonicel, J.; Vergnes, B.; Guilbert, S. *Cereal Chem.* **1999**, *76*, 361-370.

26. Domenek, S.; Morel, M. H.; Bonicel, J.; Guilbert, S. *J. Agric. Food Chem.* **2002**, *50*, 5947-5954.

27. Domenek, S.; Morel, M. H.; Redl, A.; Guilbert, S. *Macromol. Symp.* **2003**, *197*, 181-191.

28. Pommet, M.; Morel, M. H.; Redl, A.; Guilbert, S. *Polymer* **2004**, *45*, 6853-6860.

29. Redl, A.; Morel, M. H.; Bonicel, J.; Guilbert, S.; Vergnes, B. *Rheologica Acta* **1999**, *38*, 311-320.

30. Domenek, S.; Brendel, L.; Morel, M. H.; Guilbert, S. *Biomacromolecules* **2004**, *5*, 1002-1008.

31. Cuq, B.; Gontard, N.; Cuq, J. L.; Guilbert, S. *J. Agric. Food Chem.* **1996**, *44*, 1116-1122.

32. Di Gioia, L., Cuq, B., Guilbert, S. *J. Mater. Res.* 2000, *15*, 2612-2619.

33. Redl, A., ENSA, **1998**.

34. Cuq, B.; Gontard, N.; Aymard, C.; Guilbert, S. *Polym. Gels Networks* **1997**, *5*, 1-15.

35. Mangavel, C.; Barbot, J.; Gueguen, J.; Popineau, Y. *J. Agric. Food Chem.* **2003**, *51*, 1447-1452.

36. Gontard, N.; Duchez, C.; Cuq, J. L.; Guilbert, S. *Int. J. Food Sci. Technol.* **1994**, *29*, 39-50.

37. Gontard, N.; Marchesseau, S.; Cuq, J. L.; Guilbert, S. *Int. J. Food Sci. Technol.* **1995**, *30*, 49-56.

38. Gontard, N.; Thibault, R.; Cuq, B.; Guilbert, S. *J. Agric. Food Chem.* **1996**, *44*, 1064-1069.

39. MujicaPaz, H.; Gontard, N. *J. Agric. Food Chem.* 1997, *45*, 4101-4105.

40. Barron, C.; Varoquaux, P.; Guilbert, S.; Gontard, N.; Gouble, B. *J. Food Sci.* **2002**, *67*, 251-255.

41. Debeaufort, F., Voilley, A. *IAA* **1997**, *3*, 125-126.

42. Cuq, B., Redl, A. *Activité antimicrobienne des emballages*; Lavoisier, Tec. et Doc, Paris: Paris, **2000**.

43. Guilbert, S., Redl, A., Gontard, N. *Mass transport within edible and biodegradable protein based materials: Application to the design of acive biopackaging*; CRC Press: Boca Raton, **2000**.

44. Han, J. H. *Protein-based edible films and coatings carrying antimicrobial agents*; CRC Press: Boca Raton, **2002**.

45. Redl, A.; Gontard, N.; Guilbert, S. *J. Food Sci.* **1996**, *61*, 116-120.

46.   Guilbert, S., Gontard, N., Gorris, L. G. M. *Lebens. Wiss. Technol.* **1996**, *29*, 10-17.

47.   Guilbert, S. *Use of superficial edible layer to protect intermediate moisture foods: application to the protection of tropical fruit dehydrated by osmosis*; Elsevier: Londres, **1988**.

48.   Domenek S., M., M. H., Guilbert, S. *Wheat gluten based agromaterials: Environemental performance, biodegradability and physical modifications*; The Royal Society of Chemistry: Cambridge, **2004**.

49.   Patel, M., Bastioli,C., Marini, L., Würdinger, E. In *Biopolymers*; Wiley-VCH: New York, **2003**; Vol. 10.

50.   Garciarodenas, C. L.; Cuq, J. L.; Aymard, C. *Food Chem.* **1994**, *51*, 275-280.

51.   Domenek, S.; Brendel, L.; Morel, M. H.; Guilbert, S. *Cereal Chem.* **2004**, *81*, 423-428.

52.   Domenek, S.; Feuilloley, P.; Gratraud, J.; Morel, M. H.; Guilbert, S. *Chemosphere* **2004**, *54*, 551-559.

Author Index

Subject Index